Renewable Energy Options for Power Generation and Desalination

Ibrahim Dincer • Mert Temiz

Renewable Energy Options for Power Generation and Desalination

Springer

Ibrahim Dincer
Ontario Tech University
Oshawa, ON, Canada

Mert Temiz
Ontario Tech University
Oshawa, ON, Canada

ISBN 978-3-031-53436-2 ISBN 978-3-031-53437-9 (eBook)
https://doi.org/10.1007/978-3-031-53437-9

This Springer imprint is published by the registered company Springer Nature Switzerland AG
The registered company address is: Gewerbestrasse 11, 6330 Cham, Switzerland

Paper in this product is recyclable.

Preface

The subjects of power generation and desalination have always been critical for humanity since power generation means electricity which is badly needed by everyone for almost everything in our daily life and desalination means fresh water which is also terribly needed by many people in most parts of the world. We are in a technologically advanced era as we need more electricity and more fresh water with a requirement that we need to make these out of environmentally benign sources, without possibly using fossil fuel-based choices since they face challenges as these have been releasing huge amounts of carbon and making the world suffer terribly through global warming and climate changes locally and globally caused by excessive amounts of carbon dioxide. Such issues and concerns have made us consider renewable energy sources as the primary source of energy for power generation and desalination. Thus, this book is really timely and makes it focus on power generation and desalination by using renewable energy sources and systems accordingly.

The book consists of nine unique chapters, ranging fundamentals to future directions. In this regard, Chapter 1 deals with the key fundamentals of thermodynamics and its concepts, as well as energy and exergy approaches outlined for thermodynamic systems in analyzing and assessing them for practical applications. Chapter 2 is really about energy, environment, and sustainability dimensions and their interrelationships and introduces numerous environmental impact assessment and sustainable development criteria and parameters to use for power generation and desalination systems. Chapter 3 focuses on renewable energy sources and their classifications, systems, and applications. Numerous advantages and disadvantages of renewable energy systems are considered for comparative evaluation. Energy- and exergy-based assessment tools and methodologies are also presented for various systems through some case studies. Chapter 4 covers thermal energy (essentially heat) generation for power generation and thermal desalination applications. There is a prime focus made on renewables in this regard to link them to power generation and desalination in a more environmentally friendly and sustainable manner. Some key methodologies and examples are presented for better coverage. Chapter 5 is one of the most critical subjects here in this book and focuses on the methods of power generation, especially from renewable sources. There are also evaluation methods of power generation systems driven by different sources discussed for making a comparative evaluation, especially for renewables. There is further coverage of case studies and their practical importance. Chapter 6 gets into the specific methods, systems, technologies, and applications of desalination systems, especially driven by renewable energy sources. It covers historical developments and landmarks of desalination technologies and introduces the landmark-type studies on the pathways to generate fresh water by using renewables. It further covers the thermodynamic aspects of desalination within the energy and exergy points of views. Chapter 7 makes efforts to introduce traditional energy systems by covering the past and dwelling on the current applications and implications. There are global indicators and parameters regarding the traditional energy systems and their impacts presented with charts and maps in order to raise awareness. It also discusses the thermodynamic aspects of traditional energy systems discussed from energy and exergy points of view. Chapter 8 is about integrated energy systems where power generation and desalination are simultaneously covered and operated effectively and efficiently. There are about ten unique case studies with renewable-driven integrated energy systems for the generation of multiple useful outputs including desalination presented extensively for analysis and assessment using energy and exergy approaches. Chapter 9 aims to provide future directions about power generation and desalination, how the technologies will evolve, and how their roles will change.

In closing, we hope that the readers will find this book useful in understanding the concepts, systems and applications related to renewable energy-based power generation and desalination. We further hope that this will be a useful asset to their research and development activities in the field.

Ibrahim Dincer
Mert Temiz
Oshawa, Ontario, Canada

Contents

Chapter 1
Fundamentals

Nomenclature

E	total energy (kJ)
ex	specific exergy (kJ/kg)
$\dot{E}x$	exergy rate (kW)
g	gravitational acceleration (m s^{-2})
h	specific enthalpy (kJ/kg)
$\dot{m}$	mass flow rate (kg/s)
$\dot{N}$	molar flow rate (mol/s)
P	pressure (kPa)
Q	heat (kJ)
$\dot{Q}$	heat rate (kW)
s	specific entropy (kJ/kg K)
S	total entropy (kJ /K)
$\dot{S}$	entropy rate (kW/K)
T	temperature (^{o}C)
V	velocity (m/s)
w	specific work (kJ/kg)
Z	vertical position, elevation (m)

Subscripts

abs	absolute
b	boundary
C	compressor
cv	control volume
d	destruction
D	diffuser

I. Dincer, M. Temiz, *Renewable Energy Options for Power Generation and Desalination*, https://doi.org/10.1007/978-3-031-53437-9_1

DH	dehumidifier
e	exit
El	electricity
en	energy
EV	expansion valve
ex	exergy
gage	gauge
gen	generation
hp	heat pump
HX	heat exchanger
i	input
in	inlet
MC	mixing chamber
N	nozzle
o	output
out	outlet
OFH	open feedwater heater
P	pump
RO	reverse osmosis
Sep	separator
T	turbine
vac	vacuum
W	water, work
WV	three-way valve
XV	expansion valve

Greek Letters

η	efficiency
μ	mean
Δ	difference

Acronyms

EES	Engineering Equation Software
FLT	first law of thermodynamics
IPCC	Intergovernmental Panel on Climate Change
SI	International System of Units
SLT	second law of thermodynamics

1.1 Introduction

It is unthinkable for any community or society to maintain their life without energy, food, and water. These commodities have always been at the center of the agenda of communities and societies. Historically, human settlements have been placed by considering energy, food, and water factors; being close to bodies of water and sources of energy and the places suitable for agriculture, which were essential to develop countries, improve their lifestyles and make them more prosperous and wealthier communities or societies. The existence of rich energy and water sources leads to achieving countless challenges, enhances agricultural activities, increases the overall life quality, maintains the urbanization, and improves the mobility and ability for better transportation. It is also important to note that energy and water have always been two of the major reasons for the migrations of human groups, wars, conflicts, and peace. Especially after the establishment of commercial fossil fuel in 1859, the race to control energy and water sources becames more critical and the frequency of conflicts, wars as well as mass migrations substantially increased; the prime mover is shifted from human labor, animals, windmills, waterwheels to engines and turbines. Petrochemical companies and industry were considering Earth like an all-you-can-eat buffet, with no obligations, where they processed and consumed fossil-based fuels without any restrictions. The exponential increase in global fossil fuel consumption has resulted in very critical environmental issues. Combusting fossil-based fuels is the major cause of greenhouse gas (GHG) emissions, especially carbon dioxide emissions. Initially, greenhouse gases are a natural part of Earth's atmosphere; they trap heat and help to maintain the energy balance by stabilizing the global temperatures for living organisms. However, anthropogenic greenhouse gas emissions destabilize the energy balance of Earth and cause temperature rises along with extreme weather events. If the global temperature rise passes 2°C, some of the environmental impacts will be irreversible; an important portion of people are highly vulnerable to climate change, as well as some plant and animal species can face extinction [1]. Moreover, drought and water scarcity are considered other climate-related challenges that show how serious and critical the overall impact is. That is why global governments and organizations agreed to act against climate change, reduce GHG emissions, and stabilize the energy balance of Earth's atmosphere to mitigate the risks of the anthropogenic effects on the climate system. It is obvious that improving how we utilize and use energy and water will help to achieve major challenges. Ultimately, more efficient, more environmentally friendly, and more cost-effective energy and water systems will help to achieve better sustainability for humanity.

In order to develop a foundational understanding of energy systems and applications, the concepts and basics of thermodynamics are badly needed to introduce and discuss. Therefore, this particular chapter explains the fundamentals of thermodynamics and its basic concepts by presenting the main unit systems, physical quantities and thermophysical properties, thermodynamic aspects, seven-step approach of thermodynamic assessment, balance equations, and energy and exergy efficiencies. According to Dincer [2], the science of energy and exergy can be the definition of thermodynamics, in which the first law of thermodynamics brings the energy subject, and the second law of thermodynamics brings the exergy subject as two pillars of thermodynamics. Etymologically, "thermodynamics" originated from the Greek roots *therme* and *dynamis*, which represent heat and power, respectively. In order to carry out a thermodynamic analysis, energy and exergy approaches assess the system to find heat and power inputs and outputs as well as entropy generation and exergy destruction. Balance equations for mass, energy, entropy, and exergy can be derived at the beginning of thermodynamic analysis, which helps to calculate inputs and outputs according to the first and second laws of thermodynamics. Thereafter, a performance assessment can be carried out in order to determine energy and exergy efficiencies or energy and exergy coefficient of performances. To carry out these approaches, the key fundamentals and concepts of thermodynamics will be introduced in this chapter along with some necessary definitions. Furthermore, case studies and examples will help to illustrate and strengthen the understanding of thermodynamics. In addition to that, numerous approaches will be presented to help maintain the understanding process of thermodynamic concepts.

1.2 Units, Dimensions, and Properties

Units, dimensions, and properties are the basics of thermodynamics, which need to be covered in order to be adequately prepared for studies in energy systems and renewable desalination applications. Units represent the magnitudes of dimensions, in other words, physical quantities. Some physical quantities are directly measurable with simple devices, but there are also some unmeasurable quantities which need to be calculated. Measurable quantities can be referred to as primary dimensions (i.e., temperature, pressure), and unmeasurable quantities can be referred to as secondary dimensions (i.e., enthalpy, entropy).

1.2.1 Unit Systems

The unit systems include the modern form of the metric system, known as the International System of Units (SI), the British Imperial System of Units, and the United States customary units which developed from the British Imperial System of Units. Around the globe, SI units are the most widely used unit system. The current book primarily employs the SI unit.

1.2.2 Quantities and Thermophysical Properties

This subsection defines the physical quantities and thermophysical properties, especially the ones that are important for thermodynamic analysis. Mass and volume and their flow rates form the first step of the balance equations and quantify the substance amount or substance flow either for a moment or a certain amount of time. Pressure and temperature are other quantities which can be measured with a wide variety of devices. In order to determine other thermophysical properties, pressure and temperature can be used for specific enthalpy, specific entropy, specific volume, and specific exergy calculations, unless there is no phase change. If the pressure and temperature are at the phase change condition, another parameter, such as quality, is needed to calculate other thermophysical properties.

(a) *Mass, Density, and Volume*

Mass and volume are the physical quantities and form the first step of the balance equations in order to quantify the amount of thermofluid or substance input and output to the system or boundary. When the measurement is made for a certain amount of time or unit time, mass flow rate and volumetric flow rate definitions can be made. The basic SI unit of the mass is the kilogram (kg) and the basic unit time is second (s); therefore, the basic unit for the mass flow rate is kilogram per second (kg/s). Especially when the chemical substances are involved, the unit mole (mol) is commonly used along with the molecular weight (M) and it can be expressed as follows:

$$n = \frac{m}{M} \tag{1.1}$$

where the mole number is represented with n, the molecular weight is denoted with M, and the mass is denoted with m.

The relation between volume and mass can be explained with the density which quantifies the mass per unit volume as follows:

$$\rho = \frac{m}{V} \tag{1.2}$$

where the density is denoted with ρ in the basic unit of kg/m^3 in the SI unit system.

The quantity of volume per unit mass can be expressed with the specific volume as follows:

$$\nu = \frac{V}{m} \tag{1.3}$$

where the specific volume is denoted with ν in the basic unit of m^3/kg in the SI unit system.

Note that volumetric flow rate and mass flow rate are distinct from volume and mass with a dot symbol on top of the letter and the basic units are cubic meter per second (m^3/s) for the volumetric flow rate, and kilogram per second (kg/s) for the mass flow rate. Density and specific volume can be written for the flow rates of volume and mass as follows:

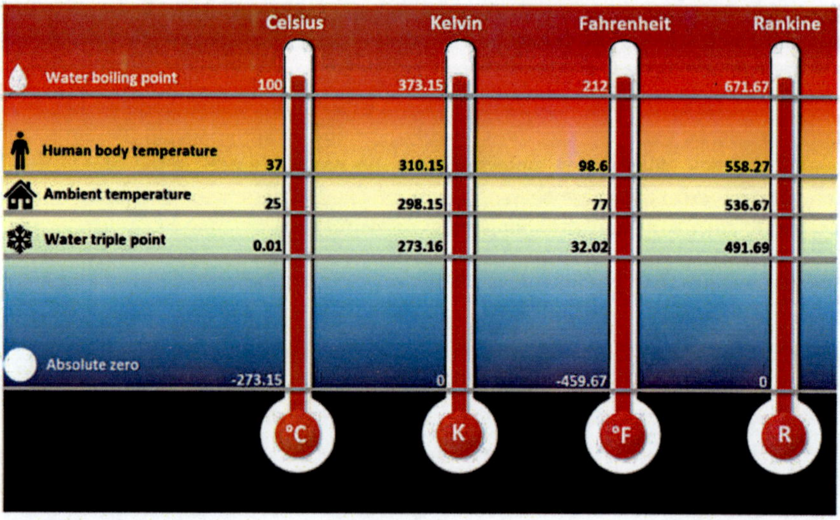

Fig. 1.1 Illustration of common temperature scales

$$\rho = \frac{\dot{m}}{\dot{V}} \tag{1.4}$$

and

$$\nu = \frac{\dot{V}}{\dot{m}} \tag{1.5}$$

where the mass flow rate is denoted with $\dot{m}$, and the volumetric flow rate is denoted with $\dot{V}$.

(b) *Temperature*

The temperature can be identified as the measure of coldness and hotness and it can be quantified with four common scales: Celsius (°C), Fahrenheit (°F), Kelvin (K), and Rankine (R). Four general temperature scales are illustrated with some of the common temperature points in Fig. 1.1.

The basic unit of temperature in the SI unit system is Celsius; however, thermodynamic analysis requires Kelvin in its calculations; therefore, Celsius and Kelvin will be commonly used in this book. Celsius and Fahrenheit scales are based on the triple point and the boiling point of the water. On the other hand, Kelvin and Rankine scales are based on the lowest possible temperature point, in other words, the absolute zero point. At the absolute zero point, molecules stop moving because of the fact that the quantification of the temperature is based on the kinetic energy of molecules.

In order to carry out a temperature conversion from one scale to another, one or more of the following relationships can be used:

$$T(\mathrm{K}) = T(^\circ\mathrm{C}) + 273.15 \tag{1.6}$$

$$T(^\circ\mathrm{F}) = 1.8\,T(^\circ\mathrm{C}) + 32 \tag{1.7}$$

$$T(\mathrm{R}) = 1.8T(\mathrm{K}) \tag{1.8}$$

$$T(\mathrm{R}) = T(^\circ\mathrm{F}) + 459.67 \tag{1.9}$$

(c) *Pressure*

Pressure is used for gases and liquids to quantify the applied force per unit area. For solids, stress can be referred to as the counterpart of pressure. Pressure can be expressed as follows:

$$P = \frac{F}{A} \tag{1.10}$$

where force is denoted with F, and unit area is denoted with A. When the mass of the body is considered, pressure can be expressed as follows:

$$P = \frac{mg}{A} \tag{1.11}$$

in which the standard acceleration due to gravity is denoted with g, and mass is denoted with m.

In the SI unit system, the unit of force is Newton (N) and the unit area is square meter (m^2); therefore, the division of force to the unit area is N/m^2 which is expressed as Pascal (Pa). There are other pressure units are also available for various applications and locations. Bar and Pa will be commonly used throughout the book. Besides these, the standard atmosphere (atm) and the pound per square inch (psi) are popular examples among pressure units. In order to convert these units into each other, Table 1.1 is given to summarize some ones for practical purposes.

There are four basic pressure measurements, namely atmospheric pressure, absolute pressure, gauge pressure, and vacuum pressure. Figure 1.2 shows the relationship between different pressure measurements.

Atmospheric pressure refers to the consideration of pressure measurement of the surrounding air on the Earth. When the atmospheric pressure is considered as the

Table 1.1 Conversion of pressure units

Unit	Atm	Bar	Psi	Pa	MPa
1 atm	1	1.01325	14.696	101325	0.101325
1 bar	1.013	1	14.5	10^5	0.1
1 psi	0.068	0.068947	1	6895	0.0068947
1 Pa	10^{-5}	10^{-5}	0.000145	1	10^{-6}

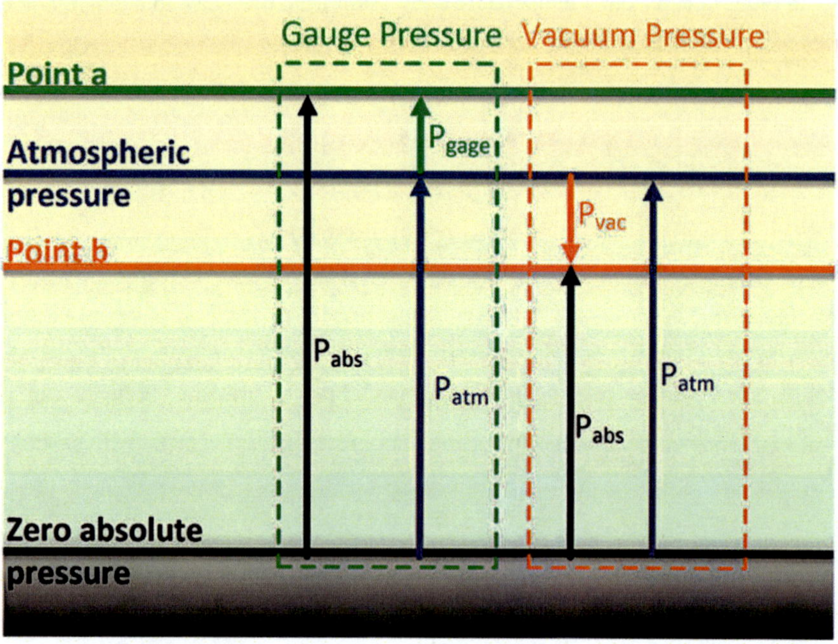

Fig. 1.2 Illustration of common temperature scales

reference, vacuum pressure and gauge pressure refer to the change from atmospheric pressure. Zero absolute pressure reference point refers to the perfect vacuum condition where there is no matter at all in the control space. According to the zero absolute pressure reference point, pressure points are measured as absolute pressure. Gauge pressure can also be considered when there is a higher pressure than the atmospheric pressure. To depict the gauge pressure, a high-pressure point (point a in Figure 1.2) can be considered. In order to measure the pressure at point a, the equation can be expressed as follows:

$$P_{\text{abs},a} = P_{\text{atm}} + P_{\text{gage}} \tag{1.12}$$

On the other hand, vacuum pressure is for the conditions when there is a lower pressure than the atmospheric pressure. According to the point b in Fig. 1.2, vacuum pressure can be expressed as follows:

$$P_{\text{abs},b} = P_{\text{atm}} - P_{\text{vac}} \tag{1.13}$$

(d) *Internal Energy*

Internal energy is the total molecular form of energy which includes molecular energies in terms of sensible, latent, chemical, nuclear, and other forms of molecular energies. Internal energy can be denoted with U in kilojoule (kJ) unit, and for the unit mass, specific internal energy can be denoted with u in kJ/kg unit. In order to formalize the total energy by considering three types of energy, internal (molecular), kinetic, and potential, it can be expressed as follows:

$$E = U + KE + PE \tag{1.14}$$

where E denotes total energy, KE denotes kinetic energy, and PE denotes potential energy. These values are actually the multiplication of specific forms of internal, kinetic, and potential energies with mass, which also can be expressed as follows:

$$E = mu + \frac{mV^2}{2} + mgz \tag{1.15}$$

where velocity is denoted by V, standard acceleration due to gravity is denoted by g which can be taken as 9.81 m/s^2, and the elevation in reference to the Earth's surface is denoted by z. Specific forms of total energy in terms of specific internal energy, specific kinetic energy, and specific potential energy can be expressed as follows:

$$e = u + ke + pe = u + \frac{V^2}{2} + gz \tag{1.16}$$

where the specific total energy is denoted by e, specific kinetic energy is denoted by ke, and specific potential energy is denoted by pe.

1.2.3 Thermodynamic Aspects

It is previously mentioned that to form a foundational understanding of energy systems and applications within the focus of desalination, understanding the concepts and basics of thermodynamics is very critical. In order to assess a system from the aspects of thermodynamics, it needs to be firstly separated from its surroundings by defining the boundaries, and then inputs, outputs, and changes in terms of thermodynamic properties need to be evaluated. A system boundary refers to the separation of the space or device from its surroundings where mass and different forms of energy can enter and leave. In order to define the types of systems, mass and energy movement through the boundary should be considered. A closed system can be defined as a system where energy can pass through the boundary, but mass is conserved. An open system refers to a system where energy and mass are allowed to pass through the boundary. A third system can be mentioned as a system where both mass and energy are conserved within the boundary.

1.2.4 Seven-Step Approach

The order of thermodynamic-based evaluation is important and needs to be followed step-by-step in order to maintain the analysis process while evaluating thermodynamic systems. A step-by-step approach for analyzing systems from thermodynamic-based energy and exergy approaches is introduced elsewhere [2], which is very helpful to facilitate the process.

Thermodynamic property can be considered as the first step of the approach. For a defined system, the entering and exiting conditions should be defined in terms of properties: mass, temperature, enthalpy, entropy, exergy, etc. [3]. Once the conditions in terms of properties are defined, the state can be formed from two or more properties. After that, a process can be formed by including at least two states. A process can be defined as a change in properties from one state to another. When there are at least two or more processes, a cycle can be formed, which means that the initial state is expected to be reached at the end of two or more processes, as depicted in Fig. 1.3.

After a cycle is formed, the first and second law analyses and performance assessments can be carried out. The fifth step can be mentioned as the first law of thermodynamics (FLT), in which mass and energy balances are evaluated in order to determine mass, heat, and power inputs and outputs to the determined boundaries. The second law of thermodynamics (SLT) follows the fifth step, which finds out the entropy generations and exergy destructions of thermodynamic systems. The final step is the performance assessment step, where energy and exergy

Fig. 1.3 Seven-step approach to studying thermodynamic systems

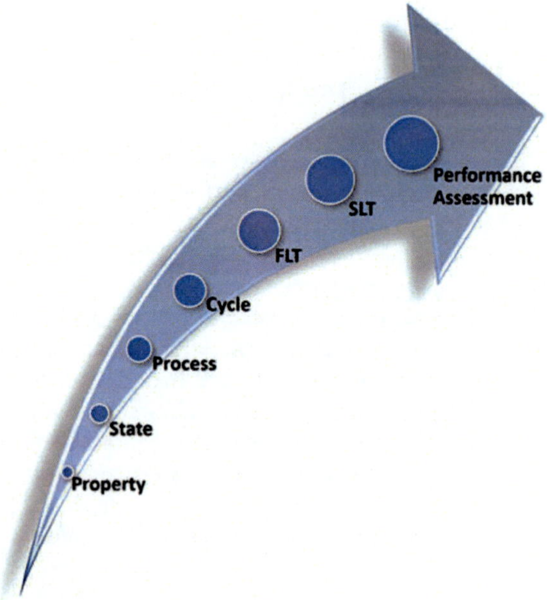

efficiencies or energy and exergy coefficient of performances can be determined. After the final step, the thermodynamic system is ready to be compared with currently available systems in terms of energy and exergy. Property is already discussed in the previous section; for this reason, the following subsections discuss the remaining six steps.

(a) *State*

State can be defined as a condition which includes two or more properties. In thermodynamics, there are also equilibrium states mentioned throughout the topics, which refer to no mechanism to create change in terms of thermal, mechanical, phase, and chemical. Thermal equilibrium refers to the state when the temperature is constant throughout accordingly and the same among the entire system. In the mechanical equilibrium, the entire system reaches the same pressure and there is no pressure change occurring. Phase equilibrium refers to the entire condition throughout the system having the same phase.

(b) *Process*

A process can be defined as the change from one state point into another one. At the end of each process, properties also change. Due to the changing and constant properties, different types of processes can be defined as follows:

- Adiabatic process: heat is conserved within the boundary.
- Isenthalpic process: enthalpy does not change during the process.
- Isentropic process: entropy does not change during the process.
- Isobaric process: pressure does not change during the process.
- Isochoric process: volume does not change during the process.
- Isothermal process: temperature does not change during the process.

(c) *Cycle*

Two or more processes can form a cycle, when the final process reaches the initial state. Commonly used power cycles, heat pump cycle, Carnot cycle, and refrigeration cycles usually consist of four or more processes. In order to form a cycle, at least two properties, two states, and two processes are needed as depicted in Fig. 1.4.

(d) *The First Law of Thermodynamics*

The FLT defines the energy conservation by stating that the energy can be converted from one form to another one; however, it cannot be created nor destroyed. In order to perform the first law of analysis, both mass and energy balances should be considered for analysis and assessment.

In order to depict the FLT, the previously mentioned Earth's energy imbalance from the introduction section can be used. When we look at the Earth's energy inputs and outputs, we can consider that a portion of the energy input is accumulated and used via the atmosphere, land, or biosphere in order to convert it in other forms of energies, such as to convert it to heat in order to keep the ambient temperature at certain levels for living organisms or convert it to chemical energy by producing

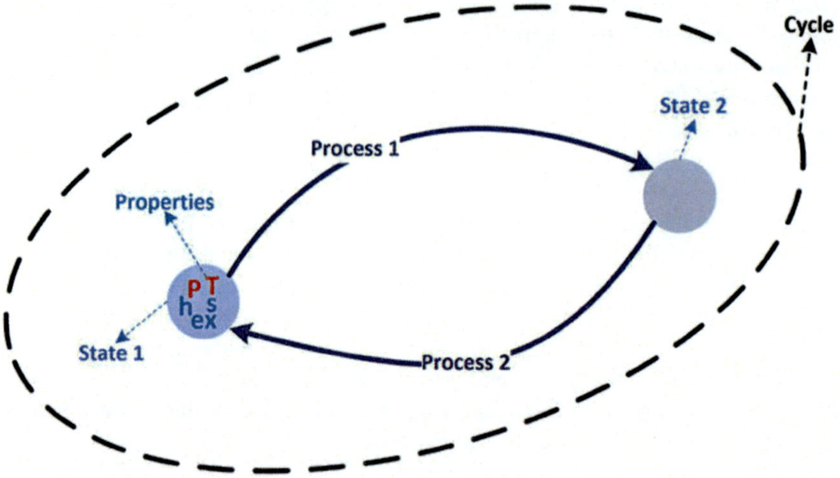

Fig. 1.4 Illustration of forming a cycle from properties, states, and processes

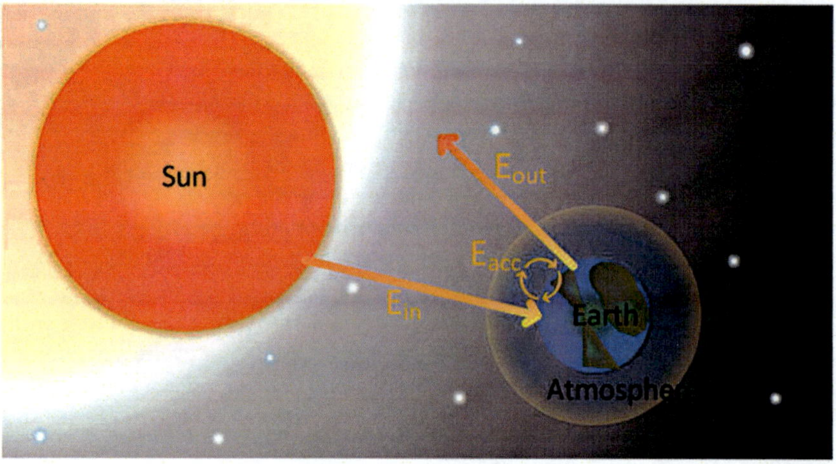

Fig. 1.5 An illustration of Earth's energy balance

glucose via the photosynthesis process. Until the industrialization, energy flow was in a balance. However, when the carbon dioxide (important heat-trapping gas) concentration in Earth's atmosphere is increased, the accumulated energy is increased, and this created imbalance which leads to climate change, global warming, extreme weather conditions, loss of species, etc.

A formulation of this macro-scale process can be expressed as the total energy input to Earth is equal to the sum of total energy released from Earth and accumulated energy as shown in Fig. 1.5. To generalize this process, it can be expressed as follows:

$$\text{Total energy input} = \text{Total energy output} + \text{Energy accumulation} \quad (1.17)$$

In order to carry out a thermodynamic analysis based on the first law of thermodynamics, the mass and energy balance equations needed to be used. For steady-state, steady flow conditions apply, general balance equations for the mass and energy balances can be expressed as follows:

$$\sum_i \dot{m}_i = \sum_o \dot{m}_o \quad (1.18)$$

$$\sum_i \dot{m}_i \left(h_i + \frac{V_i^2}{2} + gZ_i\right) + \sum_i \dot{Q}_i + \sum_i \dot{W}_i = \sum_o \dot{m}_o \left(h_o + \frac{V_o^2}{2} + gZ_o\right) + \sum_o \dot{Q}_o + \sum_o \dot{W}_o \quad (1.19)$$

where m denotes the mass, Q denotes the thermal energy, W denotes the work, h denotes the specific value of enthalpy per unit mass, V denotes the velocity, and g denotes the acceleration due to gravity.

(e) *The Second Law of Thermodynamics*

In order to carry out a more insightful analysis, the SLT analysis finds the useful energy content, or in other words, it finds the quality of the energy. The maximum amount of the available work can be quantified with exergy. In order to carry out SLT, unlike the energy analysis, a reference environment along with the reference conditions should be defined. Conversation law does not apply to exergy; therefore, exergy destruction needs to be added to the equation. A general exergy balance equation can be expressed as follows:

$$\text{Total exergy input} = \text{Total exergy output} + \text{Exergy destruction} + \text{Exergy accumulation} \quad (1.20)$$

For a certain period of time, the exergy balance equation can be written as follows:

$$\sum_i \dot{m}_i s_i + \sum_i \frac{\dot{Q}_i}{T_s} + \dot{S}_{gen} = \sum_o \dot{m}_o s_o + \sum_o \frac{\dot{Q}_o}{T_s} \quad (1.21)$$

$$\sum_i \dot{E}x_{\dot{Q}_i} + \sum_i \dot{W}_i + \sum_i \dot{m}_i ex_i = \sum_o \dot{E}x_{\dot{Q}_o} + \sum_o \dot{W}_o + \sum_o \dot{m}_o ex_o + \dot{E}x_d \quad (1.22)$$

where m denotes the mass, Q denotes the thermal energy, W denotes the work, S denotes entropy, s denotes the specific value of entropy per unit mass, Ex denotes the exergy amount, ex denotes the specific value of exergy per unit mass, h denotes the specific value of enthalpy per unit mass, V denotes the velocity and g denotes the acceleration due to gravity. In reversible processes the value of Ex_d is equal to zero, in irreversible processes it is greater than zero.

The exergy analysis is a useful tool since the energy analysis is not able to define the quality content of the energy, but the exergy analysis can provide the information of useful energy for each state point and component. Systems that are designed in an integrated fashion, have several components; therefore, the exergy analysis can provide more insightful and beneficial information about the integrated systems. During the design process, exergy analysis is the true measure to change and improve the design by checking exergy destructions and efficiencies of each component. To design the system with the exergy analysis also allows the use of less resources and generate more useful products, which are also meant to have higher exergy efficiency and lower exergy destructions or lower irreversibility.

In order to find the exergy content of the flowing stream of matter, the sum of nonflow exergy should be summed up with the stream's flow work.

$$\mathrm{Ex_{flow}} = \mathrm{Ex_{nonflow}} + (P - P_0)V \tag{1.23}$$

Or it can be written in terms of physical, chemical, kinetic, and potential exergy contents as follows:

$$\mathrm{Ex_{flow}} = \mathrm{Ex_{physical}} + \mathrm{Ex_{chemical}} + \mathrm{Ex_{kinetic}} + \mathrm{Ex_{potential}} \tag{1.24}$$

In order to find the exergy content of heat, the heat transfer across the system boundary should be considered with the reference and boundary temperatures as follows:

$$\mathrm{Ex}^{\mathrm{Q}_b} = \left(1 - \frac{T_0}{T_r}\right) Q_b \tag{1.25}$$

where b denotes the boundary.

The exergy content of the work can be considered as equal to energy as follows:

$$\mathrm{Ex}^{\mathrm{W}} = W \tag{1.26}$$

In order to find the destructed exergy content, in other words, irreversibility of the process, the following equation can be used:

$$\mathrm{Ex_d} = T_0 S_{\mathrm{gen}} \tag{1.27}$$

where S_{gen} denotes the entropy generation.

It is important to mention that the reference environment is of prime importance and must be considered in order to perform the exergy analysis properly. It is considered as an infinite system in stable equilibrium with constant temperature, pressure, and chemical potentials.

In order to carry out a thermodynamic analysis, balance equations are critical to calculate the substance, thermofluid, heat, and power inputs and outputs along with the entropy generation and exergy destruction. By considering the first and second laws of thermodynamics, the mass, energy, entropy, and exergy balance equations can be written in general form as follows:

$$\sum_{i} \dot{m}_i = \sum_{e} \dot{m}_e \tag{1.28}$$

$$\sum_{i} \dot{m}_i \left(h_i + \frac{V_i^2}{2} + gZ_i\right) + \sum_{i} \dot{Q}_i + \sum_{i} \dot{W}_i = \sum_{o} \dot{m}_o \left(h_o + \frac{V_o^2}{2} + gZ_o\right) + \sum_{o} \dot{Q}_o + \sum_{o} \dot{W}_o \tag{1.29}$$

$$\sum_{i} \dot{m}_i s_i + \sum_{i} \frac{\dot{Q}_i}{T_s} + \dot{S}_{gen} = \sum_{e} \dot{m}_e s_e + \sum_{o} \frac{\dot{Q}_o}{T_s} \tag{1.30}$$

$$\sum_{i} \dot{E}x_{\dot{Q}_i} + \sum_{i} \dot{W}_i + \sum_{i} \dot{m}_i ex_i = \sum_{o} \dot{E}x_{\dot{Q}_o} + \sum_{o} \dot{W}_o + \sum_{o} \dot{m}_o ex_o + \dot{E}x_d \tag{1.31}$$

(f) *Performance Assessment*

To finalize a thermodynamic analysis and to obtain results which can be used to compare the analyzed system with other options, a performance assessment step needs to be carried out. Comparing the systems or subsystems in terms of thermodynamic performance helps to maintain a decision-making process. One has to keep in mind that there is a strong need to follow a specific solution procedure in order to analyze a thermodynamic system. In this regard, Fig. 1.6 shows the step-by-step procedures to maintain the assessment process.

The definitions of energy and exergy efficiencies are simply the division of useful outputs to the total inputs. When the energy efficiency is calculated, useful energy outputs and total energy inputs are used. For the exergy efficiency, total exergy outputs and total exergy inputs need to be used. The energy and exergy efficiencies of a system can be expressed as follows:

$$\eta_{en} = \left(\frac{\text{Energy of useful outputs}}{\text{total energy inputs}}\right) \tag{1.32}$$

and

$$\eta_{ex} = \left(\frac{\text{Exergy of useful outputs}}{\text{total exergy inputs}}\right) \tag{1.33}$$

If a system is used to generate more than one useful output, which may be defined as cogeneration or trigeneration or quadro-generation or more than four generation systems. The efficiencies of these systems are defined as follows:

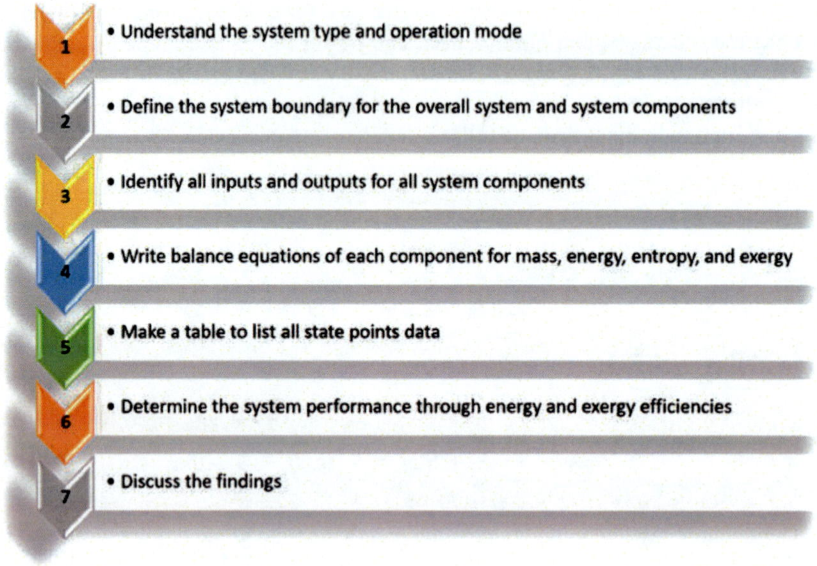

Fig. 1.6 Step-by-step procedures to solve energy systems with energy and exergy approaches

For single generation:

$$\eta_{\text{single generation}} = \frac{\text{Output}}{\text{Input}} \tag{1.34}$$

For cogeneration (double or dual generation):

$$\eta_{\text{cogeneration}} = \frac{\text{Output}_1 + \text{Output}_2}{\text{Input}} \tag{1.35}$$

For trigeneration (triple generation):

$$\eta_{\text{trigeneration}} = \frac{\text{Output}_1 + \text{Output}_2 + \text{Output}_3}{\text{Input}} \tag{1.36}$$

For quadro-generation:

$$\eta_{\text{quadgeneration}} = \frac{\text{Output}_1 + \text{Output}_2 + \text{Output}_3 + \text{Output}_4}{\text{Input}} \tag{1.37}$$

For nth generation:

$$\eta_{n-\text{generation}} = \frac{\text{Output}_1 + \text{Output}_2 + \text{Output}_3 + \text{Output}_4 + \ldots + \text{Output}_n}{\text{Input}} \tag{1.38}$$

For the specific forms of thermodynamic balance equations for mass, energy, entropy, and exergy, along with energy and exergy efficiencies for performance assessment and evaluations, Table 1.2 can be used for the commonly used components, including turbines, compressors, pumps, expansion valves, three-way valves, separators, mixing chambers, heat exchangers, open feedwater heaters, dehumidifiers, nozzles, and diffusers accordingly.

Table 1.2 Balance equations for various components

Component	Equation
Turbine 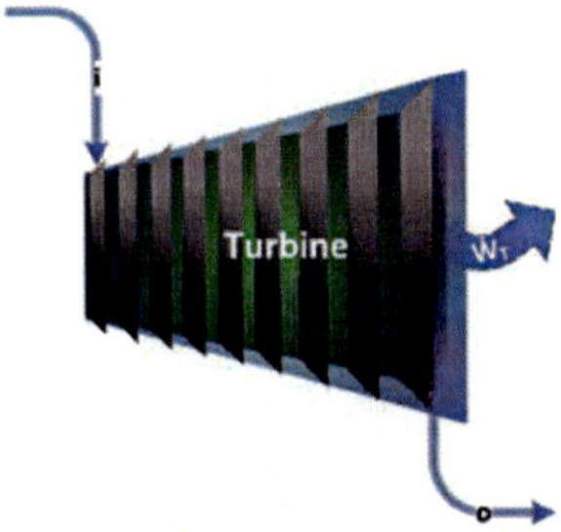	**Balance equations:** **Mass balance:** $\dot{m}_i = \dot{m}_o$ **Energy balance:** $\dot{m}_i h_i = \dot{m}_o h_o + \dot{W}_T$ **Entropy balance:** $\dot{m}_i s_i + \dot{S}_{gen,T} = \dot{m}_o s_o$ **Exergy balance:** $\dot{m}_i ex_i = \dot{m}_o ex_o + \dot{W}_T + \dot{E}x_{d,T}$ **Efficiency equations:** $\eta_{en,T} = \dot{W}_T / (\dot{m}_i h_i - \dot{m}_o h_o)$ $\eta_{ex,T} = \dot{W}_T / (\dot{m}_i ex_i - \dot{m}_o ex_o)$
Compressor 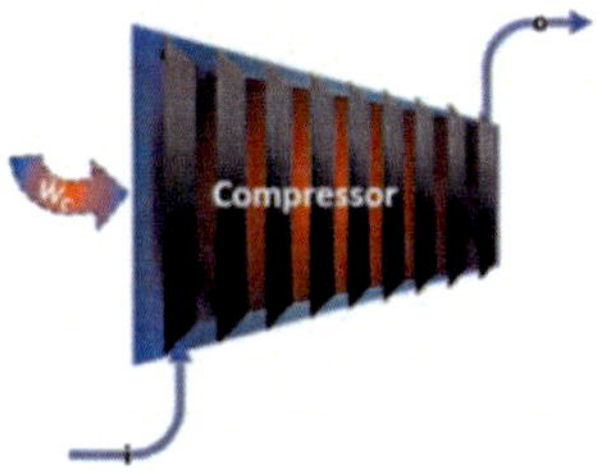	**Balance equations:** **Mass balance:** $\dot{m}_i = \dot{m}_o$ **Energy balance:** $\dot{m}_i h_i + \dot{W}_C = \dot{m}_o h_o$ **Entropy balance:** $\dot{m}_i s_i + \dot{S}_{gen,C} = \dot{m}_o s_o$ **Exergy balance:** $\dot{m}_i ex_i + \dot{W}_C = \dot{m}_o ex_o + \dot{E}x_{d,C}$ **Efficiency equations:** $\eta_{en,C} = (\dot{m}_o h_o - \dot{m}_i h_i) / \dot{W}_C$ $\eta_{ex,C} = (\dot{m}_o ex_o - \dot{m}_i ex_i) / \dot{W}_C$
Pump 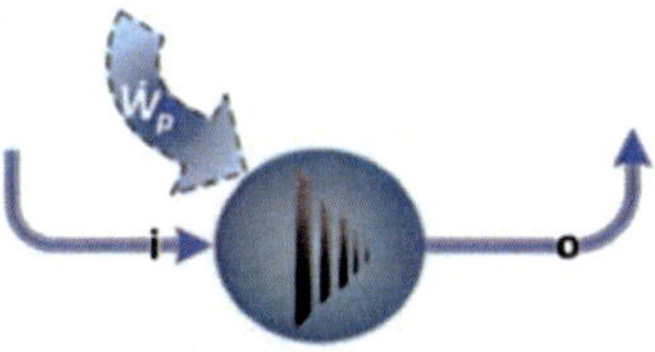	**Balance equations:** **Mass balance:** $\dot{m}_i = \dot{m}_o$ **Energy balance:** $\dot{m}_i h_i + \dot{W}_P = \dot{m}_o h_o$ **Entropy balance:** $\dot{m}_i s_i + \dot{S}_{gen,P} = \dot{m}_o s_o$ **Exergy balance:** $\dot{m}_i ex_i + \dot{W}_P = \dot{m}_o ex_o + \dot{E}x_{d,P}$ **Efficiency equations:** $\eta_{en,P} = (\dot{m}_o h_o - \dot{m}_i h_i) / \dot{W}_P$ $\eta_{ex,P} = (\dot{m}_o ex_o - \dot{m}_i ex_i) / \dot{W}_P$
Expansion valve 	**Balance equations:** **Mass balance:** $\dot{m}_i = \dot{m}_o$ **Energy balance:** $\dot{m}_i h_i = \dot{m}_o h_o$ **Entropy balance:** $\dot{m}_i s_i + \dot{S}_{gen,EV} = \dot{m}_o s_o$ **Exergy balance:** $\dot{m}_i ex_i = \dot{m}_o ex_o + \dot{E}x_{d,EV}$ **Efficiency equations:** $\eta_{en,EV} = \dot{m}_o h_o / \dot{m}_i h_i$ $\eta_{ex,EV} = \dot{m}_o ex_o / \dot{m}_i ex_i$

(continued)

Table 1.2 (continued)

Component	Equation
Three-way valve 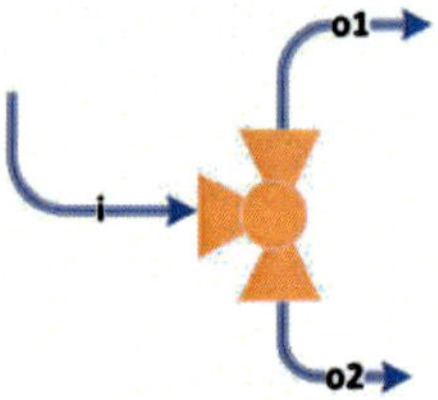 	**Balance equations:** **Mass balance:** $\dot{m}_i = \dot{m}_{o1} + \dot{m}_{o2}h_{o2}$ **Energy balance:** $\dot{m}_i h_i = \dot{m}_{o1}h_{o1} + \dot{m}_{o2}h_{o2}$ **Entropy balance:** $\dot{m}_i s_i + \dot{S}_{gen,WV} = \dot{m}_{o1}s_{o1} + \dot{m}_{o2}s_{o2}$ **Exergy balance:** $\dot{m}_i ex_i = \dot{m}_{o1}ex_{o1} + \dot{m}_{o2}ex_{o2} + \dot{E}x_{d,WV}$ **Efficiency equations:** $\eta_{en,WV} = (\dot{m}_{o1}h_{o1} + \dot{m}_{o2}h_{o2}) / \dot{m}_i h_i$ $\eta_{ex,WV} = (\dot{m}_{o1}ex_{o1} + \dot{m}_{o2}ex_{o2}) / \dot{m}_i ex_i$
Separator 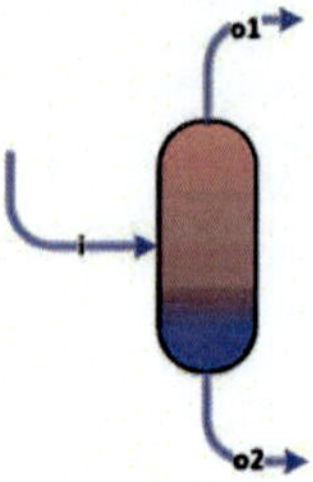 	**Balance equations:** **Mass balance:** $\dot{m}_i = \dot{m}_{o1} + \dot{m}_{o2}$ **Energy balance:** $\dot{m}_i h_i = \dot{m}_{o1}h_{o1} + \dot{m}_{o2}h_{o2}$ **Entropy balance:** $\dot{m}_i s_i + \dot{S}_{gen,Sep} = \dot{m}_{o1}s_{o1} + \dot{m}_{o2}s_{o2}$ **Exergy balance:** $\dot{m}_i ex_i = \dot{m}_{o1}ex_{o1} + \dot{m}_{o2}ex_{o2} + \dot{E}x_{d,o2Sep}$ **Efficiency equations:** $\eta_{en,Sep} = (\dot{m}_{o1}h_{o1} + \dot{m}_{o2}h_{o2}) / \dot{m}_i h_i$ $\eta_{ex,Sep} = (\dot{m}_{o1}ex_{o1} + \dot{m}_{o2}ex_{o2}) / \dot{m}_i ex_i$
Mixing chamber 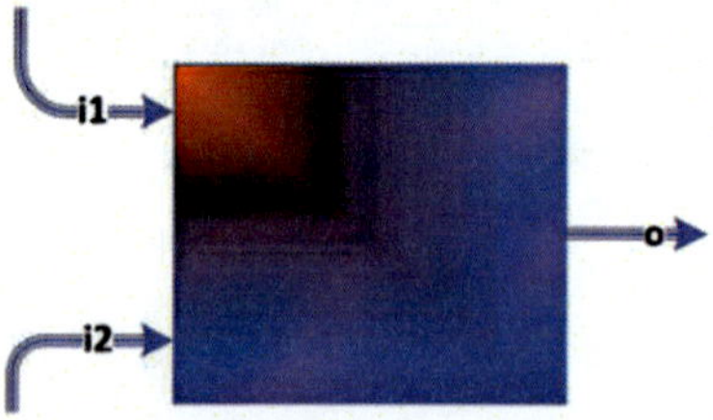 	**Balance equations:** **Mass balance:** $\dot{m}_{i1} + \dot{m}_{i2} = \dot{m}_o$ **Energy balance:** $\dot{m}_{i1}h_{i1} + \dot{m}_{i2}h_{i2} = \dot{m}_o h_o$ **Entropy balance:** $\dot{m}_{i1}s_{i1} + \dot{m}_{i2}s_{i2} + \dot{S}_{gen,MC} = \dot{m}_o s_o$ **Exergy balance:** $\dot{m}_{i1}ex_{i1} + \dot{m}_{i2}ex_{i2} = \dot{m}_o ex_o + \dot{E}x_{d,MC}$ **Efficiency equations:** $\eta_{en,MC} = \dot{m}_o h_o / (\dot{m}_{i1}h_{i1} + \dot{m}_{i2}h_{i2})$ $\eta_{ex,MC} = \dot{m}_o ex_o / (\dot{m}_{i1}ex_{i1} + \dot{m}_{i2}ex_{i2})$
Heat exchanger 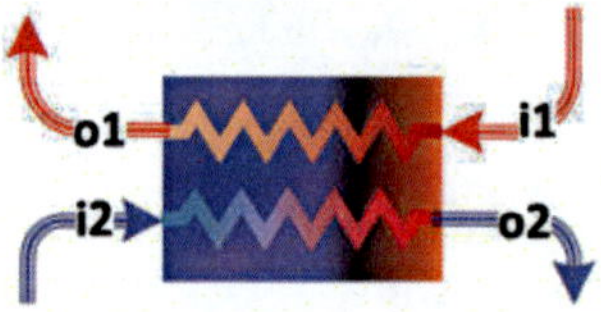 	**Balance equations:** **Mass balance:** $\dot{m}_{i1} = \dot{m}_{o1}$, $\dot{m}_{i2} = \dot{m}_{o2}$ **Energy balance:** $\dot{m}_{i1}h_{i1} + \dot{m}_{i2}h_{i2} = \dot{m}_{o1}h_{o1} + \dot{m}_{o2}h_{o2}$ **Entropy balance:** $\dot{m}_{i1}s_{i1} + \dot{m}_{i2}s_{i2} + \dot{S}_{gen,HX} = \dot{m}_{o1}s_{o1} + \dot{m}_{o2}s_{o2}$ **Exergy balance:** $\dot{m}_{i1}ex_{i1} + \dot{m}_{i2}ex_{i2} = \dot{m}_{o1}ex_{o1} + \dot{m}_{o2}ex_{o2} + \dot{E}x_{d,HX}$ **Efficiency equations:** $\eta_{en,HX} = (\dot{m}_{o1}h_{o1} + \dot{m}_{o2}h_{o2}) / (\dot{m}_{i1}h_{i1} + \dot{m}_{i2}h_{i2})$ $\eta_{ex,HX} = (\dot{m}_{o1}ex_{o1} + \dot{m}_{o2}ex_{o2}) / (\dot{m}_{i1}ex_{i1} + \dot{m}_{i2}ex_{i2})$

(continued)

Table 1.2 (continued)

Component	Equation
Open feedwater heater 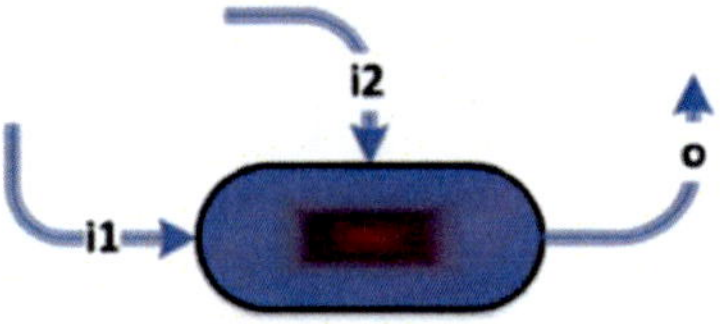	**Balance equations:** **Mass balance:** $\dot{m}_{i1}+\dot{m}_{i2}=\dot{m}_o$ **Energy balance:** $\dot{m}_{i1}h_{i1}+\dot{m}_{i2}h_{i2}=\dot{m}_oh_o$ **Entropy balance:** $\dot{m}_{i1}s_{i1}+\dot{m}_{i2}s_{i2}+\dot{S}_{gen,OFH}=\dot{m}_os_o$ **Exergy balance:** $\dot{m}_{i1}ex_{i1}+\dot{m}_{i2}ex_{i2}=\dot{m}_oex_o+\dot{E}x_{d,OFH}$ **Efficiency equations:** $\eta_{en,OFH}=\dot{m}_oh_o/(\dot{m}_{i1}h_{i1}+\dot{m}_{i2}h_{i2})$ $\eta_{ex,OFH}=\dot{m}_oex_o/(\dot{m}_{i1}ex_{i1}+\dot{m}_{i2}ex_{i2})$
Dehumidifier 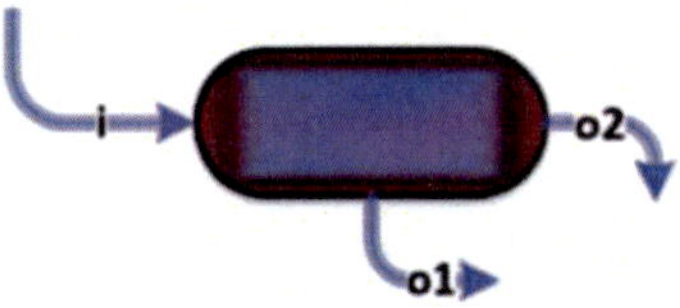	**Balance equations:** **Mass balance:** $\dot{m}_i=\dot{m}_{o1}+\dot{m}_{o2}$ **Energy balance:** $\dot{m}_ih_i=\dot{m}_{o1}h_{o1}+\dot{m}_{o2}h_{o2}$ **Entropy balance:** $\dot{m}_is_i+\dot{S}_{gen,DH}=\dot{m}_{o1}s_{o1}+\dot{m}_{o2}s_{o2}$ **Exergy balance:** $\dot{m}_iex_i=\dot{m}_{o1}ex_{o1}+\dot{m}_{o2}ex_{o2}+\dot{E}x_{d,DH}$ **Efficiency equations:** $\eta_{en,DH}=(\dot{m}_{o1}h_{o1}+\dot{m}_{o2}h_{o2})/\dot{m}_ih_i$ $\eta_{ex,DH}(\dot{m}_{o1}ex_{o1}+\dot{m}_{o2}ex_{o2})/\dot{m}_iex_i$
Reverse osmosis unit 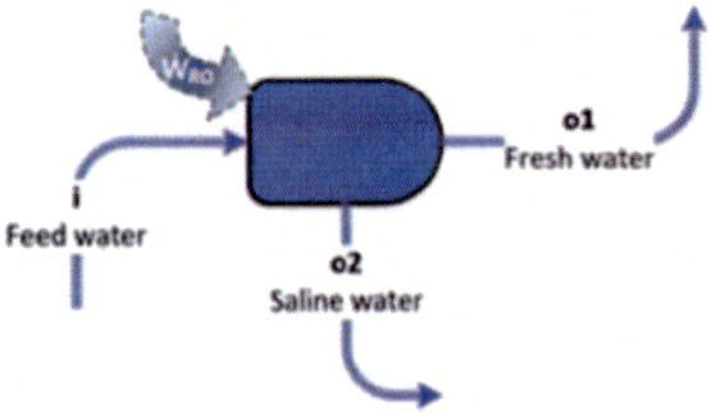	**Balance equations:** **Mass balance:** $\dot{m}_i=\dot{m}_{o1}+\dot{m}_{o2}$ **Energy balance:** $\dot{m}_ih_i+\dot{W}_{RO}=\dot{m}_{o1}h_{o1}+\dot{m}_{o2}h_{o2}$ **Entropy balance:** $\dot{m}_is_i+\dot{S}_{gen,RO}=\dot{m}_{o1}s_{o1}+\dot{m}_{o2}s_{o2}$ **Exergy balance:** $\dot{m}_iex_i+\dot{W}_{RO}=\dot{m}_{o1}ex_{o1}+\dot{m}_{o2}ex_{o2}+\dot{E}x_{d,RO}$ **Efficiency equations:** $\eta_{en,RO}=(\dot{m}_{o1}h_{o1}+\dot{m}_{o2}h_{o2}-\dot{m}_ih_i)/\dot{W}_{RO}$ $\eta_{ex,RO}=(\dot{m}_{o1}ex_{o1}+\dot{m}_{o2}ex_{o2}-\dot{m}_iex_i)/\dot{W}_{RO}$
Nozzle 	**Balance equations:** **Mass balance:** $\dot{m}_i=\dot{m}_o$ **Energy balance:** $\dot{m}_ih_i=\dot{m}_oh_o$ **Entropy balance:** $\dot{m}_is_i+\dot{S}_{gen,N}=\dot{m}_os_o$ **Exergy balance:** $\dot{m}_iex_i=\dot{m}_oex_o+\dot{E}x_{d,N}$ **Efficiency equations:** $\eta_{en,N}=\dot{m}_oh_o/\dot{m}_ih_i$ $\eta_{ex,N}=\dot{m}_oex_o/\dot{m}_iex_i$
Diffuser 	**Balance equations:** **Mass balance:** $\dot{m}_i=\dot{m}_o$ **Energy balance:** $\dot{m}_ih_i=\dot{m}_oh_o$ **Entropy balance:** $\dot{m}_is_i+\dot{S}_{gen,D}=\dot{m}_os_o$ **Exergy balance:** $\dot{m}_iex_i=\dot{m}_oex_o+\dot{E}x_{d,D}$ **Efficiency equations:** $\eta_{en,D}=\dot{m}_oh_o/\dot{m}_ih_i$ $\eta_{ex,D}=\dot{m}_oex_o/\dot{m}_iex_i$

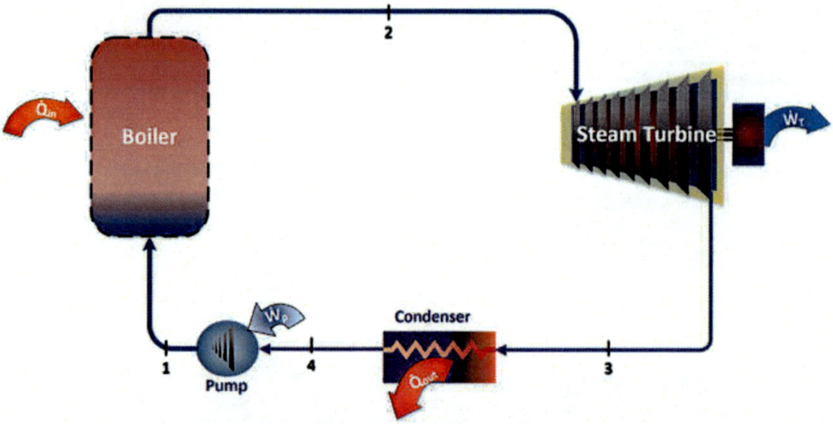

Fig. 1.7 Layout of a simple Rankine cycle power generation system

In the above sections, general balance equations for mass, energy, entropy and exergy are written and the respective energy and exergy efficiencies are listed for single-, co-, tri-, quadro-, and nth-generation options. Example 1.1 is given to strengthen the knowledge and apply step-by-step procedures.

Example 1.1 A simple Rankine cycle (as shown in Fig. 1.7) is designed to generate power by using the heat source at 950°C. Pressurized water with 16 kg/s mass flow rate at 180.6°C and 7700 kPa is heated with a boiler to generate high-temperature steam at 383.1°C. Steam enters the steam turbine, and it expands and leaves as saturated vapor at 180°C in order to generate power. The condenser rejects the heat to the surrounding until water becomes saturated liquid at state point 4.

(Take $\nu_4 = 0.001004$ m^3/kg)

(a) Illustrate and discuss the thermodynamic system.
(b) Write the mass, energy, entropy, and exergy balance equations for each component.
(c) Prepare a thermodynamic state properties table.
(d) Calculate the net power generation.
(e) Calculate the energy and exergy efficiencies.
(f) Discuss the findings and suggest improvements.

Solution 1.1
It is really important here to follow a specific solutions methodology as given in Fig. 1.6. In this regard, the system diagram is drawn to illustrate all inputs and outputs.

(a) The configuration in Fig. 1.7 shows the thermodynamic system layout for power generation. Boiler uses 950°C heat source and transfers the heat by generating superheated steam to the desired point. This power generation system leads to losses since the excess heat from the condenser is wasted and released into the atmosphere.

(b) Pump component is important to pressurize the working fluid from state point 4 to state point 1, by using power in the system. The mass, energy, entropy, and exergy balance equations of the pump can be written as follows:

$$\dot{m}_4 = \dot{m}_1$$

$$\dot{m}_4 h_4 + \dot{W}_P = \dot{m}_1 h_1$$

$$\dot{m}_4 s_4 + \dot{S}_{gen,P} = \dot{m}_1 s_1$$

$$\dot{m}_4 ex_4 + \dot{W}_{P1} = \dot{m}_1 ex_1 + \dot{E}x_{d,P}$$

After pressurized via pump, working fluid enters the boiler. Boiler superheats the water to the intended levels. Pressurized water enters concentrated solar plant to get superheated before entering steam turbine. The mass, energy, entropy, and exergy balance equations of boiler can be written as follows:

$$\dot{m}_1 = \dot{m}_2$$

$$\dot{m}_1 h_1 + \dot{Q}_{in} = \dot{m}_2 h_2$$

$$\dot{m}_1 s_1 + \frac{\dot{Q}_{in}}{T_{b,boiler}} + \dot{S}_{gen,boiler} = \dot{m}_2 s_2$$

$$\dot{m}_1 ex_1 + \dot{E}x_{Q_{in}} = \dot{m}_2 ex_2 + \dot{E}x_{d,boiler}$$

The power is generated in the turbine by expanding the superheated working fluid. The mass, energy, entropy, and exergy balance equations of the turbine can be written as follows:

$$\dot{m}_2 = \dot{m}_3$$

$$\dot{m}_2 h_2 = \dot{m}_3 h_3 + \dot{W}_T$$

$$\dot{m}_2 s_2 + \dot{S}_{gen,T} = \dot{m}_3 s_3$$

$$\dot{m}_2 ex_2 = \dot{m}_3 ex_3 + \dot{W}_T + \dot{E}x_{dest,T}$$

The condenser releases the excess heat to the surrounding ambient. The mass, energy, entropy, and exergy balance equations of the condenser can be written as follows:

$$\dot{m}_3 = \dot{m}_4$$

$$\dot{m}_3 h_3 = \dot{m}_4 h_4 + \dot{Q}_{out}$$

$$\dot{m}_3 s_3 + \dot{S}_{gen,condenser} = \dot{m}_4 s_4 + \frac{\dot{Q}_{out}}{T_{b,condenser}}$$

$$\dot{m}_3 ex_3 = \dot{m}_4 ex_4 + \dot{E}x_{Q_{out}} + \dot{E}x_{d,condenser}$$

(c) By using the given parameters and properties, thermophysical properties can be calculated, namely enthalpy, entropy, and exergy by using an Engineering Equation Solver (EES) [4]. Pump work can be found as follows:

$$w_P = \nu_4(P_1 - P_4) = 0.001004\ \frac{m^3}{kg}(7700\ kPa - 1000\ kPa) = 6.73\ \frac{kJ}{kg}$$

$$h_1 = h_4 + w_P$$

$$h_1 = 762.5 + 6.7 = 769.2\frac{kJ}{kg}$$

$$\left.\begin{array}{l} h_1 = 769.2\frac{kJ}{kg} \\ P_1 = 7700\ kPa \end{array}\right\} \begin{array}{l} T_1 = 180.6\,^\circ\mathrm{C} \\ s_2 = 2.13\ \frac{\mathrm{kJ}}{\mathrm{kg\ K}} \end{array}$$

$$P_2 = P_1 = 7700\ kPa$$

$$\left.\begin{array}{l} T_2 = 383.1\,^\circ C \\ P_2 = 7700\ kPa \end{array}\right\} \begin{array}{l} h_2 = 3098\ \frac{\mathrm{kJ}}{\mathrm{kg}} \\ s_2 = 6.56\frac{\mathrm{kJ}}{\mathrm{kg\ K}} \end{array}$$

(d) At state point 3, water is indicated as saturated vapor therefore, quality will be taken as $x_3 = 1$,

$$\left.\begin{array}{l} T_3 = 180\,^\circ\mathrm{C} \\ x_4 = 1 \end{array}\right\} \begin{array}{l} \mathrm{h}_3 = 2777\ \frac{\mathrm{kJ}}{\mathrm{kg}} \\ s_3 = 6.56\ \frac{\mathrm{kJ}}{\mathrm{kg\ K}} \end{array}$$

At state point 4, water is indicated as saturated liquid therefore, quality will be taken as $x_4 = 0$,

$$\left.\begin{array}{l} T_4 = 180\,^\circ\mathrm{C} \\ x_4 = 0 \end{array}\right\} \begin{array}{l} \mathrm{h}_4 = 762.5\ \frac{\mathrm{kJ}}{\mathrm{kg}} \\ s_4 = 2.13\ \frac{\mathrm{kJ}}{\mathrm{kg\ K}} \end{array}$$

In order to find exergy values, the following equation can be used:

$$\mathrm{ex_n = h_n - h_0 - T_0(s_n - s_0)}$$

If the reference state is considered as 25°C and 101.3 kPa, enthalpy and entropy values of water can be found as:

$$\left.\begin{array}{l} T_0 = 25\,^\circ\mathrm{C} \\ P_0 = 101.3\ kPa \end{array}\right\} \begin{array}{l} \mathrm{h_0} = 104.9\ \frac{\mathrm{kJ}}{\mathrm{kg}} \\ s_0 = 0.3672\ \frac{\mathrm{kJ}}{\mathrm{kg\,K}} \end{array}$$

Therefore, a thermophysical properties table (see Table 1.3) can be created as follows:

To find the power generation, energy balance equation of the turbine can be used as follows:

Table 1.3 Thermophysical properties of substances at each state point

State point	Substance	ṁ (kg/s)	T (°C)	P (kPa)	h (kJ/kg)	ex (kJ/kg)
1	Water	16	180.6	7700	769.2	138.7
2	Water	16	383.1	7700	3098	1218.3
3	Water	16	180	1000	2777	825.7
4	Water	16	180	1000	762.5	132

$$\dot{\mathrm{m}}_2\mathrm{h}_2 = \dot{\mathrm{m}}_1\mathrm{h}_1 + \dot{\mathrm{W}}_\mathrm{T}$$

$$\dot{\mathrm{m}}_2 = \dot{\mathrm{m}}_1$$

and the work output for the steam turbine can be found as follows:

$$\dot{\mathrm{W}}_\mathrm{T} = \dot{\mathrm{m}}_2(\mathrm{h}_2 - \mathrm{h}_1)$$

$$\dot{\mathrm{W}}_\mathrm{T} = 5136\frac{\mathrm{kJ}}{\mathrm{s}} = 5136\ kW$$

The net power generation can be calculated by subtracting pump work from the generated work as follows:

$$\dot{\mathrm{W}}_\mathrm{net} = \dot{\mathrm{W}}_{\mathrm{T}_1} - \dot{\mathrm{W}}_{\mathrm{P}_1}$$

where $\dot{\mathrm{W}}_{\mathrm{p1}}$ can be found by multiplying $\mathrm{w}_{\mathrm{P}_1}$ with mass flow rate as follows:

$$\dot{\mathrm{W}}_{\mathrm{P}_1} = \dot{\mathrm{m}}_1\mathrm{w}_{\mathrm{P}_1}$$

$$\dot{\mathrm{W}}_{\mathrm{P}_1} = \dot{\mathrm{m}}_1(\mathrm{h}_4 - \mathrm{h}_1)$$

$$\dot{\mathrm{W}}_{\mathrm{P}_1} = 107.2\frac{kJ}{s} = 107.2\ kW$$

The net electricity generation can be found as follows:

$$\dot{W}_{net} = 7715.5\frac{kJ}{s} = 5028.8\ kW$$

The heat input is calculated as follows:

$$\dot{Q}_{in} = \dot{m}_1(h_2 - h_1)$$

$$\dot{Q}_{in} = 37260.8\ kW$$

$$\dot{Ex}_{Q_{in}} = \dot{Q}_{in}\left(1 - \frac{T_0}{T_{sun}}\right) = 25566.8\ kW$$

The energy efficiency of a simple Rankine cycle system can be calculated by dividing net electricity generation of the steam Rankine cycle to total energy input as follows:

$$\eta_{en,\text{single generation}} = \frac{\dot{W}_{net}}{\dot{Q}_{in}}$$

and obtained as

$$\eta_{en,\text{single generation}} = 13.5\%$$

The exergy efficiency of simple Rankine cycle system can be calculated by dividing the net electricity generation of the steam Rankine cycle to total energy input as follows:

$$\eta_{ex,\text{single generation}} = \frac{\dot{W}_{net}}{\dot{Ex}_{Q_{in}}}$$

and found as

$$\eta_{ex,\text{single generation}} = 19.67\%$$

(e) A simple Rankine cycle assessed where the energy efficiency is found as 13.5%, and the exergy efficiency is found as 19.67%. The energy and exergy efficiencies are under the traditional power plant efficiencies. When we look at the steam turbine's exit pressure, it is quite high compared to the industry standards. The energy and exergy efficincies can be improved by reducing the steam outlet pressure or by adding a low-pressure turbine to make the cycle two stages. Moreover, additional regenerative systems or integrated systems can be used to improve the efficiencies further.

1.3 Closing Remarks

In order to design or analyze or develop any energy systems for power generation and desalination applications, a clear understanding of thermodynamic concepts, in terms of energy, and exergy, is very crucial. This particular chapter has explained the basics of thermodynamics from energy and exergy aspects, as well as presented the definitions of physical quantities, thermophysical properties. For thermodynamic analysis, a seven-step approach has been introduced extensively along with mass, energy, entropy, and exergy balance equations for commonly used components of energy systems. An illustrative example for a Rankine cycle has been included in order to practice the seven-step approach for thermodynamic assessment.

References

1. Lee H, Calvin K, Dasguta D, et al. (2023) Synthesis report of the IPCC sixth assessment report climate change 2023 (AR6). Intergovernmental Panel on Climate Change
2. Dincer I (2021) Thermodynamics: a smart approach. Wiley, Chichester
3. Dincer I, Rosen MA (2020) Exergy: energy, environment and sustainable development, 3rd edn. Elsevier Ltd. https://doi.org/10.1016/C2016-0-02067-3
4. EES (2023) Engineering Equation Solver v10.643 I F-Chart Software: Engineering Software

Chapter 2
Energy, Environment, and Sustainable Development

Nomenclature

COVID-19	Coronavirus disease 2019
GDP	gross domestic product
ICT	information and communication technologies
LCOE	levelized cost of electricity
T	temperature (°C)

2.1 Introduction

After the World War II, there has been an exponential growth in many necessities, ranging from foods to goods, particularly in energy, which has majorly been driven by the growth in population, improvement in overall life quality, technologization, industrialization, digitalization, and urbanization. Micro- to macro-scale activities have become energy-hungry processes, especially after industrialization and technologization. When we look at daily life activities, such as swiping on social media, tweeting, reading news, and streaming high-definition video, these are all data-intensive services which recently boomed energy consumption thanks to digitalization. Information and communication technologies are expected to become a major sector for energy consumption, which will represent a quarter of the global energy demand by 2030, as outlined elsewhere [1].

The key events have, on the other hand, shaped the dimensions of energy and changed the major energy sources. In a perspective article [2], it is summarized that the journey to find energy sources with a carbon/hydrogen ratio and mentioned that the COVID-19 coronavirus pandemic is closing the carbon age but opening hydrogen age. Since the control of fire by early humans, wood can be considered as the first energy source for humankind. Throughout history, wood has been replaced with

I. Dincer, M. Temiz, *Renewable Energy Options for Power Generation and Desalination*, https://doi.org/10.1007/978-3-031-53437-9_2

other energy sources which dominates the energy market globally. Coal took over the domination of wood with industrialization and the coal era started. After World War I, oil dominated the global energy market. Thereafter, natural gas started to dominate the global energy mix, especially after the oil crisis in 1973, when natural gas was used for both electricity generation and space heating to facilitate the oil crisis.

2.2 Energy

In order to maintain life, energy is fundamental. Historically, growth indications of humanity have similar patterns with energy demand or energy consumption. Population growth and economic growth require more and more energy to maintain the growing process. In order to depict the growing paces of economy, population, and energy consumption, Fig. 2.1 shows the growing trends of the global population, gross domestic product (GDP), and fossil fuel consumption.

After the industrialization, fossil fuels became the major source of energy for all kinds of energy-required activities, such as transportation, electricity generation, space heating, and industrial processes. When we look at the current energy figures, fossil fuels are still dominating the energy sector. However, there is a significant growth in renewable energy sources. Especially after the oil crisis, alternative energy sources have attracted attention. Apart from the energy security and costs, the realization of the negative impacts of greenhouse gas emissions also accelerated the renewable energy implementation. From a macro-perspective, the carbon content of fuels as energy sources tends to decrease, as mentioned in a perspective article [2]. Historically, wood was replaced by coal, and then oil took over the dominance, thereafter, natural gas became the dominant energy source. New fuel era starts with

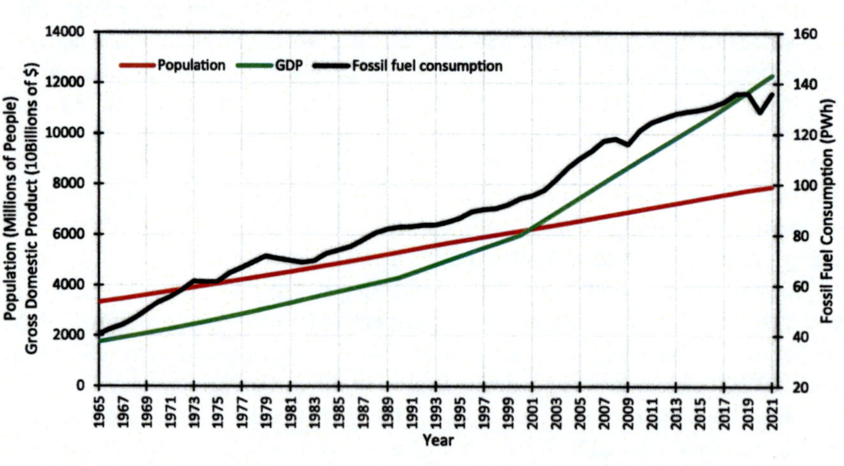

Fig. 2.1 Trends of global population, gross domestic product, and fossil fuel consumption (data from Ref. [3])

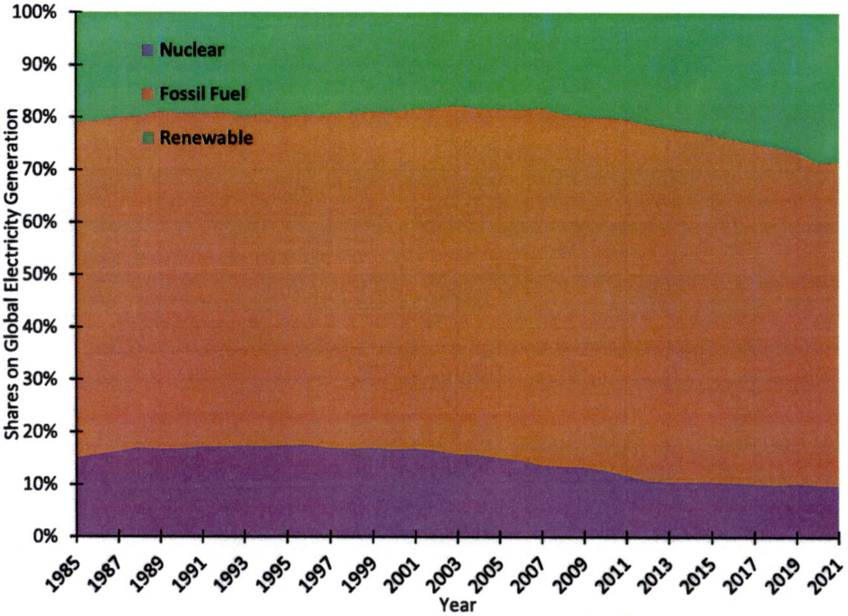

Fig. 2.2 Shares of primary energy sources for global electricity generation (data from Ref. [3])

hydrogen's dominance. In order to get rid of the carbon dioxide emissions, the hydrogen generation process should be carbon-free. Therefore, energy sources to produce hydrogen should be either renewable or nuclear. Other than fuel domination, primary energy sources need to be carbon-free in order to minimize carbon emissions. Global governments and organizations realized the importance of the transformation of energy infrastructures from fossil fuel–based energy infrastructure into renewable and carbon-free-based energy infrastructure.

Figure 2.2 shows the shares of global primary energy sources for electricity generation. Renewable electricity generation represented 20% of global electricity generation 10 years ago; however, it increased to 30% in recent years, with especially accelerated growth in wind and solar. In the next subsection, different forms of energy will be discussed and presented.

When we look at the energy sources, four main categories can be mentioned as fossil fuels, nuclear, renewables, as well as others. Currently, fossil fuels are the dominant energy source, globally, as shown in Fig. 2.2. Fossil fuels replaced wood and whale oil as the dominant energy source. During the first initiation of fossil fuels as a primary energy source, there was a misperception that the petroleum is an environmentally beneficial replacement of whale oil since it suppresses whale hunting and save whales. However, fossil fuel–driven ships were more advanced to hunt whales; therefore, the initiation of petroleum even led more whale hunting [4]. Since the initiation of petroleum, energy usage patterns have changed due to the

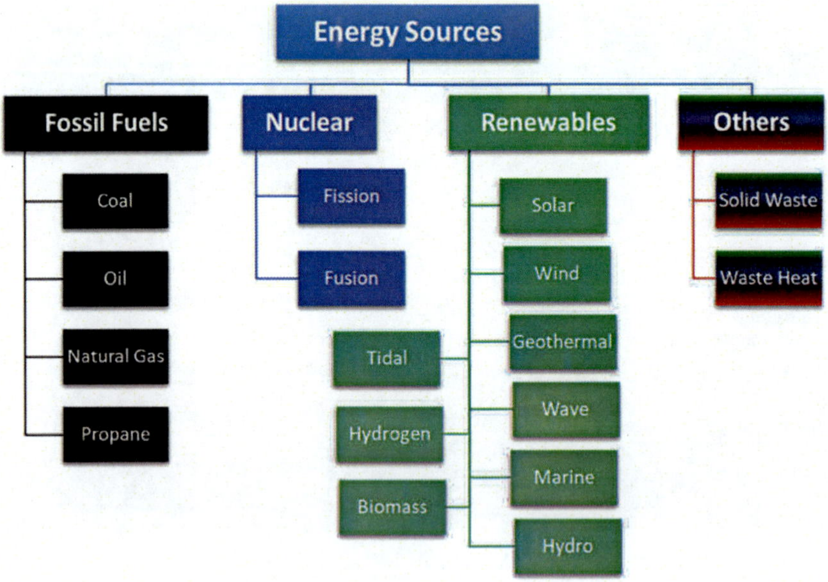

Fig. 2.3 Types of primary energy sources

advancement in numerous areas. Nuclear and renewable are two zero-carbon alternatives to fossil fuels.

Other than the climatic impacts, the development of these energy sources is critical in order to ensure the energy security. Fossil fuel–driven thermoelectric power plants in coastal areas are at risk due to rising sea levels. On the other hand, rising temperatures cause increasing losses in power transmission. On top of these, the growing demand for energy brings extra vulnerabilities to the energy sector [5]. The global energy sector changed the direction for the decarbonization of the entire energy sector, and the global energy transition started. Governments and global organizations brought strict policies and determined strategies to mitigate global carbon dioxide emissions. United Nations Framework Convention on Climate Change established environmental treaties and governments agreed to stabilize greenhouse gas concentrations to limit global temperature anomaly at 2°C [6–8]. From this aspect, the global energy transition focuses on the replacement of fossil fuel–based energy infrastructure with renewable and/or nuclear-based energy infrastructure, which are presented in Fig. 2.3.

Global energy transition is also recognized as a macro-scale engineering problem with many challenges. Even the nuclear and renewable energy sources are environmentally benign solutions; however, these are not the silver bullet to solve the entire transition problem. The intermittency of renewables is a major challenge of this transition. Availability of renewable resources is changing throughout the time since sun is not always visible in the sky and the wind is not always blowing. The continuity of the energy supply possesses critical importance. Residential areas, commercial spaces, public spaces, industrial or military areas, and hospitals are built to continuously consume energy. Therefore, an intermittent energy resource cannot be the single solution for a continuous energy supply. On the other hand, nuclear-driven energy systems can supply energy in a continuous manner; however, the flexibility is the key issue in that case since the energy demand is changing over time. Therefore, when the capacity is higher or lower, more flexible energy systems need to compensate for the changing demand. Apart from the availability and flexibility, economic reasons are also possessing important challenges. Especially at the beginning of this century, the costs of renewable energy systems were not able to compete against fossil fuel–driven energy systems.

In order to overcome challenges in the energy sector, future generation energy systems need to be smart energy systems [9]. Energy demand is exponentially growing, in order to meet the growing demand, sustainable fashion needs to be at the center of everything, in other words, sources need to be used without compromising future generations' sources as well as by keeping the environment, economy, ecology, ethics, education, and engineering (6E) in a good form for future generations. It is needed that smart energy systems should meet key expectations, as shown in Fig. 2.4.

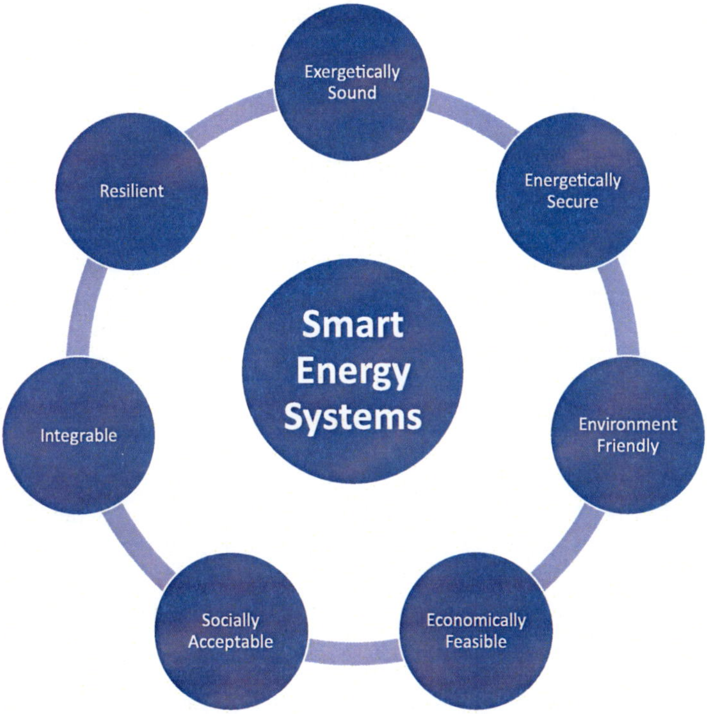

Fig. 2.4 Key expectations from smart energy systems

2.2.1 Renewables

It is widely agreed that the global energy transition is a necessity and needs to be urgently done. However, this transition requires unprecedented efforts with massive installations, infrastructure changes, and integrations never seen before. The transformation of the global energy sector is not going to be similar to the recent advancements and transformations in sectors such as biotechnology or information and communication technologies (ICT). Even the recent advancements in ICT have a tremendous influence on daily practices; there is no big physical change can be observed. There are new giant companies that have started operations in different sectors, without having any physical appearance; for instance, the largest taxi service provider is currently a technology company and owns no taxi. Similarly the largest movie and television series provider owns no physical movie theatres, but instead they use or organize currently available infrastructure or components via users to provide their services. On the other hand, the transition of the global energy infrastructure needs physical changes, which take time to build and install.

Among renewables, hydro energy started in the twenty-first century as the dominant renewable energy source. However, the growth of hydro energy has been 80% since the beginning of twenty-first century. On the other hand, massive growth is observed in wind and solar. Especially solar-based electricity generation systems have shown tremendous growth with 85,792%, where 1227.5 MWp of installed capacity in 2000 increased to 1.05 TWp at the end of 2022. Wind showed the second-largest growth and became the third-largest renewable electricity generation source. Hydro still has the largest electricity generation capacity among renewables, with 1.26 TWp. Bioenergy, geothermal, and marine are other contributors to global renewable electricity generation. Bioenergy has 148.91 GWp of installed capacities at the end of 2022. Geothermal was also showed a similar growth with hydro energy as of 2022, where 8.24 GWp of installed geothermal capacity in 2000, reached 14.88 GWp in 2022. Marine-based power generation systems have 523 MWp capacity as of 2022. In this regard, the trendlines of different types renewable energy sources, including hydro, marine, wind, solar, bioenergy, and geothermal for the period 2000–2022, are shown in Fig. 2.5. It is evident that solar and wind have been two front runners with exponential increases, providing that future increases are expected to be even more appealing.

(a) *Solar Energy*

Solar energy is the essential energy source for the species on the Earth. The fundamental energy source for primary food production of plants, which humans and animals need, also known as photosynthesis, is maintained by solar energy [11]. In order to become habitable, solar energy is key, for this reason, anywhere on Earth has solar energy at different intensities, as shown in Fig. 2.6. Solar energy exploitation by humankind started thousands of years ago. Ancient people engineered their homes by building the main façade facing south (for the northern hemisphere) in order to collect solar radiation for heating their homes.

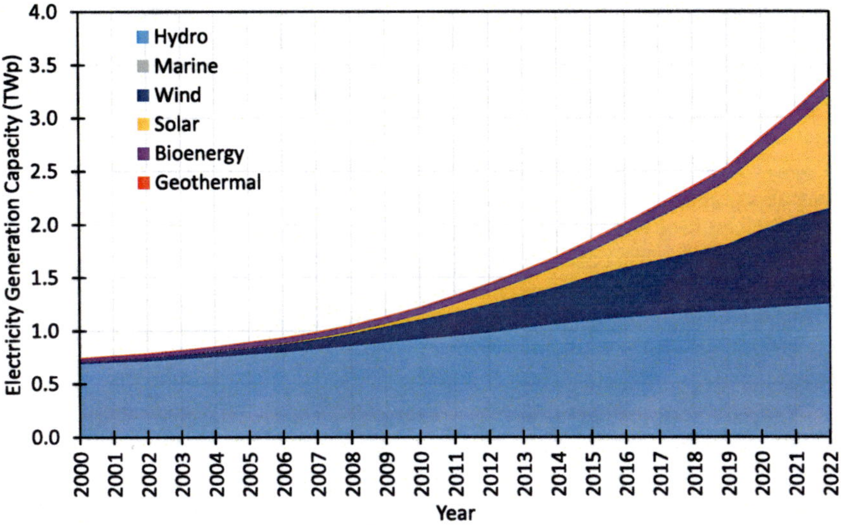

Fig. 2.5 Global renewable energy–driven electricity generation capacities (data from Ref. [10])

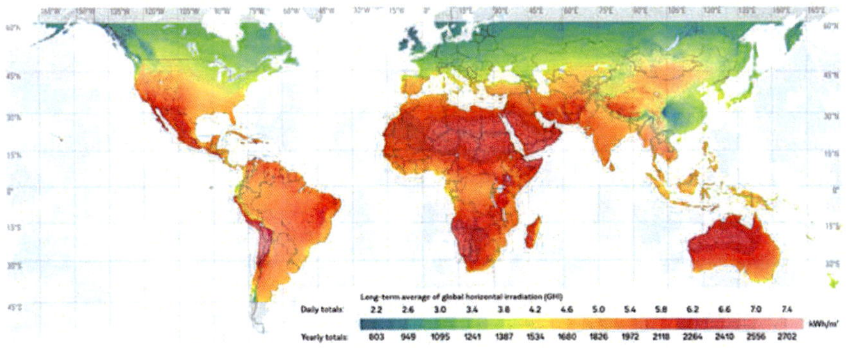

Fig. 2.6 Global horizontal irradiation map (modified from Ref. [12])

The implementation of solar energy for power generation initially started with the National Aeronautics and Space Administration's (NASA) Vanguard 1 satellite [13]. Vanguard 1 had 1 Wp of solar photovoltaic (PV) cells, which generated electricity for 6 years of operation. The cost-effectiveness of solar PVs in the 1960s was far behind today's levels. Because of the high cost, solar PV was used mainly for space operations up until 1973 and the oil crisis. With rooftop and telecom installments, solar PV technology attracted investments, and prices were dropped from US$32/Wp in 1980 to US$9/Wp in 1990. As of June 2023, solar PV module prices became less than US$0.5/Wp, which makes the technology as highly cost-effective.

Solar energy systems are implemented widely from the Arctic [14] to space [15]. Even in the less favorable areas in terms of solar energy availability, solar energy systems became attractive for investors with dropping costs. In 2018, annual solar energy system capacity additions surpassed 100 GWp. Only in 4 years, annual solar energy system net capacity additions almost doubled and reached 191.6 GWp in 2022. At the end of 2022, the global operational solar energy system power generation capacity reached 1.05 TWp [10].

Solar-based energy systems are mostly dominated by PV plants due to their better cost-effectiveness than concentrated solar power systems. Therefore, the vast majority of solar power generation systems are using solar PV technology. While 99.4% of solar power generation systems, in terms of capacity, are solar PV, 0.6% of solar power generation systems with 6.5 GWp of capacity are using concentrated solar power technology.

(b) *Wind Energy*

Wind is also one of the first energy sources that humankind exploited for their various activities. Thousands of years ago, mobility and agricultural activities were driven by wind energy. Wind is a kinetic type of energy, which is driven by solar energy since forming wind requires heating. Lands are getting heated faster than water; as proof of that, global land and sea temperature trends can be checked in Sect. 2.3. Hotter air on land starts expanding, loosens, and density becomes lower; therefore, it starts rising. On the other hand, cooler air on water is heavier and stays at lower levels of atmosphere. When lighter and hotter air on land starts rising, cooler and heavier air on sea rushes to take place of the air on land. This movement creates wind. However, during the nighttime, lands and water surfaces start cooling. Lands are also getting cooled faster, as well as air on lands. For this reason, a reverse movement starts since air on land becomes heavier and air on sea surfaces becomes lighter and rises. When we look at the movement mechanism, we can easily say wind is renewed each day.

One of the US presidents (the 16th president), Abraham Lincoln, had mentioned, "The wind is an untamed, and unharnessed force; and quite possibly one of the greatest discoveries hereafter to be made, will be taming, and harnessing of it," which is true when we look at this quote with data. Globally, more than 25.3 PWh of electricity was consumed in 2021 [17]. When we look at the available wind potential on Earth, 615 PWh of power potential can be mentioned, which is enough power to supply electricity 24.3 times global electricity consumption [18, 19].

Historically, mobility and agricultural activities can be mentioned as the first usage areas of wind. Ancient humans started to exploit wind for sailing more than 6000 years ago. In 200 BC, simple water pumps are powered by wind energy in China. When humans culturally transformed from hunting and gathering societies into agricultural societies, requirements for processing crops emerged. Therefore, wind energy is used via windmills to drive the grain grinding process in the Middle East. Windmills are used for power generation, advanced by the time, and formed wind turbines today.

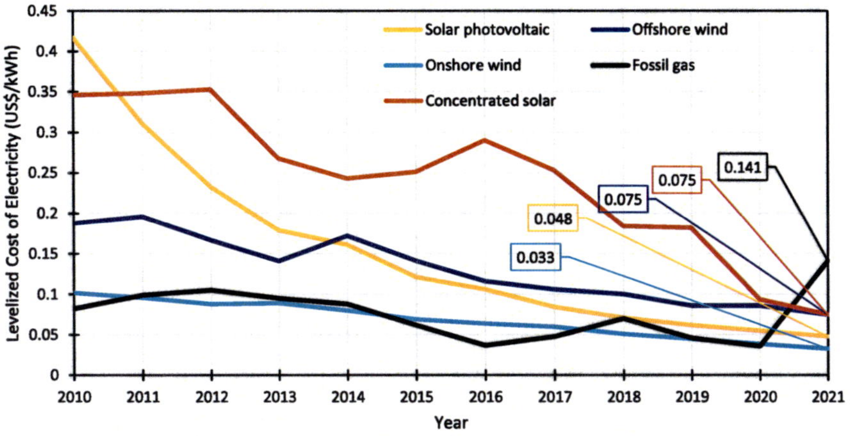

Fig. 2.7 Global levelized cost of electricity of different renewable and fossil gas–driven power generation methods (data from Ref. [16])

Currently (in 2023), wind turbines have the third largest power generation capacity among renewables after hydro and solar. Wind farms account for 898.8 GWp of power generation capacity, which is majorly formed by onshore wind farms with 835.6 GWp of power generation capacity. In 2020, more than 100 GWp of capacity additions are made annually for wind energy; however, it slowed down in 2021 and 2022, with 92.5 GWp and 74.6 GWp of annual capacity additions, respectively.

Onshore wind is the best-performing technology in terms of levelized cost of electricity (LCOE). Figure 2.7 shows the LCOE values of solar and wind technologies globally, along with fossil gas–driven power generation in Europe. Fossil gas–driven power generation was the cheapest option; however, after the fuel costs, it became one of the most expensive technologies for power generation. Onshore wind has significantly lower LCOE among power generation technologies, which makes it a popular choice for investors and decision-makers. However, a fast decline in solar PV technology costs limits the growth of wind power. Solar PVs are more flexible in terms of capacities, for this reason, small-scale investments can go through to build cost-effective solar PV plants with higher rate of returns than the minimum acceptable rate of returns. On the other hand, wind farm investments need to be at larger capacities in order to be cost-effective, which is one of the main reasons behind the solar PV's faster growth than wind farms.

(c) *Geothermal Energy*

Geothermal energy means Earth heat, which is a renewable thermal energy source that comes from the underground. It has one of the highest capacity factors among renewables. When the intermittency is a key issue for renewable technologies, especially for wind and solar, geothermal energy is continuously available to generate useful output at the required form. As shown in Fig. 2.8, it is a cost-

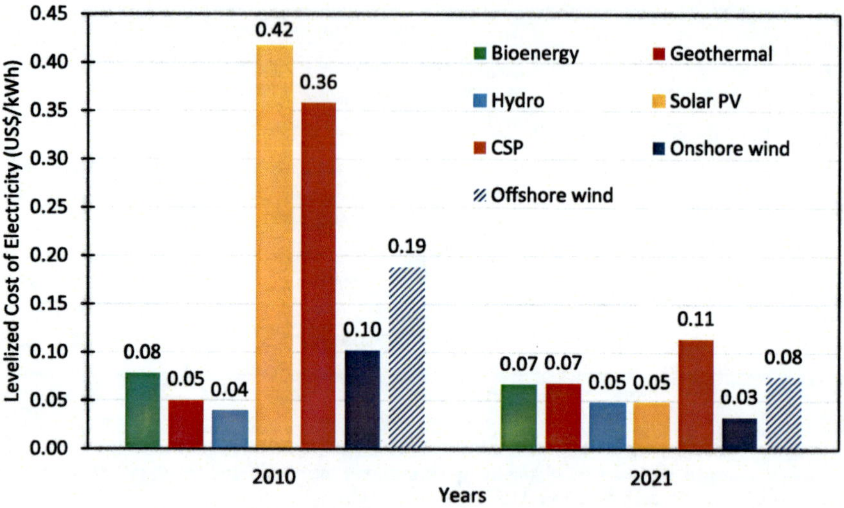

Fig. 2.8 Global levelized cost of electricity of different renewables (data from Ref. [16])

competitive source with a US$0.068/kWh of LCOE. The global capacity factor of operational geothermal power plants was 87% in 2010 and became 77% in 2021. It is now highly competitive against fossil fuel or nuclear-based power plants.

However, geothermal energy possesses more risks compared to other renewables, in terms of return on investment. Geothermal reservoir needs to consist of hot water at desired temperatures for a power generation cycle. Reservoir depth also needs to be not too deep for cost-effectiveness. Cost-effectiveness and risks keep geothermal power generation systems behind other renewables. At the beginning of the twenty-first century, global operational geothermal power plant capacity was 8.2 GWp. It almost doubled and became 14.88 GWp of global operational power generation capacity at the end of 2022. When geothermal power generation capacity additions compared to other renewables, it stays far behind since global solar power generation capacities increased 858 times and global wind power generation capacities increased 53 times.

According to the available temperatures at the reservoir, different power generation cycles can be used in geothermal power plants. Figure 2.9 shows currently operational power plants around the globe and their type of power generation technology. Dry steam, flash steam, and binary power generation cycles are the most commonly used technologies for geothermal-driven power generation processes. Dry steam and flash cycles directly use geothermal fluids from the reservoir and require higher temperatures compared to the other geothermal power generation technologies. Binary cycle, on the other hand, indirectly uses geothermal fluid and exchanges its heat with cycle's own thermofluid. Usually, dry steam and flash steam cycles require more than 180°C geothermal fluid temperatures. Binary cycle on the other hand can be executed at temperatures lower than 180°C. For lower

Fig. 2.9 Operational geothermal power plants and their power generation cycle (data from Ref. [20])

temperatures, between 90°C and 150°C, Kalina cycle and the organic Rankine cycle are two options that can be considered. Moreover, enhanced geothermal power plants inject external water into the reservoir and collect heat from the underground. Heated water then gets pumped through the surface for expansion to generate power. Enhanced geothermal systems do not require natural hot water in the geothermal reservoir. It extends the geothermal energy availability and holds a potential to implement geothermal-driven power generation systems to extended locations.

(d) *Hydro Energy*

Hydro energy has the highest power generation capacity in operation at the end of 2022. However, when currently available growing rates are considered, it is expected that solar power generation system capacities will take over the hydro energy's title in 2023 or 2024. According to the data [10], global renewable hydropower systems have 1255.5 GWp of power generation capacity, in operation.

Note that the capacity factors for hydro energy systems are usually higher than other popular renewables, such as solar and wind. Figure 2.10 shows the global power generation capacity factors in 2010 and 2022. For hydro energy systems, capacity factors are 45%, wind has 39% of capacity factor, and solar has 17% of capacity factor globally. However, seasonal changes affect the hydro energy–based power generation since it is directly connected with precipitation levels. Similar to the wind mechanism, solar energy is the main driving force of hydro energy mechanism. Hydro energy cycle starts with the evaporation of water with solar irradiation. Water vapor then rises and cumulates in the atmosphere. Once the

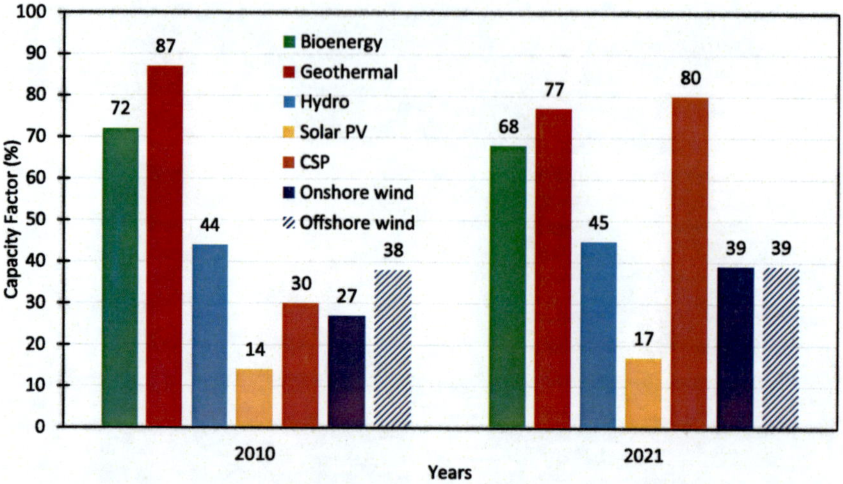

Fig. 2.10 Global capacity factors of different renewables (data from Ref. [16])

atmospheric conditions are favorable, cumulated water vapor starts precipitation on land. Natural or human-made reservoirs higher than sea levels collect water. The collected water has potential energy due to its mass and elevation difference between the upper reservoir and the lower reservoir. From the upper reservoir, water flows through the lower reservoir. Hydro turbines and generators convert the potential energy of the water into electrical energy. Because of the precipitation levels and weather conditions, hydropower capacities may vary.

(e) *Bioenergy*

Biological or organic materials such as wood, wastes from forests, gardens, farms, parks, as well as garbage contain chemical forms of stored energy. In order to exploit that chemical energy stored in organic materials, direct burning or conversion processes can be executed. Direct burning refers to the use of biomass directly to generate heat via direct combustion or gasification. On the other hand, biomass can be processed and treated in order to convert it into biofuels.

Conventional first-generation biofuels are made by using food-based biomass in order to use food-based biomass in a fermentation process. Sugarcane, corn, sweet sorghum, rapeseed, and sugar beets are some of the common biomass sources to generate biofuels such as bioethanol or biodiesel.

Advanced second-generation biofuels are made by using non-edible biomass sources in order to keep food products for the food market, and residue, by-product or waste biomass sources are used and exploited for useful fuels. Anaerobic digestion, gasification, pyrolysis, and direct combustion are the most common processes to generate second-generation biofuels or utilize non-food biomass sources.

Bioenergy-based power generation systems have 148.9 GWp of capacity globally. At the beginning of the twenty-first century, it was only 28.1 GWp. The capacity growth of global bioenergy-based power generation systems is 530% from 2000 to 2022, as shown in Fig. 2.5. When we look at the LCOE values, a slight change can be seen from US$0.078/kWh in 2010 to US$0.067/kWh in 2021, as shown in Fig. 2.8. Capacity factors for bioenergy systems are better than most of the renewables as 68% in 2022, as shown in Fig. 2.10. Overall, bioenergy systems are cost-competitive and favorable in terms of capacity factor. Especially when non-food products are used, bioenergy systems both prevent the production of waste and exploit those by-product sources to generate useful outputs.

In terms of emissions, biomass generates carbon dioxide; however, that carbon dioxide is already a product of the photosynthesis process of the biomass source. Therefore, it can be considered that net carbon dioxide emissions are in balance and no extra carbon dioxide emissions occur because of biomass usage.

(f) *Marine Energy*

Tidal, tidal currents, waves, salinity gradient, and thermal gradient–type renewable energy sources can be considered under marine energy.

Tidal energy systems use a structure called a barrage in order to control water levels. Between tidal basin and sea or ocean, barrage structure uses gates and forms a potential energy difference between the basin and the main water body. By using a turbine and generator, potential energy difference can be converted into electrical energy.

Tidal current energy systems similarly use tidal, however, kinetic energy associated with tidal current is exploited this time. There are various technologies that can exploit the tidal current; usually, turbines are used underwater to utilize the kinetic energy of tidal currents.

For wave energy systems, various concepts and technologies are available. Both kinetic and potential energies of waves are utilized by these components. The rise and fall movement of waves creates potential energy, and waves' horizontal movement brings kinetic energy to the equation; both these energies can be converted into mechanical or electrical energy with suitable components.

Temperature gradient–based energy systems can be considered under marine energy. Ocean thermal energy conversion (OTEC) systems exploit the temperature difference between deep water and the surface to run a power cycle. Sunlight, as well as solar radiation, cannot travel more than around 1 km deep. Therefore, deep water stays cooler than the surface water. This temperature difference can be exploited via OTEC systems to generate power. Another way to exploit temperature difference is suggested as a reverse OTEC cycle [14]. Especially in Arctic regions, ambient air temperatures go very low, such as −40°C, while sea temperatures stay around 0°C. The temperature difference between ambient air and water body can be exploited by using suitable thermofluids.

Salinity gradient energy is based on the difference in salt concentration of different fluids. A membrane divides two fluids which can pass water but not salt. Water moves from the freshwater reservoir into the saltwater reservoir. Saltwater reservoirs become more pressurized or become higher due to the type of the salinity gradient power generation system. Movement of the system boundary can be exploited as mechanical or electrical power.

Overall, marine energy systems possess 90 PWh annual power generation potential. Globally, more than 25.3 PWh of electricity was consumed in 2021 [17]. Therefore, potential marine energy systems can meet current global power needs, more than three times.

(g) *Hydrogen Energy*

Hydrogen energy systems are potential alternatives to fossil fuel–based energy systems due to the many critical advantages of hydrogen. In order to replace fossil fuels with a carbon-free solution, renewable energy sources cannot be sufficient as stand-alone systems. Fossil fuels are energy carrier mediums which can be carried anywhere and can be used anytime. Renewables need to be available at a specific location and time in order to utilize it. Hydrogen is also an energy carrier medium which can be used anytime and anywhere. In contrast to fossil fuels, hydrogen does not contain carbon; therefore, hydrogen-based power generation does not emit any greenhouse gases. Monoatomic hydrogen simply contains one proton and one electron. Since monoatomic hydrogen is unstable, diatomic hydrogen (H_2) can be found in nature, and when hydrogen is mentioned, it refers to H_2 throughout this book.

Current global hydrogen demand majorly comes from the chemical industry, where hydrogen is used as a chemical substance such as a processing agent. On the other hand, hydrogen is a promising alternative to fossil fuels and completes renewable energy systems from many aspects. One of the most important challenges of renewables is intermittency. Hydrogen energy systems can be used as an energy storage system, which can compensate for intermittencies and also help flexibilities. On the other hand, hydrogen can be used as a fuel for mobility; hydrogen vehicles consume hydrogen via hydrogen fuel cells and generate mechanical power for mobility. Another advantage of hydrogen is its logistics; hydrogen can be transported in liquid form or with nitrogen in ammonia form, from country to country or locally. Hydrogen may be able to use similar infrastructure with natural gas pipelines to be transferred and diverted from one location to another as long as it is blended with natural in some acceptable volumetric ratios.

(h) *Energy Storage*

Renewable energy sources are only half of the picture for the energy transition. Ultimately, a complete transition into carbon-free societies requires energy storage mediums. Fossil fuel–based energy systems are very flexible, whenever the customer needs energy, fossil fuel–based systems can generate the required energy. When there is no demand from customers, fossil fuels can be stored as is without converting fossil fuels to any other form of energy since they are already stored energy in chemical form as a fuel. On the other hand, renewables need to be utilized and used once they are available.

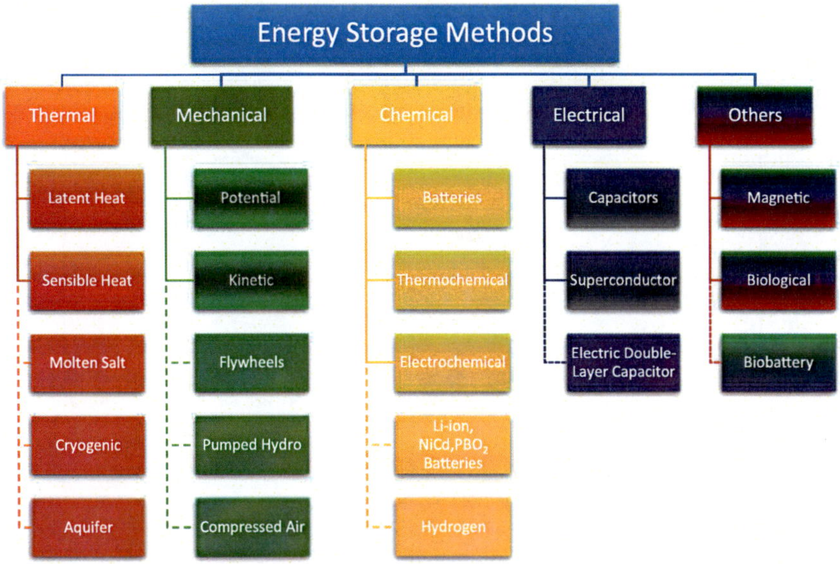

Fig. 2.11 Energy storage methods and examples

Energy storage methods are given in Fig. 2.11. Energy storage can be done both between source and system or system and service. If there is an unutilized excess source, it can be stored before processing via the energy system. If the energy system has already processed the source and generated a new form of energy, it can be stored as well. Thermal energy sources, such as geothermal energy, can be stored via thermal energy storage systems. Mechanical, chemical, and electrical forms of storage are usually used between system and service. Moreover, storage mediums between system and service can be directly used when needed or they can be processed again by the system. Renewable energy sources will be presented with details and examples in Chap. 3.

2.3 Environment

Energy richness with oil domination helped society to become more advanced with many developments in different technological avenues. Improved life quality and advanced societies grew faster and became more and more hungry for energy. Fossil fuel consumption and population growth happened at a similar pace.

However, advanced societies, improved life qualities, and a larger population brought a by-product, which is anthropogenic greenhouse emissions. In order to utilize fossil fuels to generate energy in the desired form, fossil fuels need to be combusted. The combustion process of fossil fuels is the major contributor to

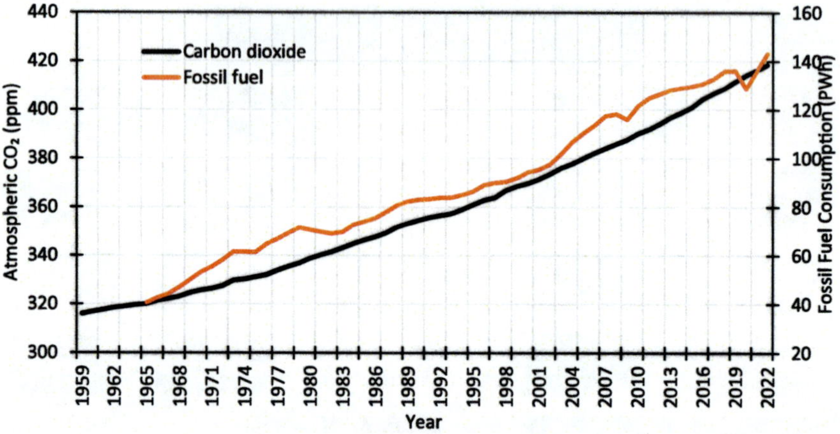

Fig. 2.12 Atmospheric carbon dioxide annual average at Mauna Loa Observatory (data from Ref. [21]) and fossil fuel consumption trends (data from Ref. [3])

anthropogenic greenhouse gas emissions. As previously discussed in Chap. 1, Earth's energy balance is changing due to the increasing carbon dioxide particles in the atmospheric gas mixture. Figure 2.12 shows how atmospheric CO_2 is increasing throughout the year. Since carbon dioxide has the ability to trap and accumulate heat that is reached from the sun via radiation heat transfer, accumulated heat is increasing, which leads to increasing temperatures. Manua Loa Observatory in Hawaii started to record atmospheric carbon dioxide levels in 1958. The annual average carbon dioxide level in the atmosphere was recorded as 316 ppm in 1959. By that time, greenhouse gas emissions tremendously increased, and the annual average carbon dioxide level in the atmosphere reached 418.6 ppm in 2022. One may ask whether there are any historical highs and lows. During the ice age and interglacial periods, there were peaks and valleys; however, the previous historical peak was around 300 ppm [22]. When we look at the current levels, they are 40% higher than the historical peaks.

The growing pace of fossil fuel consumption and atmospheric carbon dioxide increases the chance to cause drastic changes on Earth. Arctic and Antarctic ice can be considered as the backbone of the balance on Earth which is one of the driving factors of jest streams. However, changing global temperatures affect the thickness and extent of the sea ice both in the north and in the south poles. The weakened sea ice can lead to changes in jet streams. Jet streams, in response to those changes, can lead to global catastrophic events.

When we look at the global temperature changes in water bodies such as oceans and seas, after the 1980s, there was a significant positive anomaly when 1901 and 2000 temperatures taken as a base point. Water bodies are getting heated slower than land; therefore, global water body temperature anomalies are lower than land temperature anomalies. Figure 2.13a shows the global sea temperature anomalies for each average month of 10 years' period. Especially after 2010, global sea

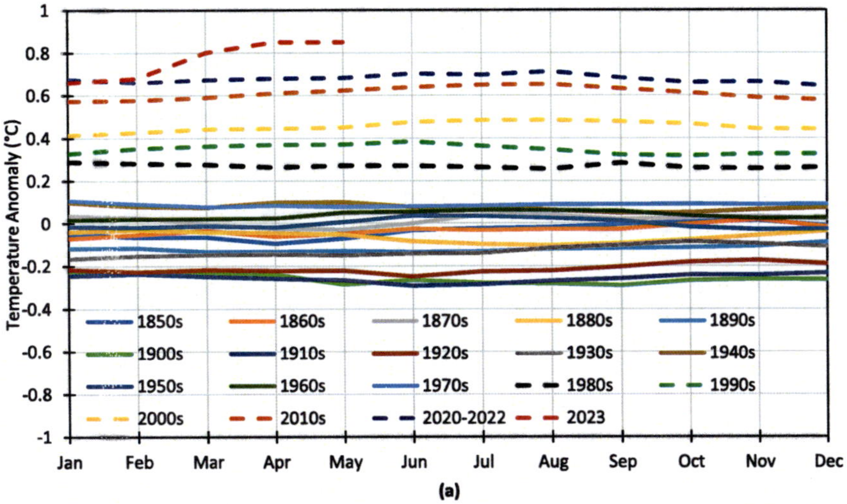

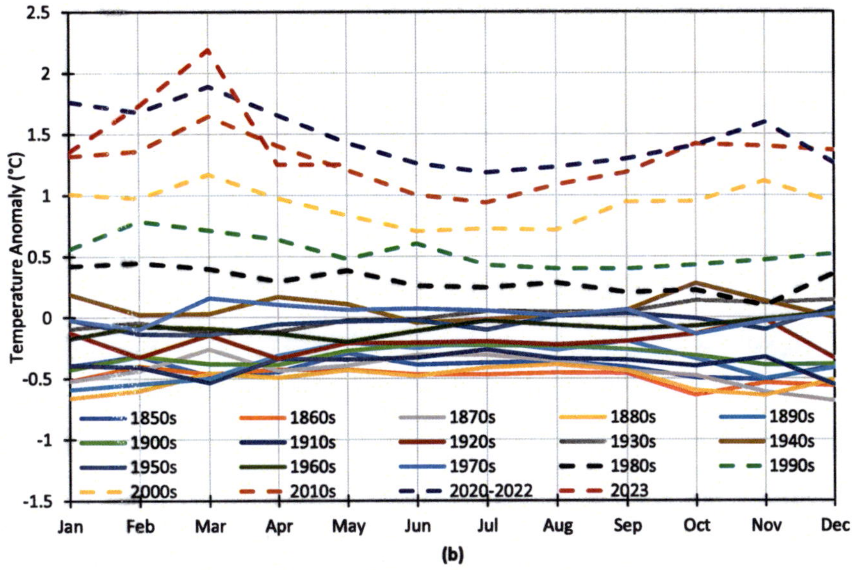

Fig. 2.13 Global sea (a) and land (b) temperature anomalies according to 1901–2000 base point (data from Ref. [23])

temperature anomalies have been observed in a way surpassing 0.5°C and reached above 0.85°C recently.

Water has almost four times higher heat capacity than land. Also, water bodies on Earth account for two-thirds of the Earth's surface. This phenomenon also brings some climatic differences between coastal and inland cities. Coastal cities experience temperature changes in a less drastic way since the water body softens the

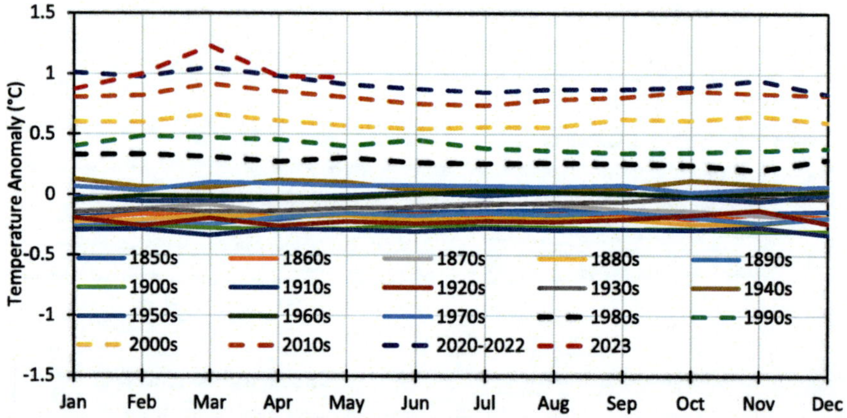

Fig. 2.14 Global temperature anomalies according to 1901–2000 base point (data from Ref. [23])

change in temperature while cooling the ambient during hotter periods and warming the ambient during cooler periods. On the other hand, inland cities have more drastic changes in temperatures, while they feel colder winters and hotter summers than coastal cities. Moisture is another parameter that affects temperature and contributes to the difference between coastal and inland cities. It is now further observed that water bodies evaporate during hot days, which absorbs some portion of the heat for a phase change. As a thermodynamic knowledge, during a phase change of a thermofluid, temperature change cannot be observed. Therefore, the evaporation of water bodies uses heat to change the phase, and temperatures can be kept more stable. For these reasons, there is a significant change between global water bodies and land temperatures.

Global land temperature anomalies are already passed 1.5°C and rising exponentially, as reflected in Fig. 2.13b. Lands are getting heated faster than water bodies; therefore, temperature anomalies are happening at a larger pace on lands. At the beginning of the twenty-first century, land temperature anomaly pushed 1°C and surpassed it in upcoming years. In the last decade, land temperature anomaly surpassed 1.5°C and even exceeded 2°C in March 2023.

Figure 2.14 shows overall global temperature anomalies by 1901 and 2000 base points. Overall, global temperature anomalies are already pushing 1°C and rising to unprecedented levels. Only about 30 years before 2020s, the temperature anomalies observed were about 0.5°C in 1990s. For a 0.5°C temperature rise, 30 years was enough. For this reason, global governments and organizations promised to act urgently to limit temperature anomaly at 2°C. A bigger image can be depicted in Fig. 2.15, which shows how related temperature change and carbon dioxide levels as well as the urgency to act against climate change.

There are global consequences of changing carbon dioxide concentrations in the atmosphere and warming global temperatures. There is a major risk of species losses for more than 30,000 species including birds, mammals, reptiles, fish, invertebrates, corals, and cephalopods. Moreover, there are serious risks to human health; besides the

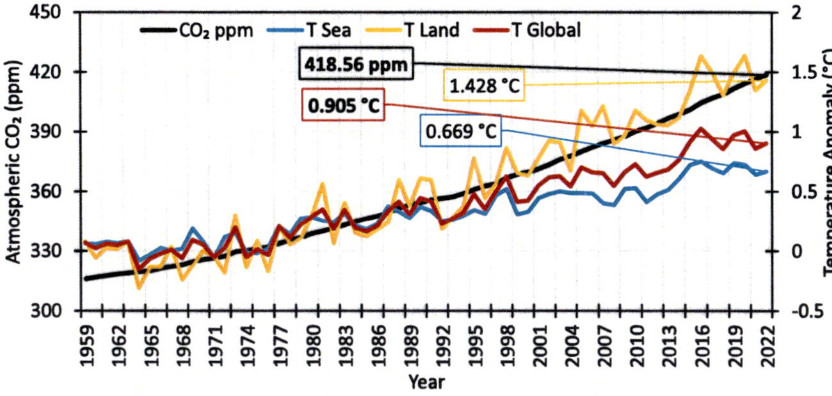

Fig. 2.15 Global temperature scales (data from Refs. [3] and [23])

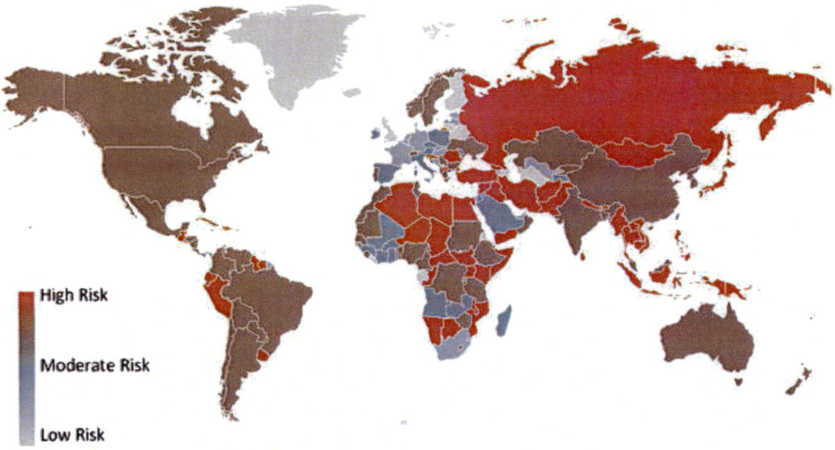

Fig. 2.16 Countries risk map for climate-related catastrophic and extreme weather events (data from Ref. [24])

air quality and respiratory impacts, high-temperature and high-humidity conditions, and having those conditions more frequently and impactfully, possess serious risks of developing hypothermic conditions, which leads to a higher risk of mortality, as depicted in Fig. 2.16.

Global warming effects and temperature anomalies can be observed more easily with more changes in the northern hemisphere than in the southern hemisphere since land coverage is double in the northern hemisphere. Especially in polar circles, warming is happening almost four times faster than the globe according to Rantanen et al. [25]. The reason behind faster Arctic warming is the feedback loop where warming melts sea ice and more areas of the sea exposes to sunlight which leads to more warming and more melting of the sea ice, as shown in Fig. 2.17. This feedback

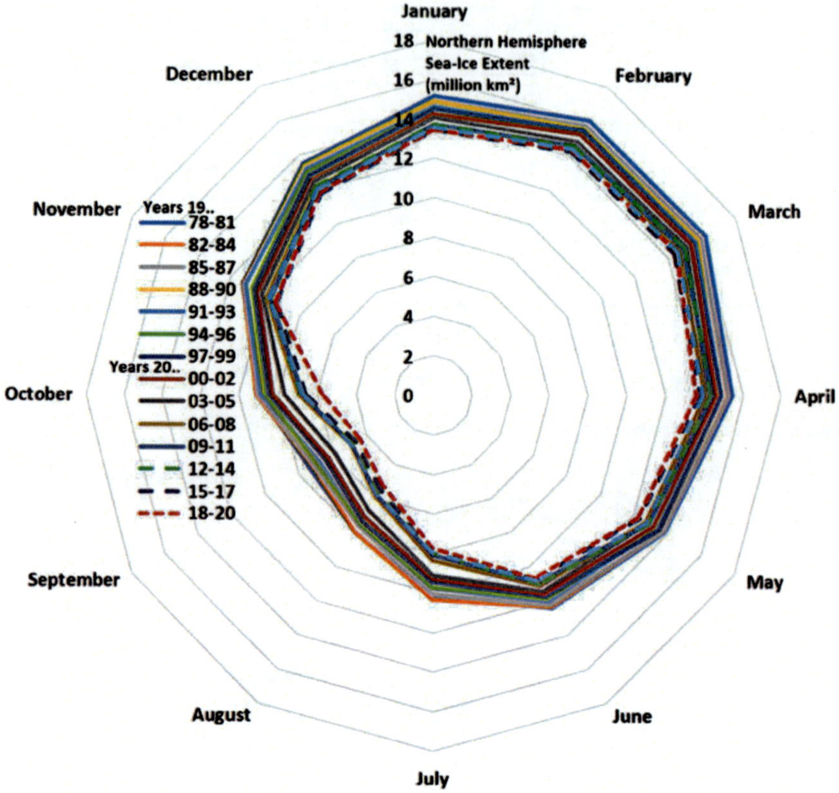

Fig. 2.17 Northern hemisphere sea-ice extent by months (data from Ref. [26])

loop process is called the Arctic amplification. The respective data also prove that the sea-ice extent is significantly smaller than historical levels. Some historical records confirmed various observations in September 2020, for example, where the sea-ice extent shrank to 4 km^2.

Historically, the annual sea-ice extent average was around 12 million km^2 in the 1980s. Recently, it shrank around 10 million km^2 which is the historical annual average for the Arctic sea-ice extent. Figure 2.18 shows the annual average of the northern hemisphere's sea-ice extent, where shrinking is significantly observed.

On the other side of the Earth, Antarctic sea ice also shrank to levels, which exceptionally small. Figure 2.19 shows the Antarctic sea-ice extent anomalies from 2010 to 2023, where the baseline is the average between 1978 and 2023. The red line shows 2023 levels which are exceptionally lower than the baseline. On June 12, 2023, a historical minimum record was observed by satellite data where coverage was 10.7 million km^2 for Antarctic sea-ice extent [27]. When compared to the average between 1981 and 2010, 2.3 million km^2 is lower and is a historical record for that date, which is larger than the land area of Saudi Arabia.

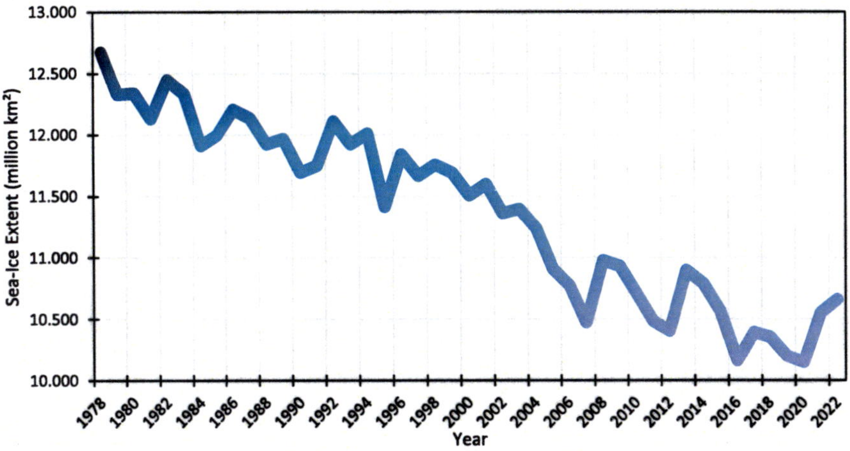

Fig. 2.18 Annual average of Arctic sea-ice extent (data from Ref. [26])

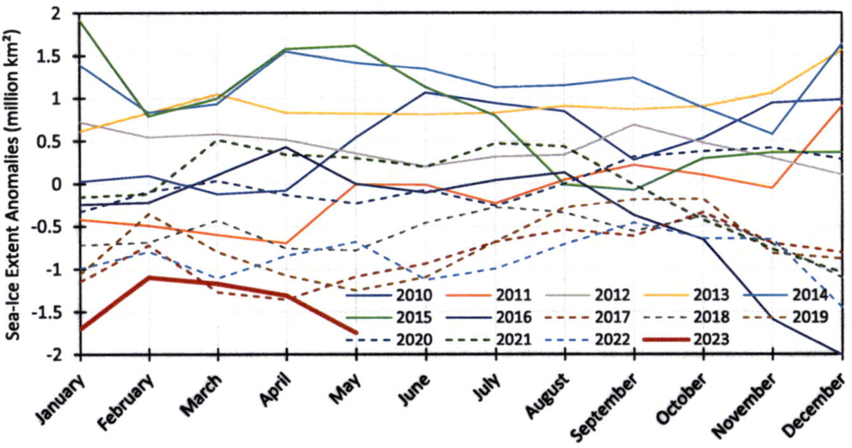

Fig. 2.19 Antarctic sea-ice extent by months (data from Ref. [26])

2.4 Sustainable Development

It is crucial to use energy while ensuring that it does not compromise future generations' sources and to keep the environment in good form. The definition of sustainable development by the Brundtland Commission is "development that meets the needs of the present without compromising the ability of future generations to meet their own needs" [28]. Sustainable development principles can be applied from micro-scale to macro-scale applications. It preserves the resources both for the

Fig. 2.20 Six key dimensions of sustainability

current generation and for the future generations while ensuring the advancement of society with technological applications in various areas. Six key dimensions can be mentioned for sustainability, namely energy, environment, education, economy, engineering, and ethics (6E), as shown in Fig. 2.20.

Energy and environment are two critical pillars of sustainability, which are also interconnected, as mentioned in previous sections. It is important to understand future demands and carefully plan smart energy applications by using energy sources efficiently. From this aspect, renewables and desalination systems can be highlighted along with hydrogen when designed in an integrated and smart fashion with high source utilization.

Ethics, which is proposed as a key shelter to even cover social, judicial, and moral sub-dimensions, on the other hand, is another primary dimension where the guiding principles come together and harmonize with integrity. It is critical to build a system based on moral principles and codes, systematically.

Economy is another important pillar of sustainability since it is one of the major driving forces for the private sector. It is critical to establish important economic policies that ensure the protection of the environment while keeping systems

Fig. 2.21 Essential needs of the humankind

profitable. It directly interconnects with other pillars and possess constraints for other pillars.

Education is recognized as one of the key pillars that can raise awareness for sustainability. It is crucial to improve knowledge and innovate in order to achieve key challenges of sustainability. Only education can improve knowledge and maintain the innovation process to overcome challenges.

Engineering is critical to come up with innovative solutions to challenges, as well as to improve currently available methods and processes in order to minimize the waste and ensure better utilization of available sources.

Essential needs are the most critical ones before proceeding with further advancements. These can be summarized in four categories as clean air, clean water, clean food, and clean energy, which is shown in Figure 2.21. These four needs are directly or indirectly related to energy. As mentioned previously, clean air and clean energy can be ensured with renewable or zero-carbon energy options.

From this context, clean energy and clean water are intricately connected in many aspects. Water is important for many energy processes, either as a coolant, as a thermofluid, or in some cases as a source, such as hydro energy, marine energy, or hydrogen energy systems. In energy storage processes, water can also be used as an energy storage medium. On the other hand, when there is a need for clean water, energy systems can produce desalinated water from saline water. In a smart manner,

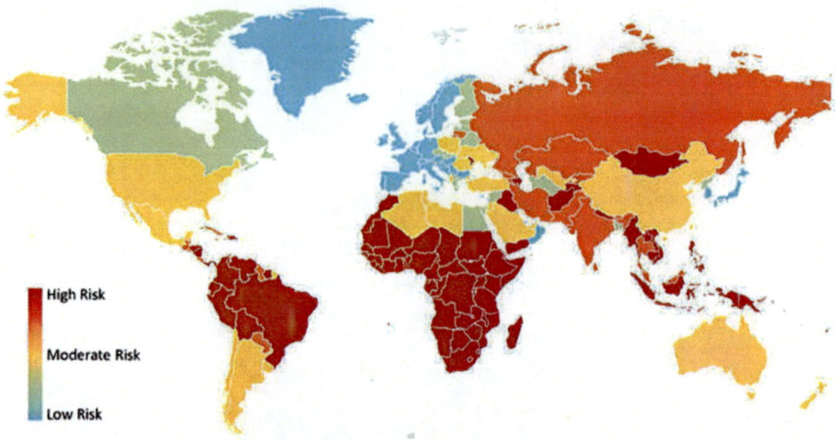

Fig. 2.22 Risk map of countries for water stress (data from Ref. [24])

some required processes can be integrated with the desalination process in order to obtain clean water. For instance, the cooling process of power generation cycles achieved in condensers is actually wasting heat unless it is recovered to use for something, where the desalination process can easily be integrated with, and waste heat can then be exploited to produce something useful.

Water scarcity is one of the major challenges that humanity is facing. Especially with the increasing influence of global warming, water scarcity is becoming a critical issue for many countries. Besides environmental reasons, socioeconomic reasons are also important aspects that affect the risk factors for water stress. Figure 2.22 shows the risk levels of countries for water stress. As mentioned above, water and energy have an intricate connection; growing population and improved quality of life trigger energy hunger as well as water use since currently available fossil fuel–driven power generation processes require water. From this aspect, fossil fuel usage, growing population, and global warming have combined effects on water stress. Therefore, it is very critical to address water-related issues by considering energy and the environment. Currently, more than 733 million people live in high-risk locations for water stress. Directly or indirectly, 2.3 billion people are facing water stress [29]. Desalination is a lifeblood for some countries; however, supportive infrastructure is required for the implementation.

In order to address sustainability and assess related parameters, exergy is a critical tool. Source utilization is key for sustainability since preserving needed sources for future generations and giving a habitable planet to them requires efficient utilization of clean sources. From this aspect, energy and its tools are limited to exactly understand what is the useful energy content in an energy source and what will be the useful energy content when it is processed. Exergy tools, on the other hand, can give plentiful critical information which can be referred to as X-ray images of energy systems. Figure 2.23 summarizes how exergy efficiency is interconnected with dimensions of sustainability.

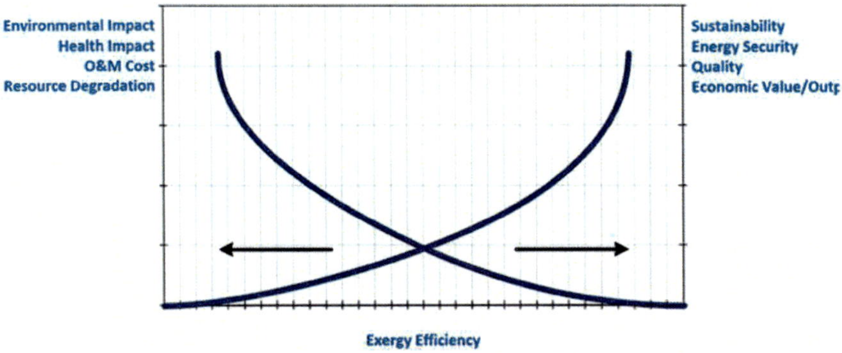

Fig. 2.23 Qualitative representation of the relation between exergy efficiency and critical dimension

2.5 Closing Remarks

It is essential to gain consciousness of sustainability by raising awareness of its dimensions from energy to environmental issues. The current chapter comprehensively presents the current status of the environmental issues with insightful data and depicts those with summarizing charts. It is obvious that environmental issues are at a crucial stage where each indicator shows catastrophic changes are very near if the world does not decide to change its path from a carbon society into a carbon-free society. From this aspect, the pathway to a carbon-free society includes using renewable or carbon-free energy sources, improving source utilization efficiency, implementing smart energy systems, and using correct, complete, and consistent assessment methods.

References

1. Andrae ASG, Edler T (2015) On global electricity usage of communication technology: trends to 2030. Challenges 6:117–157. https://doi.org/10.3390/CHALLE6010117
2. Dincer I (2020) Covid-19 coronavirus: closing carbon age, but opening hydrogen age. Int J Energy Res 44:6093–6097. https://doi.org/10.1002/er.5569
3. Ritchie H, Roser M, Rosado P (2022) Energy data. Our World in Data
4. York R, Bell SE (2019) Energy transitions or additions?: why a transition from fossil fuels requires more than the growth of renewable energy. Energy Res Soc Sci 51:40–43. https://doi.org/10.1016/J.ERSS.2019.01.008
5. Ballard JR (2015) U.S. energy infrastructure: climate change vulnerabilities and adaptation efforts
6. United Nations Climate Change – COP 27 (2022) COP27 reaches breakthrough agreement. United Nations Framework Convention on Climate Change, Egypt
7. Sharma A (2021) COP 26 president designate reflections note. In: Sessional proceedings of United Nations COP26 Climate Change Conference 2021, Scotland, United Kingdom

8. United Nations (2015) Paris agreement – UNFCCC
9. Dincer I, Acar C (2017) Smart energy systems for a sustainable future. Appl Energy 194:225–235. https://doi.org/10.1016/j.apenergy.2016.12.058
10. International Renewable Energy Agency (IRENA) (2023) renewable capacity statistics 2023. IRENA, Abu Dhabi
11. Smil V (2017) Energy and civilization: a history. The MIT Press. https://doi.org/10.1353/tech.2018.0067
12. [Data/information/map] obtained from the "Global Solar Atlas 2.0, a free, web-based application is developed and operated by the company Solargis s.r.o. on behalf of the World Bank Group, utilizing Solargis data, with funding provided by the Energy Sector n.d.
13. Easton RL, Votaw MJ (1959) Vanguard I IGY Satellite (1958 Beta). Rev Sci Instrum 30:70–75. https://doi.org/10.1063/1.1716492
14. Temiz M, Dincer I (2022) A unique ocean and solar based multigenerational system with hydrogen production and thermal energy storage for Arctic communities. Energy 239:122126. https://doi.org/10.1016/j.energy.2021.122126
15. Fan G, Duan B, Zhang Y, Li X, Ji X (2021) Full-spectrum selective thin film based photonic cooler for solar cells of space solar power station. Acta Astronaut. 180:196–204. https://doi.org/10.1016/J.ACTAASTRO.2020.12.035
16. IRENA (2023) Global LCOE and auction values. International Renewable Energy Agency, Abu Dhabi
17. Energy Data System (2023) International electricity data. Energy Information Administration (EIA). https://www.eia.gov/international/data/world/electricity/electricity-consumption
18. Lu X, McElroy MB, Kiviluoma J (2009) Global potential for wind-generated electricity. Proc Natl Acad Sci USA 106:10933–10938. https://doi.org/10.1073/PNAS.0904101106/ASSET/3AEA0C51-DE74-467D-ABD4-6A7302F45F30/ASSETS/GRAPHIC/ZPQ9990983150007.JPEG
19. Archer CL, Jacobson MZ (2005) Evaluation of global wind power. J Geophys Res Atmos 110: 1–20. https://doi.org/10.1029/2004JD005462
20. Uihlein A (2018) JRC geothermal power plant dataset. European Commission, Joint Research Centre (JRC) [Dataset] PID: http://data.europa.eu/89h/jrc-10128-10001. n.d.
21. Global Monitoring Laboratory (2023) Trends in Atmospheric Carbon Dioxide. Earth System Research Laboratories, National Oceanic and Atmospheric Administration
22. Lüthi D, Le Floch M, Bereiter B, Blunier T, Barnola JM, Siegenthaler U et al (2008) High-resolution carbon dioxide concentration record 650,000–800,000 years before present. Nature 453:379–382. https://doi.org/10.1038/nature06949
23. National Centers for Environmental Information (2023) Anomalies and index data | Global surface temperature anomalies. National Oceanic and Atmospheric Administration
24. Institute for economics and peace. Ecological threat report 2022: Countries facing high ecological threats 2023. https://www.visionofhumanity.org/maps/ecological-threat-report/#/. Accessed 4 July 2023
25. Rantanen M, Karpechko AY, Lipponen A, Nordling K, Hyvärinen O, Ruosteenoja K et al (2022) The Arctic has warmed nearly four times faster than the globe since 1979. Commun Earth Environ 3:1–10. https://doi.org/10.1038/s43247-022-00498-3
26. National Snow and Ice Data Center (2023) Sea ice data and analysis tools | Arctic sea ice news and analysis. National Aeronautics and Space Administration
27. Brodzik MJ, Stewart JS (2016) Near-real-time SSM/I-SSMIS EASE-Grid daily global ice concentration and snow extent, Version 5. 2016. NASA National Snow and Ice Data Center Distributed Active Archive Center
28. United Nations General Assembly (UNGA) (1987) Report of the world commission on environment and development: our common future, United Nations General Assembly, Oslo
29. United Nations (2021) Summary progress update 2021 – SDG 6 – water and sanitation for all. Geneva, Switzerland

Chapter 3
Renewable Energy Sources

Nomenclature

A	area (m^2)
CA	California
COD	chemical oxygen demand
DD	decimal degrees
DNI	direct normal irradiance
EBE	energy balance equation
EnBE	entropy balance equation
ExBE	exergy balance equation
ex	specific exergy (kJ/kg)
$\dot{E}x$	exergy rate (kW)
g	gravitational acceleration (m s^{-2})
GHI	global horizontal irradiation
h	specific enthalpy (kJ/kg)
H	heating
LF	linear Fresnel
LHV	lower heating value (kJ/kg)
$\dot{m}$	mass flow rate (kg/s)
MBE	mass balance equation
$\dot{N}$	molar flow rate (mol/s)
NREL	National Renewable Energy Laboratory
OTEC	ocean thermal energy conversion
P	pressure (kPa)
PTC	parabolic trough collector
PV	photovoltaic
PVGIS	Photovoltaic Geographical Information System
Q	heat (kJ)
$\dot{Q}$	heat rate (kW)

I. Dincer, M. Temiz, *Renewable Energy Options for Power Generation and Desalination*, https://doi.org/10.1007/978-3-031-53437-9_3

LCOE	levelized cost of electricity
S	total entropy (kJ /K)
s	specific entropy (kJ/kg K), source
$\dot{S}$	entropy rate (kW/K)
T	temperature (°C)
SAM	System Advisor Model
TS	total solids
TVS	total volatile solids
V	velocity (m/s)
VFA	volatile fatty acid
$\dot{W}$	power (kW)
X	mass ratio of component
Z	vertical position

3.1 Introduction

We have been in an era where fossil fuels, primarily with an order of coal, oil and natural gas, have played a dominant role since the Industrial Revolution. The use of these fossil fuels has helped a lot in speeding up technological development and societal progress. We have now reached a level where we face local and global challenges in terms of energy, environment, and economy and need to make a shift from fossil fuels to something else which are more environmentally benign and more sustainable. So, the true address has been identified as renewables. Renewable energy sources, namely solar, wind, hydro, geothermal, biomass, and ocean, are considered promising candidates to replace the entire energy sourcing. In order to proceed with an effective and fast energy transition by including the entirety of global energy infrastructure; organizations, governments, corporations, and individuals play critical roles.

Renewables are now truly recognized as alternative energy sources to replace the use of fossil fuels. At the beginning of the century, renewable energy systems were very expensive alternatives to fossil fuel combustion and other fossil fuel–driven energy systems. Over time, renewable energies have gained attention from organizations, government agencies, and corporations.

There are many reasons behind the driving forces of the growing need for renewable energy. Among all the reasons behind the renewable energy transition, one reason possesses critical importance and requires urgent action, which is climate change. Unfortunately, the latest indicators show that the Earth will become an uninhabitable place if no significant actions are taken until the mid-twenty-first century. The environmental dimensions are comprehensively discussed in the previous chapters. It is important to note that the energy transition needs to be done by replacing all fossil-based energy sources with renewables and low-carbon alternatives in order to achieve net zero or carbon-neutral goals. Before proceeding with the

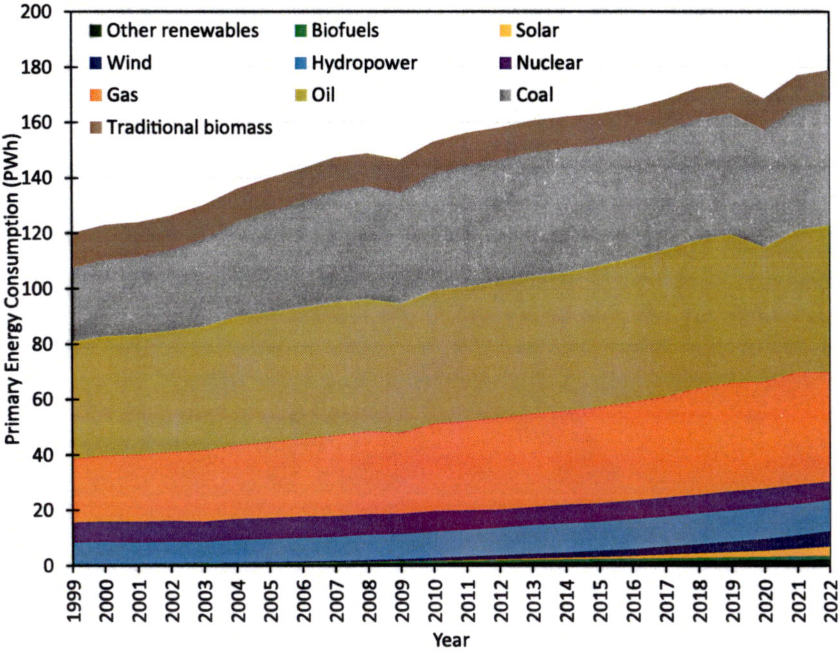

Fig. 3.1 Global primary energy consumption from different energy sources (data from Ref. [1])

particular renewable energy solutions, it would be beneficial to look into the current picture of energy sources in order to understand what needs to be done.

There are two different pictures of global energy consumption. Power generation is one of them which is shown in previous sections. However, it only includes a portion of fuel consumption. In order to see the bigger picture, primary energy consumption is the one to look at it. The growing concerns about fossil fuel consumption and its environmental impact need to be addressed by considering the primary energy consumption. From this aspect, reliance on hydrocarbons is at very critical levels as shown in Fig. 3.1. Fossil fuels, namely coal, gas, and oil, represent 76.71% of global primary energy consumption. Among fossil fuels, oil is the most dominant fuel on the Earth as of 2022. Oil represents 29.61% of global primary energy consumption. Coal and gas represent almost half of the global primary energy consumption at 47.1%, which is equal to 84,267.08 TWh of energy. A special note needs to be given here regarding the traditional biomass. Biomass can be mainly categorized into two types: traditional biomass and modern biomass. Modern biomass is the biomass obtained and utilized with sustainable methods. On the other hand, traditional biomass is unsustainably obtained and utilized biomass, which is mainly used for cooking and heating, especially in developing countries. When untraditional biomass is included with unsustainable fuels, 82.92% of the world's primary energy consumption comes from unsustainable fossil fuels or traditional biomass, which is also the major reason for anthropogenic greenhouse gas emissions (Fig. 3.2).

It is obvious that global energy sourcing highly relies on hydrocarbons. Less than 20% of primary energy consumption is sourced from low-carbon fuels such as nuclear and renewables. Although the current global energy picture is highly carbon-dependent, there is a global push for renewable energy transition. The global push for renewable energy transition leads to a global race for countries and regions to become future renewable energy and clean fuel hubs on local and global levels. Some countries have more sustainable energy sourcing as shown in Fig. 3.3.

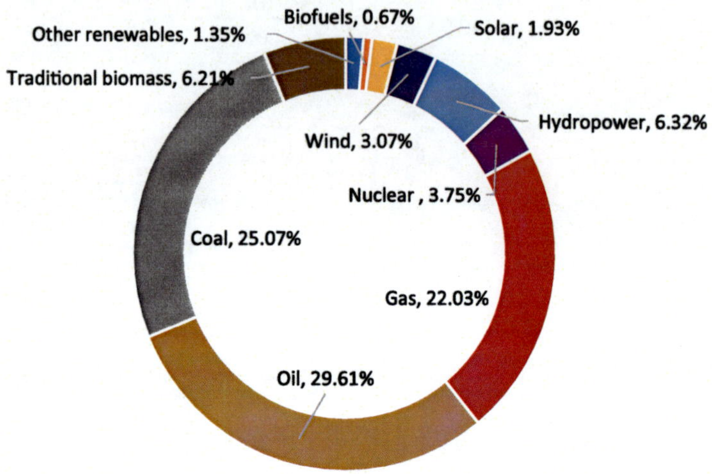

Fig. 3.2 Global primary energy consumption shares in 2022 (data from Ref. [1])

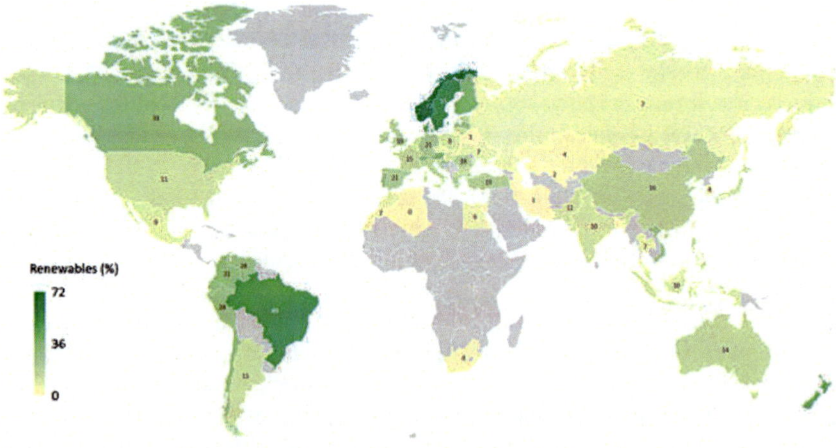

Fig. 3.3 Shares of renewables in primary energy consumption (data from Ref. [1])

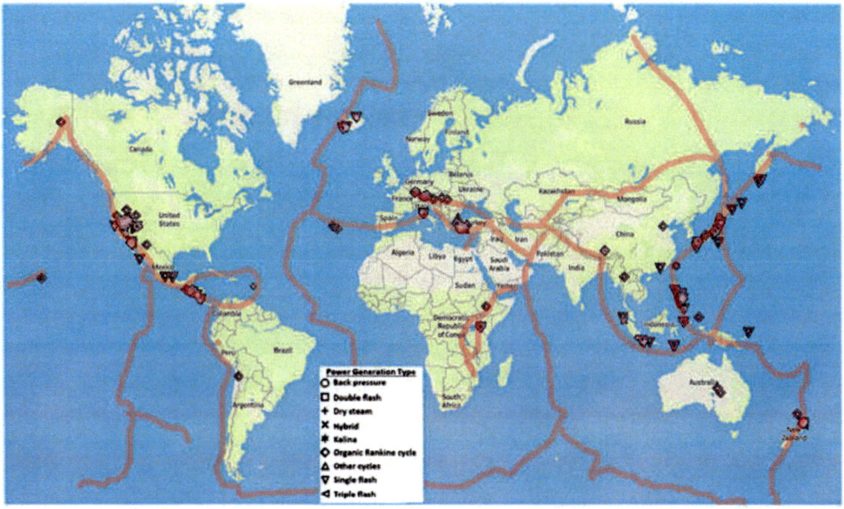

Fig. 3.4 Tectonic plates boundaries and geothermal power plants (data from Ref. [1])

When it comes to renewables in general, the local characteristics, in terms of source availability and needs for specific services, play a critical role. Note that the previously given "*5S approach*" for better sustainability is crucial to apply in order to build a complete net zero infrastructure successfully. Source is the first step of this 5S approach, and some of the sources vary on different locations. Renewable energy source availability highly varies from location to location; for instance, geothermal energy is mostly available around the world's tectonic plate boundaries as shown in Fig. 3.4. Wind, solar, hydro, and marine are other renewable energy sources that have different local availabilities. It is almost impossible to find a renewable energy solution for a location with different feasibilities. Richness in terms of renewable energy sources helps bring a high degree of feasibility for a specific application or project and hence results in a low payback period (Fig. 3.5).

There are many countries with high shares of renewables, which are encouraging future efforts for renewable energy transition. Iceland is an important example and model for other countries when it comes to renewables. Iceland's electricity grid is powered by 100% renewable sources, majorly hydro and geothermal, which is why Iceland got its nickname as "the land of ice and fire" [2]. When it is specifically analyzed, Iceland is located between major tectonic plates (Eurasian and North American plates). One can complain about being located in an Earthquake zone, active volcanic zone, or surrounded by glaciers; however, Icelandic people exploited those various natural phenomena to heat their homes and to generate power for their needs. When glaciers melt, running rivers from mountains or high reservoirs to the ocean or low reservoirs are exploited to generate power via hydropower plants. Underground reservoirs of geothermal energy are especially useful for space heating, which is a critical need for a country in the Arctic Circle.

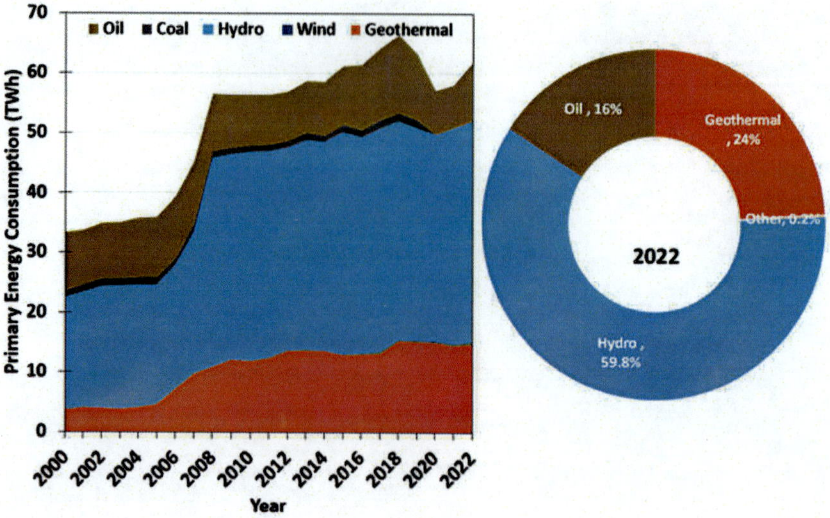

Fig. 3.5 Shares of primary energy sources of Iceland (data from Ref. [1])

The major driving force for the renewable energy transition of Iceland was not environmental but economic. Once the exported fossil fuels became unaffordable for Iceland, local people, especially farmers, looked into ways to use natural sources to heat their homes and farms as well as to generate power. The government also incentivized these efforts and supported farmers and the public who seek the renewable energy transition. The joint efforts from the government and the public from the beginning of Iceland's energy transition created a very successful renewable energy transition example.

The renewable energy transition in Iceland showed many inspirations for the rest of the world. One of them shows that funding a country-wide renewable energy transition does not require to be a rich and developed country, which can be globally observed in Fig. 3.6: since 1985, upper-middle-income countries have had larger renewable shares than high-income countries. From this aspect, cost cannot be an excuse to persist the reliance on fossil fuels; the reality is quite the reverse, cost or economic reasons should be the driver of renewable energy transition, which can be easily proven with the feasibility studies if they consider longer terms.

When it comes to continent-wide success stories, South America is an important model with its significantly higher renewable share among continents, which can be seen in Fig. 3.7. There are multiple reasons behind the high renewable shares of South American energy infrastructure. One of them is the massive investment in hydropower from the 1960s. On the other hand, energy source for transportation is also met by biofuels. There are diversified renewable energy sources which also bring complementarity to the energy picture of South America.

New Zealand is another good example of a renewable energy transition, as summarized in Fig 3.8. Hydropower meets more than half of the electricity needs

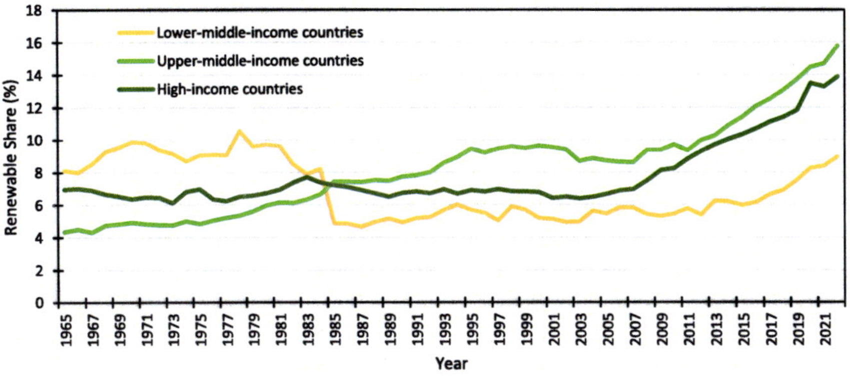

Fig. 3.6 Shares of renewables in countries with different income (data from [1])

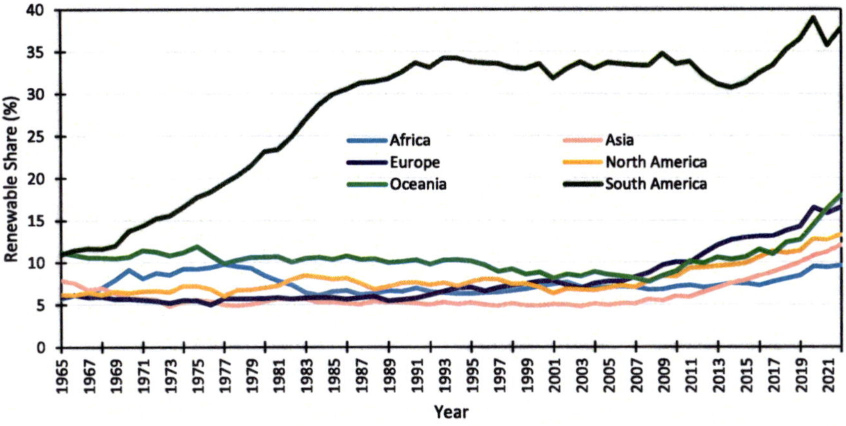

Fig. 3.7 Shares of renewables of continents (data from Ref. [1])

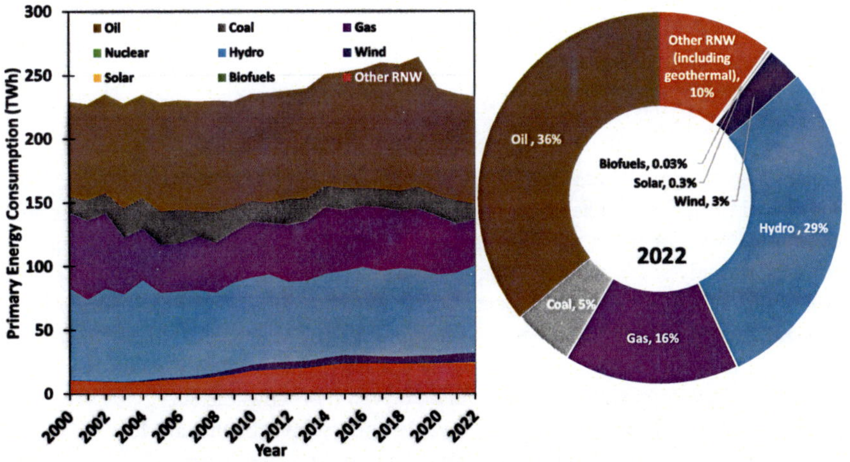

Fig. 3.8 Shares of primary energy sources of New Zealand (data from Ref. [1])

of New Zealand. Since the first hydropower investment was made in 1885, the country has been massively investing in hydropower and other renewables such as geothermal. On the other hand, geothermal meets more than 17% of the country's electricity needs. New Zealand is another example where being located in an Earthquake zone, active volcanic zone, or surrounded by glaciers. Such natural phenomena are exploited to help cover the energy needs of New Zealand.

After making a comprehensive overview of energy sources historically and the need for energy transition to renewable energy options to provide better sustainable development, it is a necessity to develop a solid understanding of what those renewables are and how they work. This chapter presents facts with the most recent data and depicts them with critical parameters how they are interconnected and to what degree the globe is affected. The past, present, and future of renewables and their impacts are comprehensively presented. In addition, the technical aspects of each of these renewable systems are addressed with insightful details. The current state of the art of renewables and their roles are discussed with case studies where each component of a renewable energy system is presented in detail. Furthermore, this chapter aims to help identify goals, objectives, and which renewable energy solutions make better sense to the particular circumstances.

The abovementioned regional examples show that the renewable energy transition is a unique process for each country. The reason behind that is every country has different resources and different needs. Local source conditions along with local needs determine which renewable energy systems would be feasible. It is also crucial to develop a solid understanding of renewable energy systems in order to execute a successful renewable energy transition. The most feasible solution is usually an integration of different energy systems. Also, a unique energy system can be the most feasible solution which has never been developed before. From this aspect, the following sections aim to provide an understanding of renewable energy systems and further to show how to assess these renewable energy systems and applications by considering the primary tools of thermodynamics through the first and second laws of thermodynamics.

3.2 Renewable Energy Sources

Renewable energy sources are the most promising future energy sources to take over fossil fuel dominance. There are a variety of types that can generate different forms of energy which can be used in processes for useful outputs. Renewable energy sources, namely solar, wind, hydro, geothermal, biomass, and marine, can be used when and where available to meet the needs of people. Figure 3.9 presents the classification of renewable energy systems with subcategories, which makes it a categorized illustration.

Any energy-demanding process can be driven by renewable energy sources with a correct design. The majority of renewable energy investments are focused on the power generation sector, which is crucial and needs to be decarbonized entirely to

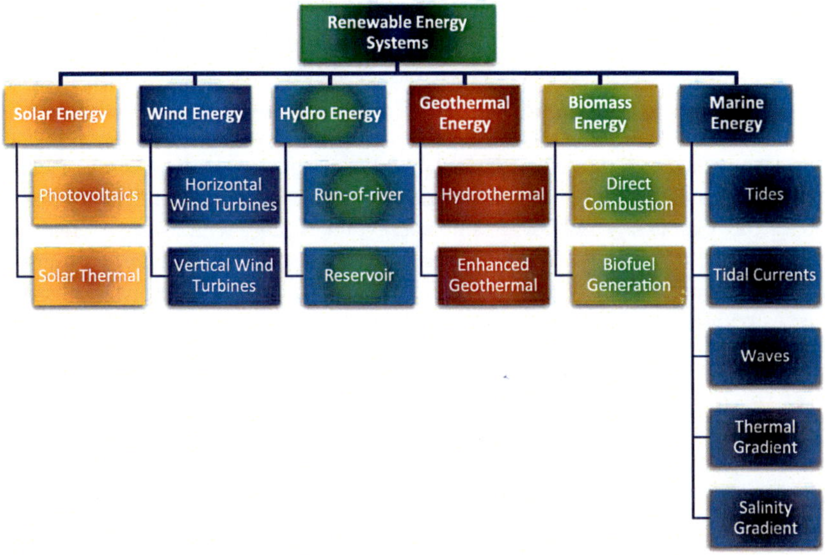

Fig. 3.9 Classification of renewable energy sources

meet the climate goals. However, there are other sectors such as space heating and cooling, transportation, agriculture, information and communication technologies, and industries, which also need to be decarbonized by energizing these sectors via renewable energy sources. However, environmental problems with increased pollution, emissions, and population growth bring many challenges and issues that need to be fixed. Energy and water insecurities are getting more serious with the continuous and growing use of fossil fuels. Energy scarcities and water scarcities are getting more serious and critical with unsustainable anthropogenic activities.

The transition of unsustainable fossil fuel–driven anthropogenic activities with renewable energy–driven alternatives is the most effective way to fight against energy and water insecurities. When we look at the water insecurities itself, major anthropogenic changes are caused by many events as follows:

- Climate change
- Population growth
- Economic growth
- Air pollution
- Water pollution
- Solid wastes
- Unsustainable agriculture
- Unsustainable energy production
- Unsustainable industrial activities

These are directly or indirectly linked with fossil fuel usage. Fossil fuel usage can be mentioned as the root cause of climate change, unsustainable energy production, and unsustainable industrial activities. Water insecurities driven by population growth, economic growth, water pollution, and unsustainable agriculture can be mitigated by using sustainable and smart solutions which are energized by renewable energy sources. From this aspect, renewable energy sources are the key to overcome the most critical challenges. But it is important to note that, renewables need to be supported by other key technologies. In energy, renewables with energy storage and hydrogen; in water, renewables with desalination, are the key combinations to meet the ultimate sustainability goal. In the following subsections, renewable energy sources with case studies will be presented.

3.2.1 Solar Energy

Solar energy is the main energy source on Earth. The process in the sun is called a thermonuclear fusion reaction, where hydrogen atoms get fused to form helium. The radiation from this process reaches Earth and can be harvested by natural or man-made systems. In the biosphere, all living animals and plants, relies on solar energy.

The first human-made solar energy harvesting systems were used to make fire back in the seventh century BC, when a magnifying glass was used to concentrate sun's rays. Then houses built for passive solar heating by facing their Façade to the sun. The first solar collector was made by Swiss scientist Horace de Saussure in 1767 and was used by Sir John Herschel for cooking purposes. The discovery of the photovoltaic effect by a French scientist, Edmond Becquerel, was in 1839. In 1883, solar cells based on selenium wafers were described theoretically by American inventor Charles Fritts. Einstein [3] published a paper about the photoelectric effect in 1904 and then received the Nobel Prize for his theories about the photoelectric effect in 1921. The first silicon-based photovoltaic cell was developed by Bell Labs in 1954. The first application area for the photovoltaic cells was a satellite named Vanguard 1 which launched with its 1 W solar panel in 1958. The first concentrated solar system was built in 1968 by Professor Giovanni Francia near Genoa, Italy. In 1973, the first residential application "Solar One" was implemented by the University of Delaware. The first megawatt-scale application was built in California in 1982. The University of Wales achieved 20% of energy efficiency for solar PV cells in 1985. Recently, solar has boomed with the effect of cost drops which led to surpassing 100 GW of cumulative solar PV capacity in 2012. Applications are varied in various sectors, a solar-powered plane was flown around the world in 2016. When it comes to 2018, solar PV installations reached 100 GW of annual capacity additions. During 2022, 1300 TWh of electricity is produced by solar PV systems. Figure 3.10 summarizes the history of solar energy.

As mentioned previously, solar energy is available at different intensities and changes with time because of the Earth and Sun's relative movement and the

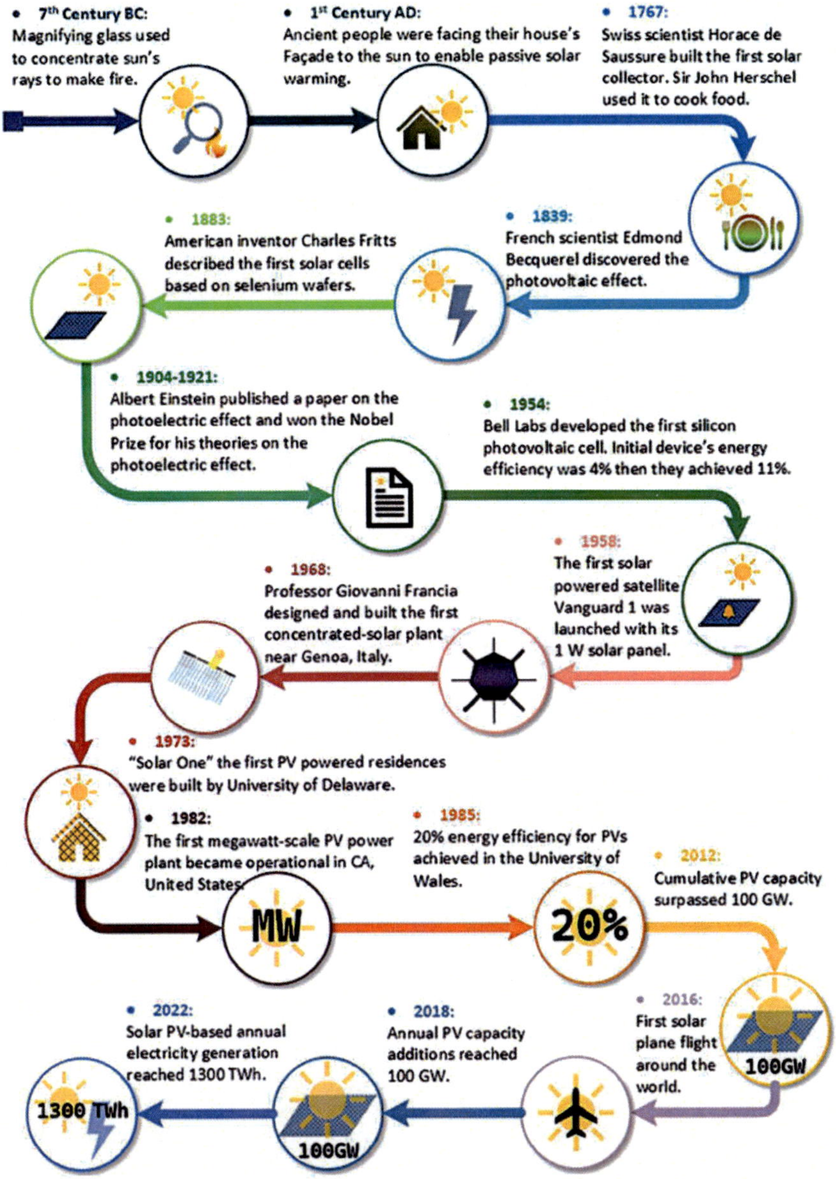

Fig. 3.10 History of solar energy (data from Ref. [1])

atmospheric conditions. When it is considering the application area and type of service need, the suitable type of the solar energy system varies. There are two main technologies available to harness solar energy: solar thermal and photovoltaics as shown in Fig. 3.11. The main difference between those types lies in the conversion

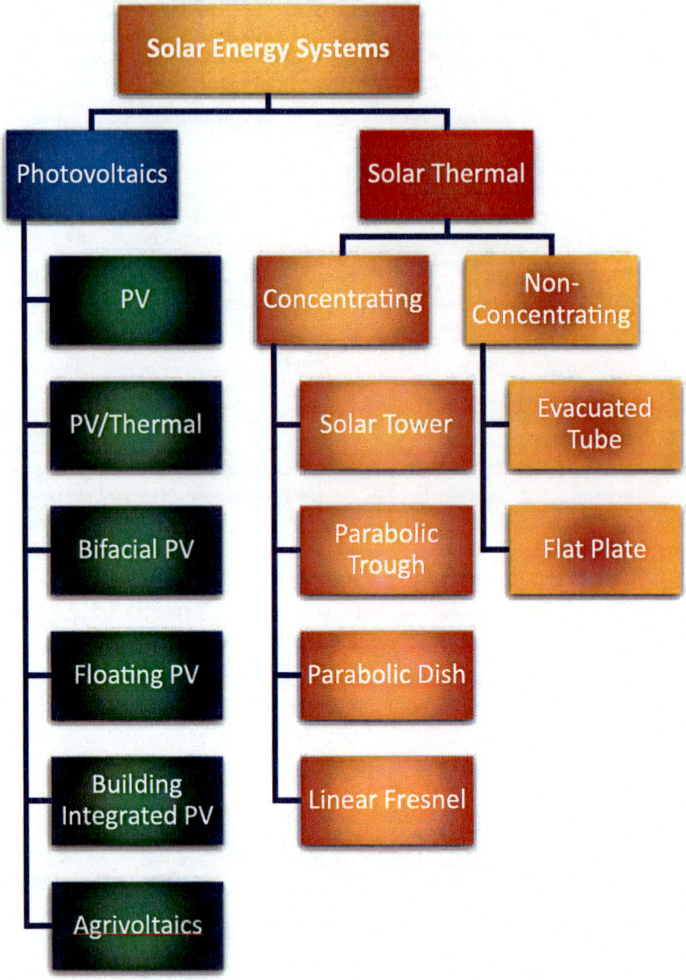

Fig. 3.11 Types of solar energy systems

of solar energy. Solar thermal systems convert solar energy into heat, which can be either directly used or converted into different types of useful outputs such as power. On the other hand, photovoltaics directly turns solar energy into DC electricity. The following subsections aim to comprehensively discuss solar thermal systems and photovoltaics for electricity generation.

(a) *Solar Thermal Systems.*
Solar thermal systems harness solar energy by absorbing the solar radiation to heat a heat transfer fluid, the so-called thermofluid. Later on, a heat transfer medium can be used for heating purposes directly or converted as process heat

for various processes to generate different useful outputs. Solar thermal systems can be investigated in two categories: non-concentrating and concentrating solar thermal systems.

Solar thermal systems are considered one of the unique solutions to harness renewable energy to produce a variety of useful outputs in a very promising way. One advantage of solar thermal systems is their uniqueness to achieve very high temperatures. Among renewables, solar- and geothermal-driven systems can achieve more than 300°C, but the current state of the art enables solar thermal systems, namely solar tower, parabolic dish, linear Fresnel, and parabolic trough collector, to produce more than 500°C heat. It is very crucial to achieve these temperatures in order to run unique processes. A notable example of high-temperature processes is the thermochemical hydrogen production process, which usually requires higher temperatures. Its importance relies on source utilization performance to generate hydrogen due to the direct conversion of heat into chemical energy as hydrogen. Photovoltaics are one option to start harnessing solar energy to generate hydrogen. However, its performance is far behind thermochemical cycles when source-to-service energy and exergy efficiencies are compared.

Moreover, solar thermal systems technically offer a potential solution for hard-to-electrify sectors. High-temperature heat can also be used in a variety of industrial processes. One advantage to having process heat within the system is that there is the potential to exploit the heat until ambient temperatures. From high-grade heat to ambient temperatures, there are numerous processes that can be used according to the service needs of the specific application. A higher number of processes to generate multiple useful outputs leads to better utilization of energy sources.

(b) *Non-concentrating Solar Thermal Systems.*
Non-concentrating solar thermal systems are alternative solutions to generate heat by harvesting solar energy primarily for space and water heating purposes. Compared to concentrating solar thermal systems, non-concentrating solar thermal systems are cheaper solutions and require less complex equipment. However, heat generation cannot reach high temperatures; therefore, it limits the variety of processes. It is possible to generate power via organic Rankine cycle or ocean thermal conversion processes with non-concentrating solar systems, however, concentrating solar collectors are significantly more suitable for power generation.

There are two main types of non-concentrating solar collectors: evacuated tubes and flat plate solar collectors. Evacuated tubes collect solar radiation via their absorber and transfer it to the water in most cases. The inside of the tube is de-pressurized to create a vacuum condition, which can further help to insulate the collectors to not lose heat outside. Multiple tubes connect and form a system according to the capacity requirement. Flat plate solar collectors consist of a heat-absorbing backplate and multiple connected tubes placed above the backplate. The backplate collects solar irradiation and transfers the heat to the tubes where water is circulated.

Non-concentrating systems are simple systems which are used primarily for heating purposes. Typically, non-concentrating collectors are integrated with domestic hot water tanks for heat storage purposes. An illustrative example of non-concentrating solar thermal systems will be given in the next chapter.

(c) *Concentrating Solar Thermal Systems.*
Concentrating solar thermal systems are more advanced technologies compared to non-concentrating systems. By virtue of the high-temperature heat generation ability of concentrating solar thermal systems, there is a rich option for potential integrations. Suitable processes vary from power generation to hydrogen production.

Solar towers and parabolic trough systems are mature technologies which are commercially proven with plants currently operating around the world. Linear Fresnel technology follows parabolic trough and solar tower technologies with early commercial applications. Dish-Stirling technology has demonstration projects which still seek commercial proof.

Process heat can be generated at high temperatures. Linear Fresnel usually produces mid-temperatures, typically around 250–350°C. Parabolic trough technology, on the other hand, is suitable for applications which need around 500°C heat. Typically, parabolic trough systems use process heat around 350–400°C, moreover, can produce up to 550°C. Solar towers can go higher temperatures compared to the parabolic trough and linear Fresnel. A wider range of process heat temperatures is suitable for solar tower applications, such as 250–600°C. Dish-Stirling technology has the highest temperature among concentrating solar energy technologies, which can generate process heat typically between 550 and 750°C, respectively.

(d) *Solar Tower.*
Solar tower collector-type systems typically can go higher temperatures compared to the parabolic trough and linear Fresnel. A wider range of process heat temperatures are suitable, typically from 250 to 600°C. Collector concentration can be equal to more than 1000 suns, which enables high temperatures.

Solar tower collectors concentrate sun via heliostat mirrors, which reflect the sunlight to a target point. Heliostat mirrors use both horizontal and vertical axis trackers in order to reflect the sun on target anytime during the daytime. Depending on the design, thermofluid (typically water or liquid sodium) is heated at the target on the tower where sun rays are focused by heliostat. Thermofluid is used to operate specific processes integrated with the solar tower. Currently available commercial solar tower systems are developed for power generation purposes. However, many other processes can be run to produce a variety of useful outputs.

Case Study 3.1 A case study is carried out by considering weather data for Daggett, CA, United States at 34.85, −116.78, where the average global horizontal irradiation is 5.83 kWh/m^2/day and average ambient temperature is 17°C. A solar tower-based power generation system is designed for the selected location in order to generate 104 MW_e of electricity at the design point. The proposed design is simulated by using the System Advisor Model (SAM) from the National Renewable Energy

Laboratory (NREL). It is important to note that, solar tower systems are usually coupled with thermal energy storage systems, typically, molten salt storage systems, in order to improve the flexibility and operate the system with less intermittencies. The current case study used a direct solar tower system without energy storage, in order to present distinct characteristics of the solar tower system.

Solar tower system consists of a heliostat field, tower, receiver, and power cycle blocks. Each block needs to be designed in a compatible manner in order to run the system in harmony. The designed system collects solar irradiance via heliostats, which track the sun with its dual-axis solar tracker. Sun's rays are concentrated on the receiver which is at the top of the tower. Thermofluid enters the receiver and absorbs the heat. Since there is no energy storage designed, thermofluid goes into the heat exchanger to heat water to get superheated steam. Superheated steam enters the turbine and expands there to generate power. The condenser condenses the steam until it becomes liquid water. The pump pressurizes the water and sends it to the heat exchanger. Once the thermofluid in the heat exchanger rejects its heat to the water, the thermofluid proceeds to the solar tower again to absorb the heat. The solar tower power generation system and heliostat field are illustrated in Fig. 3.12.

In order to generate 104 MW_e of electricity at the design point, a solar tower is designed at 193.5 m of height, which holds a receiver with 21.6 m of height and 17.65 m of diameter. The receiver used stainless AISI316, which has 0.88 of coating emittance, and 0.94 of coating absorbance. At the design point, the receiver generates 669.9 MW_t heat at 574°C when 290°C of thermofluid enters the receiver.

According to the simulation, the designed system generated 304.16 GWh of electricity in a typical meteorological year, with a 33.5% capacity factor. 79,157 m^3 of water is used for cooling in the condenser. Figure 3.13 shows power generation and global horizontal irradiation for average hours of each month in a typical meteorological year.

Case Study 3.2 As mentioned previously, renewable energy transition success stories have their unique cases. The reason behind this is the source availability and service needs are different at different locations. For this reason, the solar tower is also can be used as a unique solution for unique cases. A recent study by Sorgulu and Dincer [4], develops a unique solar tower-based integrated system in order to generate power, hydrogen, fresh water, and heat. Their newly developed system, as illustrated in Fig. 3.14, uses a solar tower to generate high-temperature process heat, which is firstly used to generate hydrogen via a heavy-element halide thermochemical cycle. The hydrogen is produced and blended with natural gas, which is used as a fuel for residential heating and appliances, as well as for power generation via a gas turbine. Generated power and excess heat are used to generate fresh water via a multi-effect thermal desalination system, as well as with a reverse osmosis unit.

The important usage of high-temperature heat in this system is to generate hydrogen via a four-step thermochemical cycle. The reaction steps of this specific thermochemical cycle require 300°C heat to run. To run a thermochemical cycle at 300°C, a solar tower is a good solution to generate high-temperature heat to exchange the heat with it. On the other hand, they use a gas turbine and residential fuel pipeline, which are currently available energy infrastructures in both developed

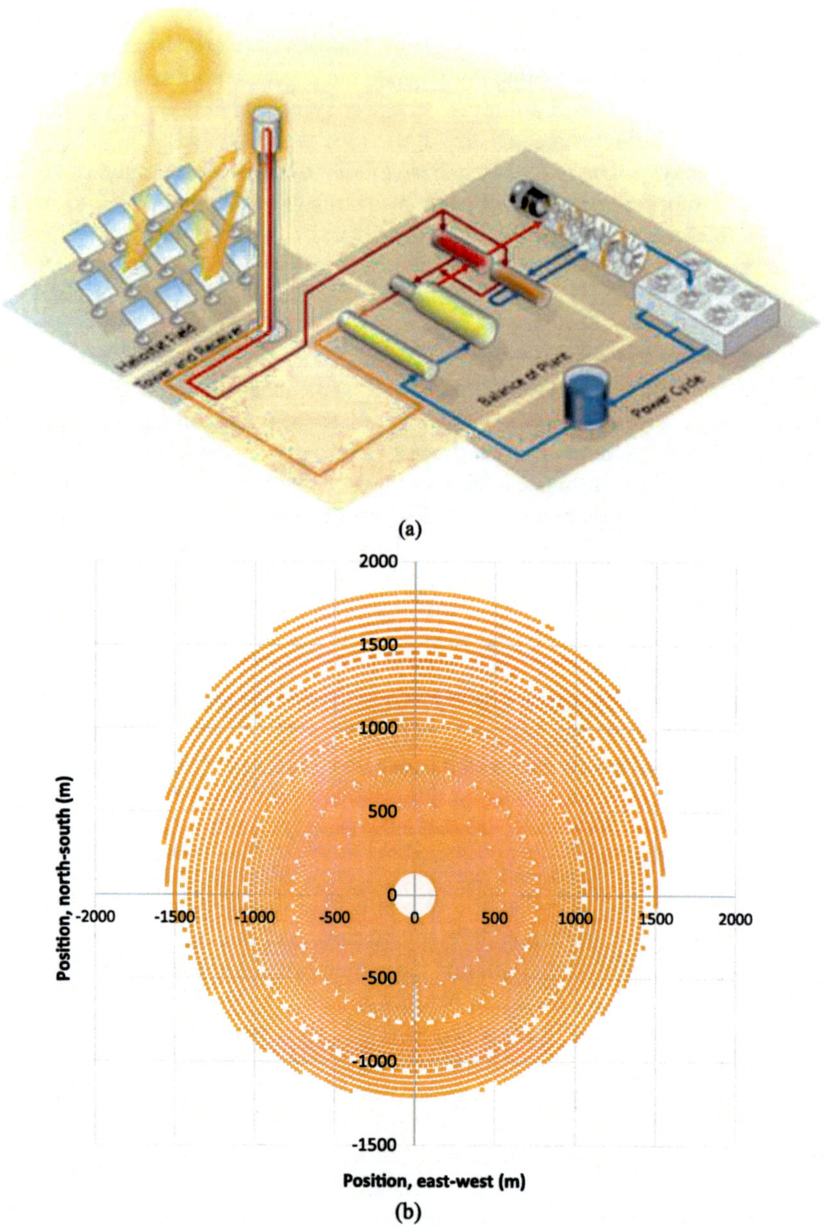

Fig. 3.12 (**a**) Illustration of a solar tower system (modified from Ref. [4]) and (**b**) heliostat positions on the field

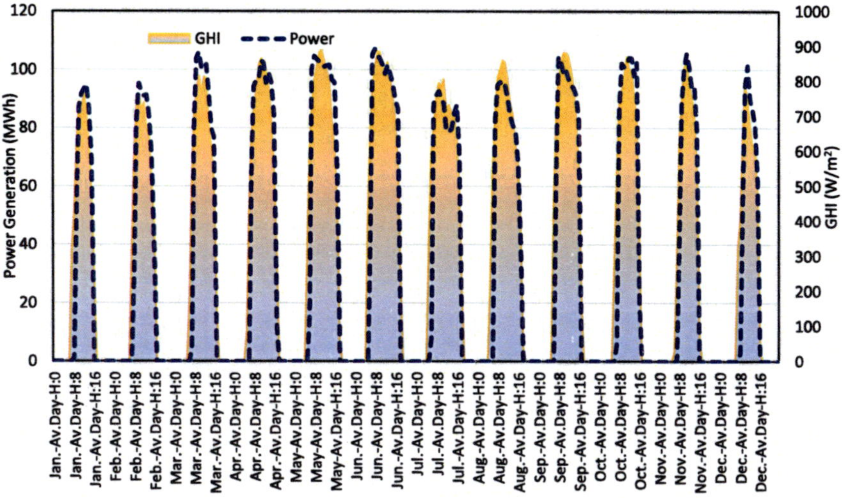

Fig. 3.13 Power generation and global horizontal irradiation for the solar tower system

and developing countries. From this aspect, their proposed system does not require major infrastructure change after the hydrogen and natural gas blend. Additional desalination units are very critical in order to mitigate water scarcity by exploiting the waste or excess energy.

They evaluated their new system according to the first and second laws of thermodynamics from energy and exergy aspects. Solar data is used from the Photovoltaic Geographical Information System (PVGIS) for Istanbul, Turkey location [5]. They calculated thermophysical properties at each state point via the mass balance equation (MBE), energy balance equation (EBE), entropy balance equation (EnBE), and exergy balance equation (ExBE), which can generally be expressed as follows:

$$\text{MBE}: \sum_{\mathrm{i}} \dot{m}_{\mathrm{i}} = \sum_{\mathrm{o}} \dot{m}_{\mathrm{o}}$$

$$\text{EBE}: \sum_{\mathrm{i}} \dot{Q}_{\mathrm{i}} + \sum_{\mathrm{i}} \dot{W}_{\mathrm{i}} + \sum_{\mathrm{i}} \dot{m}_{\mathrm{i}} \left(h_{\mathrm{i}} + \frac{V_{\mathrm{i}}^2}{2} + gZ_{\mathrm{i}} \right) = \sum_{\mathrm{o}} \dot{Q}_{\mathrm{o}} + \sum_{\mathrm{o}} \dot{W}_{\mathrm{o}} + \sum_{\mathrm{o}} \dot{m}_{\mathrm{o}} \left(h_{\mathrm{o}} + \frac{V_{\mathrm{o}}^2}{2} + gZ_{\mathrm{o}} \right)$$

$$\text{EnBE}: \sum_{\mathrm{i}} \dot{m}_{\mathrm{i}} s_{\mathrm{i}} + \sum_{\mathrm{net}} \frac{\dot{Q}_{\mathrm{net}}}{T_S} + \dot{S}_{\mathrm{gen}} = + \sum_{\mathrm{o}} \dot{m}_{\mathrm{o}} s_{\mathrm{o}}$$

$$\text{ExBE}: \sum_{\mathrm{i}} \dot{E}x_{\dot{Q}_{\mathrm{i}}} + \sum_{\mathrm{i}} \dot{W}_{\mathrm{i}} + \sum_{\mathrm{i}} \dot{m}_{\mathrm{i}} ex_{\mathrm{i}} = \sum_{\mathrm{o}} \dot{E}x_{\dot{Q}_{\mathrm{o}}} + \sum_{\mathrm{o}} \dot{W}_{\mathrm{o}} + \sum_{\mathrm{o}} \dot{m}_{\mathrm{o}} ex_{\mathrm{o}} + \dot{E}x_{\mathrm{d}}$$

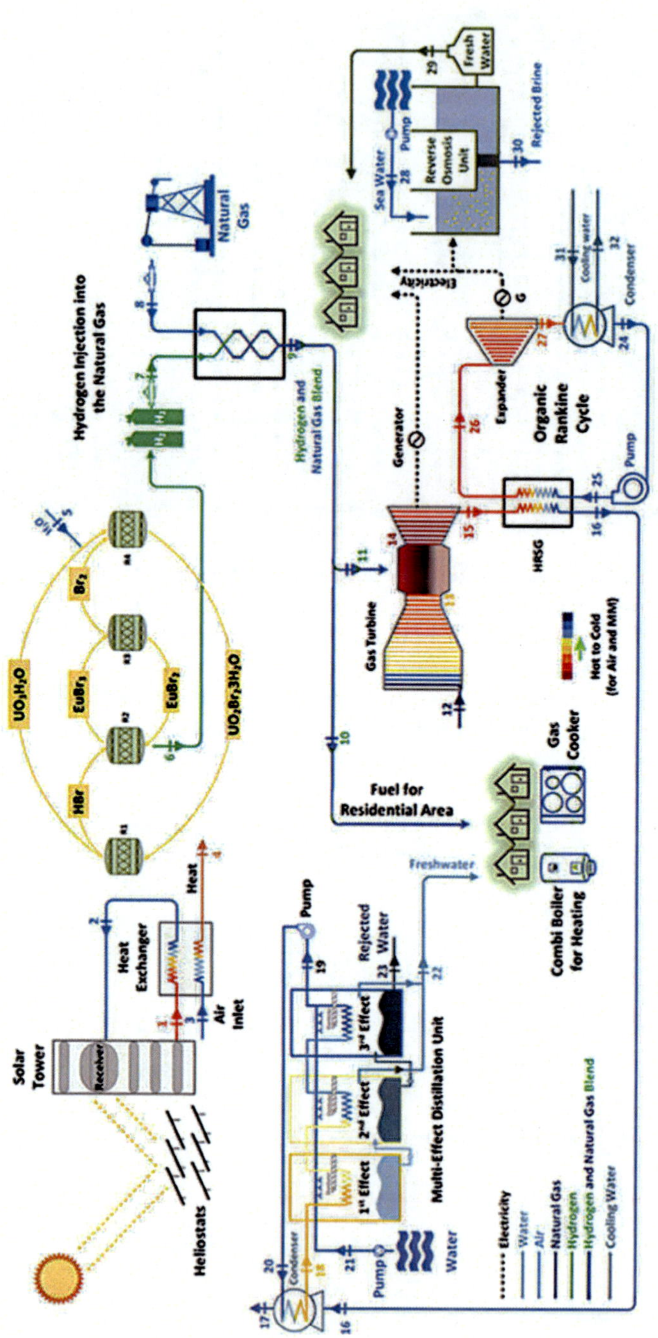

Fig. 3.14 An illustration of solar tower–based integrated system (adapted from Ref. [4])

In order to find how the system utilizes the source, the overall energy and exergy efficiencies are then calculated as follows:

$$\eta_{en} = \frac{\dot{W}_{net} + \dot{Q}_H + \dot{m}_{fw}h_{fw}}{\dot{Q}_{solar} + \dot{m}_{LNG}h_{LNG} + \dot{m}_{sw}h_{sw}}$$

and

$$\eta_{ex} = \frac{\dot{W}_{net} + \dot{E}x_{Q_H} + \dot{m}_{fw}ex_{fw}}{\dot{E}x_{Q_{solar}} + \dot{m}_{LNG}ex_{LNG} + \dot{m}_{sw}ex_{sw}}$$

Their calculations showed that the integrated community energy system for 10,000 households in Istanbul, Turkey, would need hydrogen at 0.011 kg/s. In order to produce that required hydrogen, their proposed thermochemical cycle consumes 13.5 MW of heat. These calculations include blending natural gas. If pure hydrogen is used in the system, 200 MW of heat would be required to produce hydrogen via their proposed thermochemical cycle.

When we look at the temperatures from solar tower system, they presented that the solar tower exit has a 563°C temperature, and 513°C heat is supplied to the thermochemical cycle in order to produce the hydrogen. Without considering the gas turbine cycle, the highest temperature in the system were found for the solar tower system. They used Hitec solar salt to carry out heat, which is operatable between 238 and 593°C [6]. Temperatures can be seen in Table 3.1 with other thermophysical properties for each state point.

At steady-state conditions, they found the net power generation as 4.14 MW. They supplied 0.36 MW of power and 0.38 MW of heat to run a reverse osmosis and a multi-effect distillation desalination unit. According to their calculations, they found 26.1% of energy and 21.28% of exergy efficiencies, as shown in Fig. 3.15. These results show the importance of developing unique solutions when it comes to renewables. The process of producing hydrogen by using solar energy is a process with lower efficiencies, such as 12%. When we look at the PV-based hydrogen production, from solar energy to hydrogen efficiency, it can be calculated by multiplying typical solar PV solar radiation to electricity efficiency as 20% and a typical proton exchange membrane electrolyzer efficiency as 60%, which equals 12%. The approach in their system was to collect solar energy and keep it in heat form, which enables better source utilization (solar radiation to electricity for PVs typically 20%, solar radiation to heat for concentrating solar systems typically 60%). Thereafter, they used heat in a thermochemical cycle, where high-temperature heat is the energy source rather than electricity. Although the energy and exergy efficiencies of the heavy element halide thermochemical cycle, found in their study, are 9.5 and 10.5%, their solution enables the exploitation of the wastes. Further processes can use the remaining heat for their needs. It is important to note that solar-driven power generation systems usually have higher exergy efficiencies due to the exergy of

Table 3.1 The thermophysical properties of each state point of the overall integrated system [4]

State point	T (K)	P (kPa)	ex (kJ/kg)	h (kJ/kg)	s (kJ/kg)
1	836.2	101.3	472.2	866.9	1.578
2	563.2	101.3	229.7	432.9	0.94
3	300	101.3	0	300.4	5.702
4	786.2	101.3	207.7	807.1	6.698
5	573.2	101.3	616.5	3075	8.211
6	305	101.3	0.5903	4030	53.7
7	305	101.3	0.5903	4030	53.7
8	305	101.3	0.09246	14.35	0.049
9	305	101.3	0.1123	126.6	1.548
10	305	101.3	0.1123	126.6	1.548
11	305	101.3	0.1123	126.6	1.548
12	300	121.6	15.7	300.4	5.65
13	693.5	2018	405.7	706.7	5.704
14	1573	1938	1139	1725	6.655
15	868.2	116.7	289.2	915.3	6.787
16	368.2	110.9	14.53	369.1	5.882
17	288.2	105.3	3.592	288.5	5.65
18	333.2	117.3	7.174	251.5	0.832
19	303	107.7	0.0687	125.2	0.435
20	303	121.3	0.0824	125.2	0.435
21	300	106.4	0.0051	112.7	0.393
22	300	104.3	0.003	112.7	0.393
23	300	102.2	0.0009	112.7	0.393
24	343.7	38.4	5.682	336.4	1.443
25	344.3	1708	8.017	338.7	1.443
26	563	1640	212.2	933.7	2.746
27	506.5	40	119.6	843.9	2.755
28	300.1	6000	5.911	118.6	0.393
29	298.9	180	0.0875	108.1	0.378
30	302.2	5100	10.64	131.4	0.42
31	300	101.3	0	112.7	0.393
32	340	101.3	10.25	279.9	0.917

electricity is equal to the energy of electricity, however, exergy of solar radiation as a denominator will be lower than the energy of solar radiation.

The key about solar towers, it enables high-temperature processes, which possesses a critical importance especially when seeking solutions for hard-to-electrify sectors and processes. It is further proven in [4], where high-temperature thermochemical cycle is enabled with solar tower integration. A unique solution was integrated into their system, where they blended hydrogen with natural gas to use currently available infrastructure. Their approach is a more environmentally friendly solution compared to the pure natural gas-driven processes. Also, it enables to

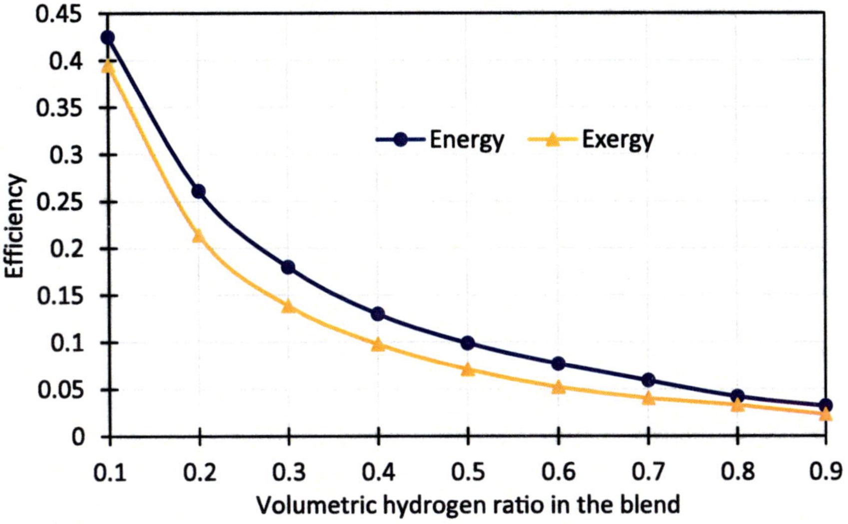

Fig. 3.15 Overall energy and exergy efficiencies of the solar-driven integrated system by volumetric hydrogen ratio in the blend (data from [4])

further understand, test, and assess the hydrogen-based systems with lower costs since the proposed system can use currently available natural gas infrastructures.

(e) *Parabolic Trough Collector.*
Parabolic trough systems are also mature technologies which are commercially proven with currently operating parabolic trough collector-type concentrated solar power plants around the globe. A parabolic trough collector consists of a curved mirror, which is typically made of polished metal in order to focus the sunlight on a focal line where the absorber tube stands. Thermofluid flows through the absorber tube to absorb the heat. A variety of thermofluids can be used, such as thermal oil, molten salt, etc. It is a typical practice to integrate parabolic trough collectors via thermal energy storage systems, typically molten salt heat storage tanks. Thermal energy storage systems enable the parabolic trough collector-type concentrated solar systems to operate in a more stable manner, since thermal energy storage can compensate hours without sunlight according to its storage capacity (Table 3.2).

One may note that parabolic trough collector systems can generate high-temperature heat up to 550°C, which is suitable for a variety of processes, including thermochemical cycle and power generation. Typically, parabolic trough systems use process heat around 350–400°C. Currently available parabolic trough systems are operating to generate power via a power cycle. A thermofluid is used in a circuit sis connected to the second circuit via a heat exchanger. The second circuit uses water as a thermofluid. The heat exchanger exchanges heat between the first and

Table 3.2 The parabolic trough–type concentrated solar power plant design parameters [4]

Parameter	Value
Heat absorption capacity	624 MW_t
Collector exit temperature	391°C
Collector inlet temperature	293°C
Number of collector loops	181
Power generation capacity	111 MW_e (gross), 99.9 MW_e (net)
Heat transfer fluid	Therminol VP-1
Location	Daggett, CA, United States
Latitude	34.85 DD
Longitude	−116.78 DD
Elevation	561 m
Meteorological data	National Solar Radiation Database
Weather station ID	91,486
Meteorological data type	Typical meteorological year
Annual global horizontal irradiance	2127.95 kWh/m^2
Annual direct normal irradiance	2799.55 kWh/m^2
Annual diffuse horizontal irradiance	456.25 kWh/m^2
Row spacing	15 m
Stow angle	170°
Deploy angle	10°
Washes per year	63
Total aperture reflective area	949,888 m^2
Solar field area	587 acres
Total land area	822 acres
Tracking power	181 kW
Pump efficiency	0.85
Piping thermal loss coefficient	0.45 W/m^2 K
Collector	SkyFuel sky trough
Collector reflective aperture area	656 m^2
Aperture width	6 m
Length of collector assembly	115 m
Number of modules per assembly	8
Receiver	Schott PTR80
Absorber tube inner diameter	0.076 m
Absorber tube outer diameter	0.08 m
Power cycle	Rankine cycle
Boiler pressure	100 bar
Condenser type	Air-cooled
Thermal energy storage	Not available

second circuits to transfer the heat from the first thermofluid (usually a molten salt) to the water. A Rankine cycle runs to generate power, where superheated steam goes into the turbine and expands. Thereafter, it gets cooled and pressurized via a pump. It enters the heat exchanger to become a superheated steam, and the cycle continues

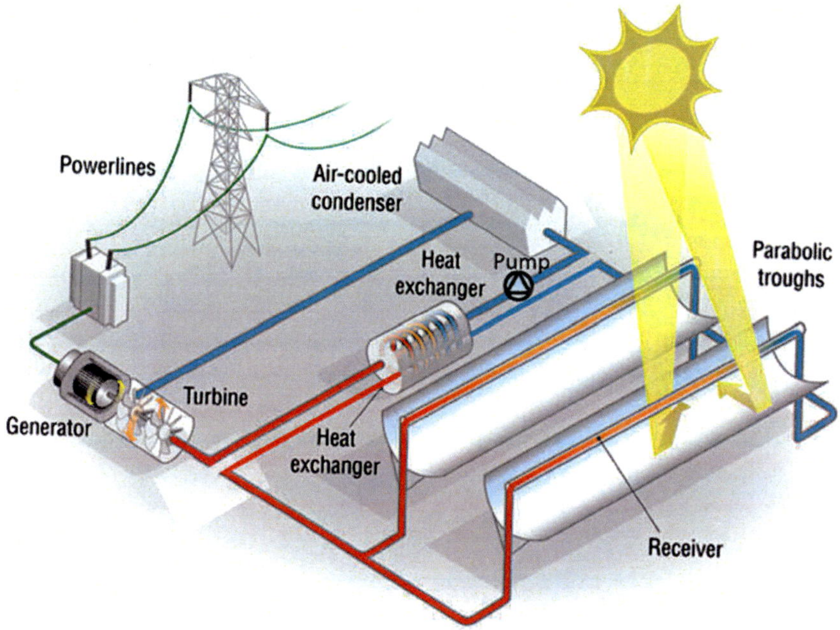

Fig. 3.16 Illustration of a parabolic trough concentrated solar power plant (modified from Ref. [4])

while generating power. In most cases, a thermal energy storage system is integrated between two circuits, which maintains the heat transfer in order to produce power in a more stable manner, and also compensates for the intermittencies of solar energy.

Case Study 3.3 For realistic analyses of renewable energy systems, it is important to use actual or calculated weather data for a specific location. It is also crucial to use the data with high resolution of time-steps. A case study is carried out to analyze a parabolic trough concentrated solar power plant, which is illustrated in Fig. 3.16, by considering an actual weather data with 60 min time-steps in a typical meteorological year for a specific location. Note that the SAM from NREL is used to carry out a time-dependent case study. The location is selected as Daggett, CA, United States at 34.85, −116.78. According to the weather data, average global horizontal irradiation is 5.83 kWh/m^2/day, and the average ambient temperature is 17°C.

In the case study, the first and second laws of thermodynamics using the energy and exergy analysis methods is used to evaluate the system. The SAM simulates the system for each hour in a typical meteorological year to find comprehensive parameters, such as mass flow rates, pumping power, thermal power, condenser operation, etc. However, the SAM only applies the first law of thermodynamics. For the second law analysis, additional calculations needed. The SAM results can be imported into a spreadsheet software for the second law analysis from exergy aspect.

In order to find the solar irradiation reaching the surface area, parabolic trough collector's aperture area needs to be multiplied by the direct normal irradiance:

$$\dot{q}_{\mathrm{DNI,PTC}} \times A_{\mathrm{DNI,PTC}} = \dot{Q}_{\mathrm{DNI,PTC}}$$

Where $A_{\mathrm{DNI,PTC}}$ denotes the total aperture area of parabolic trough collector, $\dot{q}_{\mathrm{DNI,PTC}}$ denotes the direct normal irradiance reaching the surface in selected location for unit area, and $\dot{Q}_{\mathrm{DNI,PTC}}$ denotes the direct normal solar irradiance reaching the parabolic trough collector.

For the second-law analysis, exergy of direct normal irradiance needs to be calculated. Exergy of direct normal irradiance has a correlation with both reference and source temperatures. It represents the available useful solar energy within the direct normal irradiance by considering the reference and source temperatures. The exergy of direct normal irradiance can be calculated as follows:

$$\dot{E}x_{Q_{\mathrm{DNI,PTC}}} = \dot{Q}_{\mathrm{DNI,PTC}} \times \left(1 - \left(\frac{T_0}{T_s}\right)\right)$$

In the second circuit, a Rankine cycle runs to generate power by using the heat source. Some portion of the heat can be transferred to the second circuit, remaining of the heat is wasted.

It is important to find how the source is utilized. With better source utilization, sources can be exploited for more useful outputs. Source utilization can be found by using the energy and exergy efficiencies. For the specific system, solar energy is used to generate useful heat and then power. The overall energy and exergy efficiencies of the parabolic trough collector-type concentrated solar power plant are defined as follows:

$$\eta_{\mathrm{en}} = \frac{\dot{W}_{\mathrm{net}}}{\dot{Q}_{\mathrm{DNI,PTC}}}$$

and

$$\eta_{\mathrm{ex}} = \frac{\dot{W}_{\mathrm{net}}}{\dot{E}x_{Q_{\mathrm{DNI,PTC}}}}$$

According to the calculations, 2658.3 GWh of solar energy reaches the parabolic trough collectors' 949,888 m^2 of total reflective aperture area. However, the heat transfer fluid is only able to carry 31.88% of the total solar energy to the second circuit as 847.5 GW of heat. When we look at the exergy of total direct normal irradiance, 2519.6 GWh of solar exergy reaches the parabolic trough collectors in the case study. The heat transfer fluid carries 18.38% of solar exergy to the second circuit. Figure 3.17 shows the direct normal irradiance and heat generation of the parabolic trough system, along with its exergy contents.

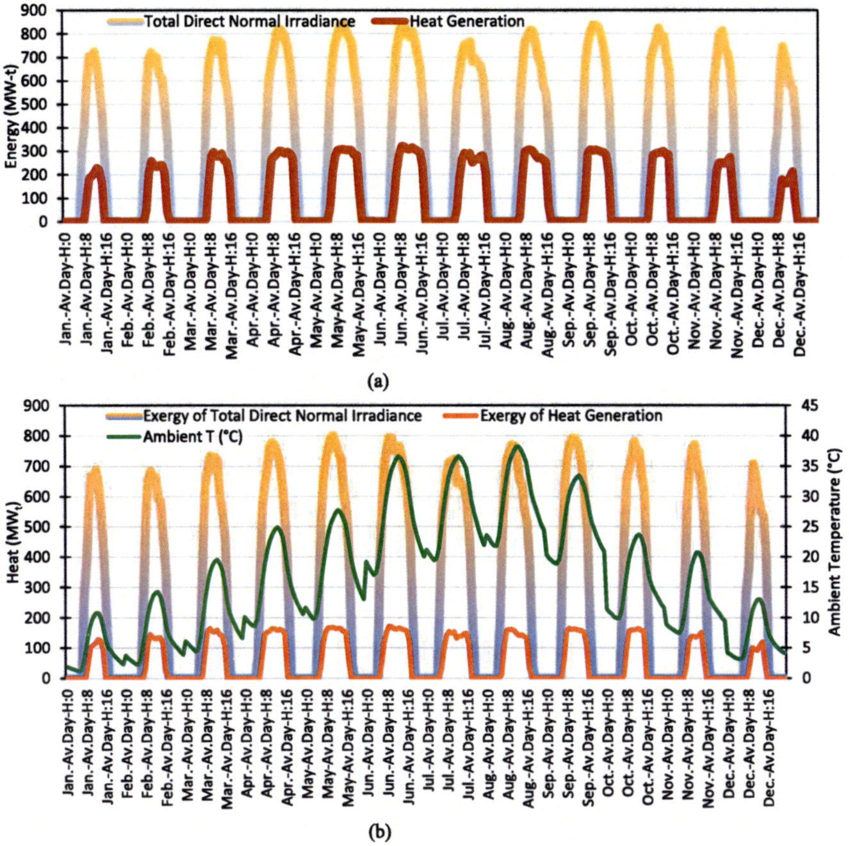

Fig. 3.17 (**a**) Direct normal irradiance and heat generation of the parabolic trough system in the case study and (**b**) their exergy contents

In the second circuit, heat from the first circuit is used to generate superheated steam. The turbine expands the superheated steam and generates power. In a typical meteorological year, the system in the case study can generate 274.9 GWh of power, as shown in Fig. 3.18. Since the objective of the given system is to generate power, the overall energy and exergy efficiencies can be calculated by using this data. The overall energy and exergy efficiencies are calculated as 10.34% and 10.91% for a typical meteorological year, as shown in Fig. 3.19.

When the system in the case study is compared with typical commercial systems, the efficiencies are lower for the system in the case study. The main reason is the instability of the given system, which is designed without heat storage to directly see the basic concentrated solar plant. The heat storage system can maintain the process and keep the temperatures at suitable levels for more stable and more efficient operations.

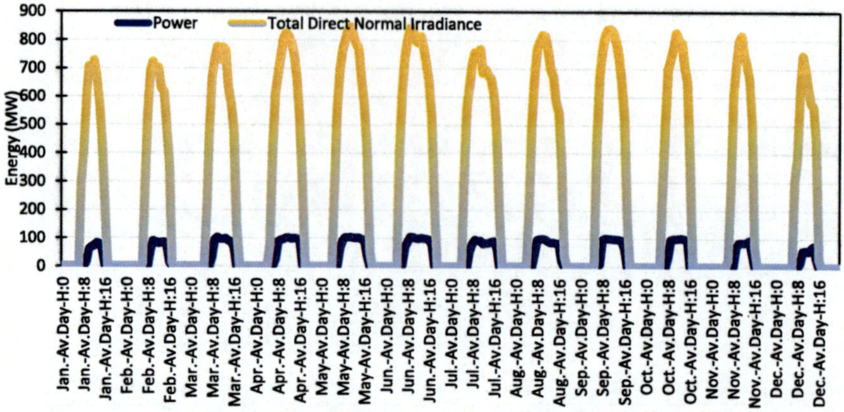

Fig. 3.18 Power generation from total irradiance for parabolic trough collector–type system for each hour in a typical meteorological year

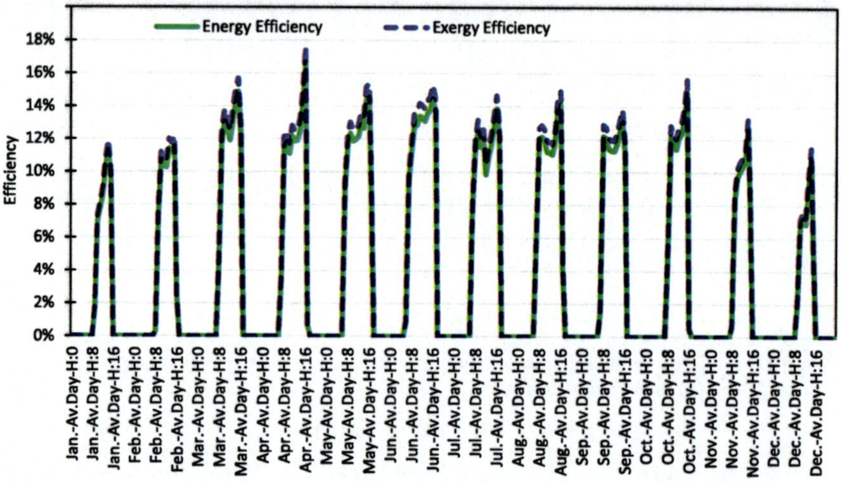

Fig. 3.19 The overall energy and exergy efficiencies of a parabolic trough collector–type system for each hour in a typical meteorological year

(f) *Linear Fresnel Collector.*

Linear Fresnel collector-type concentrated solar energy systems consist of flat mirrors which reflect the solar rays to the receiver tube. According to the design, multiple long mirrors align horizontally and face to the top where the receiver tube is placed. During the daytime, mirrors do the tracking movement on a single-axis horizontal tracker in order to keep the sunlight on the receiver tube.

Linear Fresnel technology is not mature like solar towers or parabolic trough-type concentrated solar plants. However, with recent pushes to renewable energy, linear

Table 3.3 Some linear Fresnel projects across the world [7]

Field	Location	Power capacity	Field outlet temperature	Commissioning date
Augustin Fresnel 1	Occitanie, France	300 kW	300°C	2012
Dhursar	Rajasthan, India	125 MW	510.1°C	2014
eLLO	Occitanie, France	9 MW	285°C	2019
Huaqiang	Hebei, China	15 MW	N/A	2018
Lanzhou Dacheng Dunhuang	Gansu, China	10 MW	N/A	2016
Lanzhou Dacheng Dunhuang	Gansu, China	50 MW	550°C	2019
Puerto Errado 2	Murcia, Spain	30 MW	270°C	2012

Fresnel projects are becoming more popular. Table 3.1 shows the operational linear Fresnel projects around the world as of 2023. When we look at the operational projects, the process heat temperature highly varies from 270°C to 550°C. High process heat temperatures can unlock many new opportunities for linear Fresnel systems since the variety of potential integrations enrich with increasing process heat temperatures (Table 3.3).

Case Study 3.4 A case study is carried out for a linear Fresnel collector-type concentrated solar power system. The same power cycle is used with the parabolic trough type; however, heat exchanger in Fig. 3.13 is replaced with recirculated boiler loop of a linear Fresnel as shown in Fig. 3.20.

In order to find the solar irradiation reaching the surface area, linear Fresnel collector's aperture area needs to be multiplied with the direct normal irradiance:

$$\dot{q}_{DNI,LF} \times A_{DNI,LF} = \dot{Q}_{DNI,LF}$$

where $A_{DNI,LF}$ denotes the total aperture area of linear Fresnel collector, $\dot{q}_{DNI,LF}$ denotes the direct normal irradiance reaching the surface in selected location for unit area, $\dot{Q}_{DNI,LF}$ denotes the direct normal solar irradiance reaching the linear Fresnel.

Exergy analysis is critical to get more insightful results. To find exergy value of parameters, the second law analysis is needed. The exergy of direct normal irradiance for linear Fresnel can be calculated as follows:

$$\dot{E}x_{QDNI,LF} = \dot{Q}_{DNI,LF} \times \left(1 - \left(\frac{T_0}{T_s}\right)\right)$$

Linear Fresnel–type concentrated solar power systems utilize solar energy to generate power. In order to find efficiencies, solar energy and exergy can be considered as inputs and the power is the useful output. The overall energy and exergy efficiencies of the linear Fresnel collector-type concentrated solar power plant can be calculated as follows:

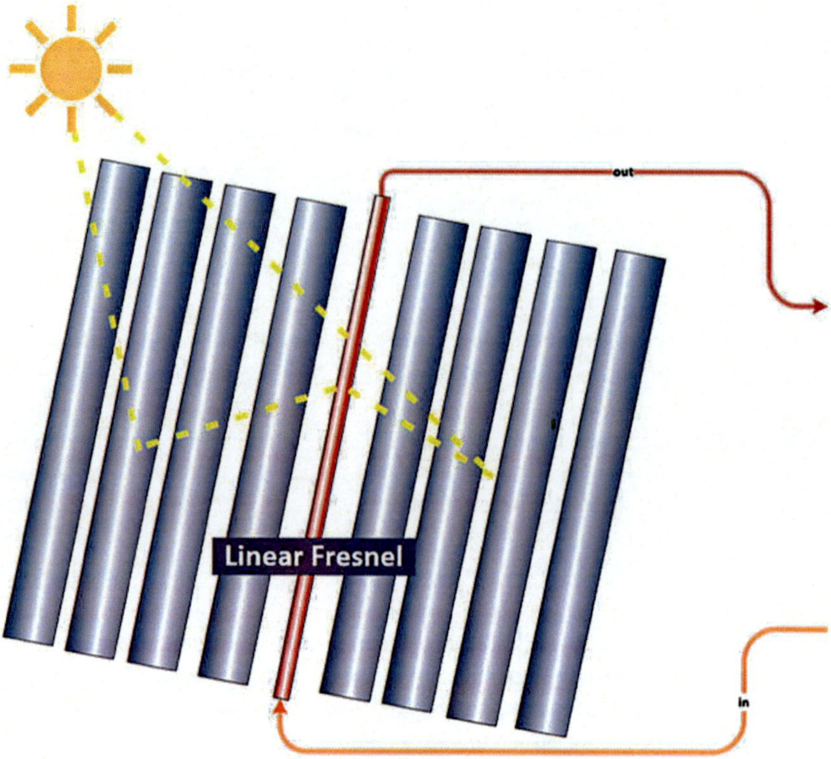

Fig. 3.20 Illustration of a linear Fresnel–type collector

$$\eta_{en} = \frac{\dot{W}_{net}}{\dot{Q}_{DNI,LF}}$$

and

$$\eta_{ex} = \frac{\dot{W}_{net}}{\dot{E}x_{Q_{DNI,LF}}}$$

To generate power from solar energy, during the heat conversion and then power generation processes, energy being wasted. Figure 3.21 clearly shows from solar energy, how the conversion is completed to generate heat and power.

For the case study, 842,848 m^2 of total reflective field aperture area is considered for Daggett, CA, United States. In a typical meteorological year, 2460.7 GWh of solar energy (direct normal irradiance) reaches the mirrors. Heat transfer fluid is able to absorb and carry 742.6 GWh of heat. The power generation cycle generated 248 GWh of net electricity in a typical meteorological year for the selected location.

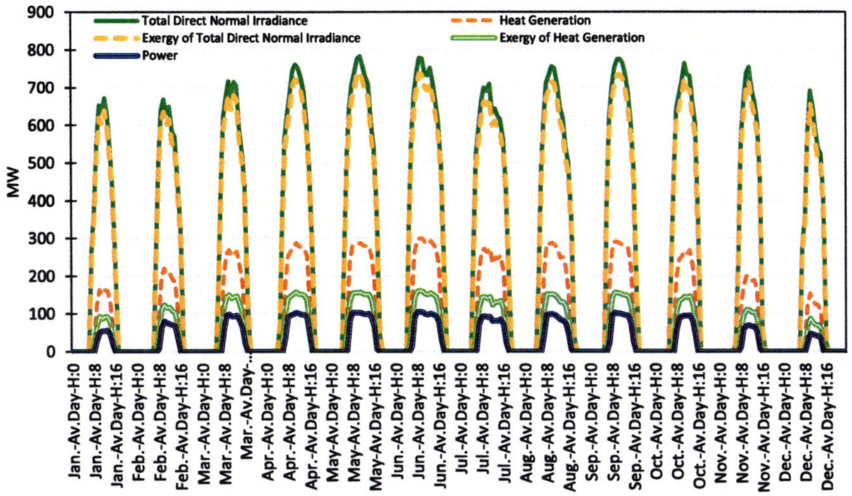

Fig. 3.21 Energy and exergy capacities of direct normal irradiance, heat, and power for linear Fresnel collector–type concentrated solar power plant

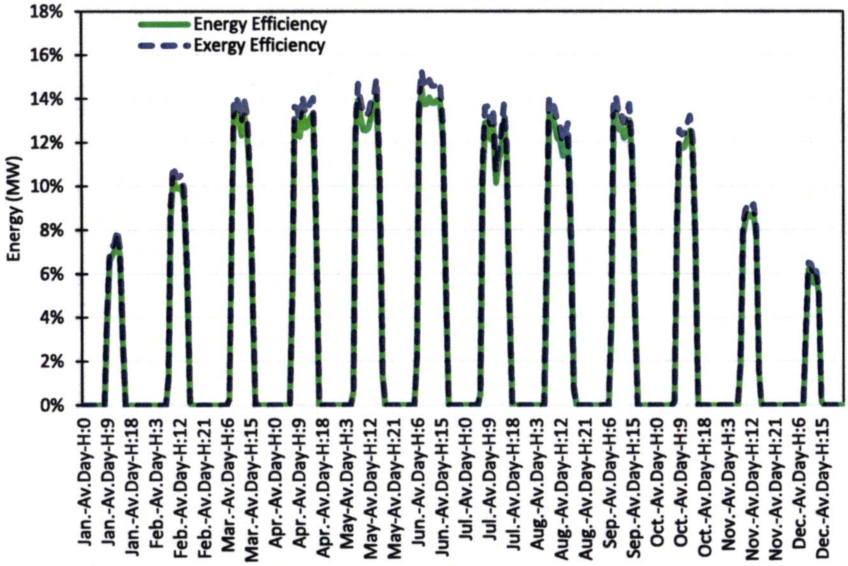

Fig. 3.22 The overall energy and exergy efficiencies of a linear Fresnel collector-type system for each hour in a typical meteorological year

The overall energy and exergy efficiencies are calculated as 10.08% and 10.63%. Figure 3.22 shows the energy and exergy efficiencies of the system in a typical meteorological year. The results are slightly lower than the previous case study with parabolic trough collectors; however, specific cases can result differently when design and conditions change.

(g) *Photovoltaics.*

Photovoltaics are one of the greatest inventions in energy area, which is developed starting in the nineteenth century and has become more efficient and more cost-effective recently. In 2012, 100 GW cumulative capacity is surpassed, and in 2018, annual capacity additions reached 100 GW. In 2022, 1300 TWh of electricity generated by solar PV technology without emitting greenhouse gases, thanks to the cost-effectiveness and increased performance of solar PV technology.

Solar cells are made of semiconductor material, they are connected in chains to form a more powerful unit called PV modules or PV panels. Each solar cell has p- and n-type layers, where p-type layer has excess electrons and n-type layer has positively charged holes. When the solar cell is exposed to the sunlight, solar irradiance transfers its energy to the electrons, which enables electrons to move into the positively charged holes in the n-type layer. This electron movement creates the electrical current, and it can be extracted via conductive material. Solar cell chains connect each other to form PV modules or panels; modules connect each other in series, then series form strings. According to the design, as an example in grid-connected designs, strings connect to the inverters and inverters connect to the transformers to connect to the electrical grid.

For the cell material, mono-crystalline, poly-crystalline, amorphous silicon, cadmium telluride/cadmium sulphide, copper indium gallium diselenide, copper zinc tin sulphide, dye-sensitized, organic, perovskite, polymer, and quantum dot types can be used. Among the cell material types, mono- and poly-crystalline types are currently popular technologies. However, for applications, such as building integrated PV, more flexible materials can be advantageous such as amorphous silicon cells. For different cases, a unique cell material can be more suitable, which needs to be assessed to find the feasibility for specific conditions.

In a solar PV plant, there are many components that form the entire system. PV modules and inverters are the major components of a PV plant. Additionally, when a solar PV plant investment is in the decision-making process, Fig. 3.23 shows the breakdown of each product or service that needs to be considered.

Two major challenges can be mentioned that block the widespread implementation of solar PV systems. First is the intermittency, which is a major challenge for many renewables, such as wind and solar. Energy storage or source shifting are two potential solutions to the intermittencies. There are many alternatives for source shifting and energy storage. In order to create a successful complementarity between service, storage, and sources, the optimum system will be unique for different locations and applications, since the service needs, and source availabilities are different for different cases. Potentially, compressed air energy storage [8], pumped hydro [9], compressed air [10], hydrogen [11], and battery [12] are some of the alternative energy storage solutions to couple with solar PV systems. Typically, thermal energy storage options are not feasible solutions for solar PVs, since solar PVs generate power, which will be significantly lost once power to thermal energy and then thermal energy to power conversions are carried out. On the other hand,

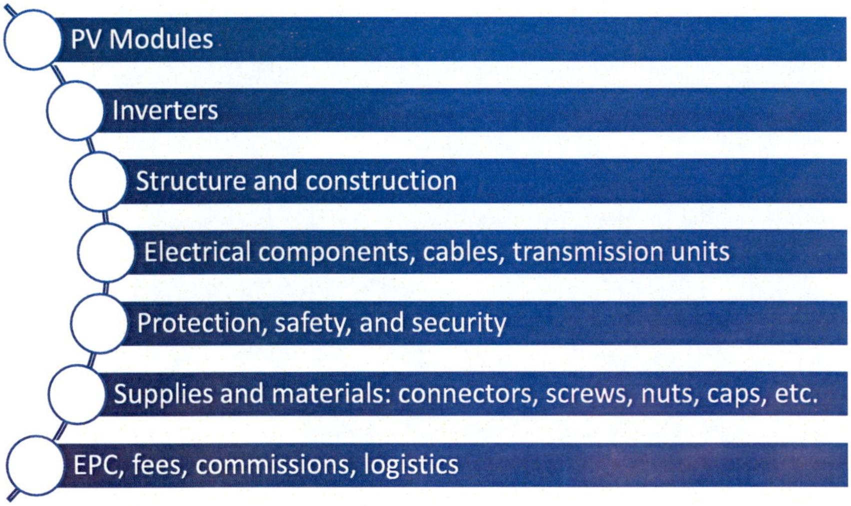

Fig. 3.23 Breakdown of components for a solar PV plant investment

mechanical energy, potential energy, or chemical energy forms are more suitable storage options, which will lead less conversion losses. Source shifting is another alternative which helps to compensate for intermittencies. Solar PV systems can be integrated with other systems which use wind [13], geothermal [14], or hydro [15] energy sources. Multiple sources can improve the system reliability.

The second major challenge can be mentioned as the land footprint of solar PV plants. Depending on the design and technology, a MW of solar PV plant typically needs 4–8 acres of land. Land scarcity is a critical problem for many countries, especially in future it will be more problematic with growing populations and economies. A potential solution to overcome this challenge is to use the land for dual or even more reasons. Unexploited surfaces can be exploited such as rivers, oceans, or lake surfaces with floating PV systems [16, 17]. If transparency is arranged properly, Agrivoltaics can be installed on agricultural lands with higher structures and wider pitch distances compared to regular PV plants [18]. In the transportation industry, solar PVs are also popular, such as solar PV-driven aircraft [19], locomotives [20], railways [21], and highways [22], where unexploited surfaces of transportation vehicles or transportation ways are used to install PV modules.

(h) *Bifacial PVs.*
Bifacial PV technology is a module design where two sides of PV modules are able to collect the sun's rays as shown in Fig. 3.24. When the components of the sun's rays are considered, there are many available possibilities to reach the back surface of solar PV modules. The geometric orientation of the PV plant is very critical to exploit the potential of bifaciality. Typically, higher structures and more reflective surfaces with high albedo are two important boosters of bifacial gain.

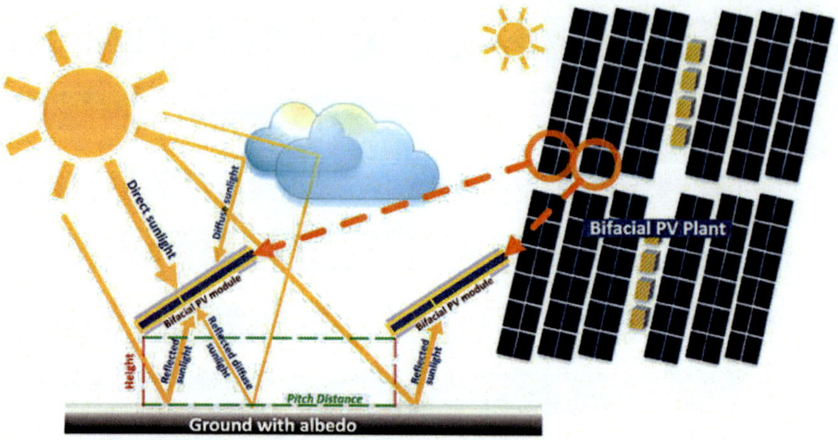

Fig. 3.24 Illustration of bifacial PV modules

The ground albedo is one of the important parameters which significantly effects the bifacial gain. The Albedo comes from the Latin word *Albus*, which means *white*. For example, fresh snow or whiter terrains have higher albedo, and darker terrains have lower albedo. A typical albedo can be considered as 0.2, which means 20% of the light can be reflected from the surface. There are many techniques that can be used to improve ground albedo. Chipping stones are one of the examples, which is a low-cost bifacial gain booster. Moreover, chipping stones can be mixed and painted with reflective paints such as road paints, which can potentially improve the ground albedo to around 80%. When the cost analysis is made, improving ground albedo is typically very effective to improve the levelized cost of electricity.

Combining high-ground albedo with a suitable design can synergetically boost the bifacial gain. Height and pitch distance are two important geometric orientation parameters; when PV modules are mounted at higher distances to the ground as well as longer distances to other PV modules, the sun's rays will find more surface to reflect. Horizontal single-axis tracker systems are also a good match for bifacial PV plants in order to improve bifacial gains. In the literature, 20% [23–25], 29% [26], and 25% [27] bifacial gains were found in specific case studies.

(i) *Floating PV.*

Floating PV plants use unexploited, unutilized water bodies to be constructed. Floating PV plants consist of pontoons, rafts or any floating object as their structural components. As mentioned previously, the optimum renewable energy system for each region and each case is unique. Floating PVs become very advantageous in different cases. Land footprint is a major challenge for PV plants. Especially for regions with land scarcity, floating PV plants are a potential solution to use solar energy via PVs.

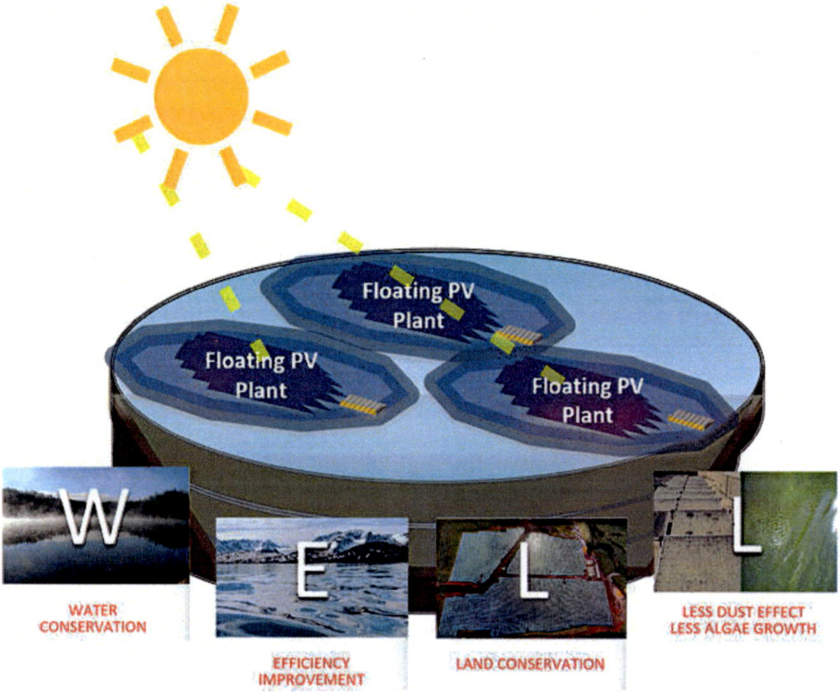

Fig. 3.25 Floating PV plant and its advantages

There are many advantages to use floating PVs, some advantages are summarized in Fig. 3.25. Land conservation is the main advantage of floating PVs: 4–8 acres of land for a MW of PV plant can be used for other useful activities, such as to build residential areas, agricultural activities, mining, tourism, etc. Water conservation is another advantage, which is important for irrigation dams or freshwater reservoirs. Floating structures cover the water body and minimize the interaction with sun's rays as well as wind, which leads to water conservation due to the lower evaporation. Temperature coefficients are important for solar PV power generation performance. Higher temperatures lead to lower power generation for PV modules. Water bodies have cooler surfaces due to the water evaporation, which decreases the temperatures; therefore, higher power generation leads to efficiency improvement. Another advantage to minimize the interaction of sun's rays with the water body is to minimize algae growth. Since algae and some microorganisms require sunlight to grow, it is unwanted for water bodies since it possesses public health concerns. Moreover, water bodies appear to have almost no dust content compared to the lands. Dust is an important problem for solar PVs, which requires periodic washes of modules. Less dust effect on water bodies can minimize losses regarding dusty PV module surfaces.

On the other hand, the effect of floating PVs on sea creatures is still not well examined. It needs to be clearly examined for specific cases, in order to eliminate environmental risks and impacts.

Case Study 3.5 A case study is carried out with the data obtained from [28] for bifacial and floating PVs for the Cochrane, Ontario location. A MW_p of floating PV plant and a MW_p of bifacial PV plants are compared for the same meteorological data. In a typical meteorological year, 1735 kWh/m^2 global horizontal irradiance is available for unit area in Cochrane, Ontario. Floating PV plant is designed on a fixed floating pontoon, horizontally placed two PV modules in a row facing to south with 34° of tilt angle. The pitch distance is given as 8 m. Bifacial PV plant is designed on terrestrial land with a horizontal single-axis tracker structure. 8 m of pitch distance is given between tracker columns. A MW_p of Floating PV plant design produced 1432.2 MWh of electricity in a year, while a MW_p of bifacial PV plant produced 1763.8 MWh of electricity, as shown in Fig. 3.26. During the months of November, December, January, and February, floating PV plant produced more electricity than the bifacial PV plant. The main reason of bifacial PV plants are designed with horizontal axis which has 0° of tilt angle, in other words, bifacial PV modules are parallel to the ground at midday. On the other hand, a floating PV plant uses 34° of tilt angle, which is more advantageous, especially during winter months due to the sun's angle.

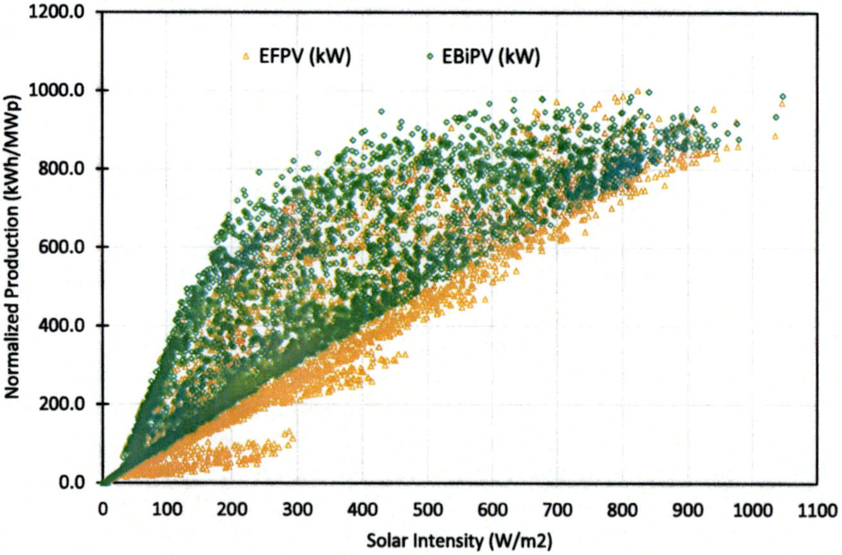

Fig. 3.26 Normalized electricity production of FPV and BiPV by solar intensity (data from Ref. [28])

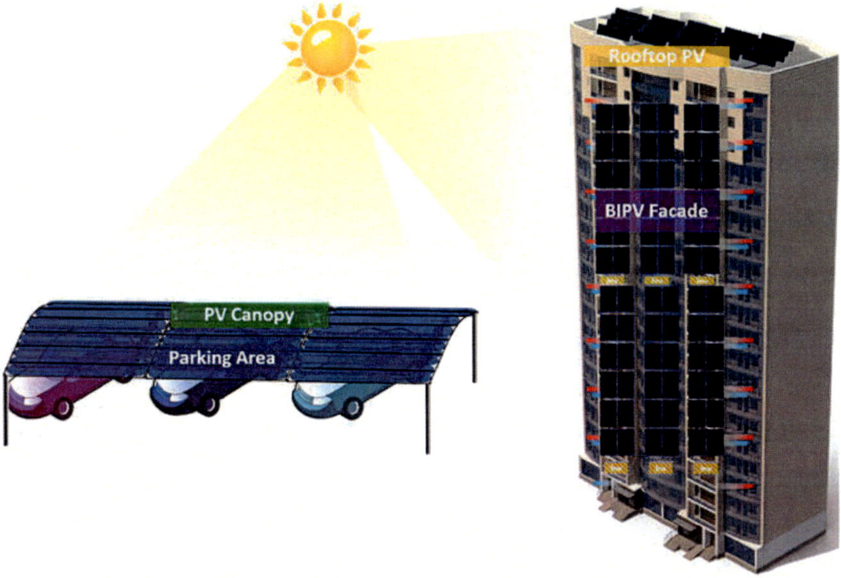

Fig. 3.27 Building-integrated PVs on parking area, rooftop, and façade

(j) *Building Integrated Photovoltaics.*
In order to overcome the land footprint challenge of solar PVs, using residential areas is another alternative. Using residential areas to install PV modules is one of the solutions, in which parking areas, building façades, or/and rooftop areas can be used to install PV modules as shown in Fig. 3.27.

When the residential area is densely populated, façades typically have more shade than usual. In those cases, roofs are usually the best option to use for densely populated areas. On the other hand, especially if the sun-facing façade has no blocking structure between the sun and the façade, it might possibly be an advantageous case for using the building façade to mount PV modules. Parking areas are also potential areas which can be exploited for PV usage. A new legislation is approved in France which mandates parking areas with more than 80 vehicle capacity needs to be covered by solar panels. There are many advantages to cover parking areas or buildings with solar PVs. Besides the land footprint, building or parking area-integrated PVs can help to decrease cooling loads for cars and buildings. On the other hand, the extra weight of solar PVs on buildings can give extra stress to the building and the ability of buildings to resist and react to earthquakes and other catastrophic events can be affected. Therefore, structural analysis should be carefully carried out for building integrated PVs.

Case Study 3.6 A case study is carried out in order to investigate the performance of different residential PV applications, namely parking area PV canopy, building-integrated PV façade, and rooftop PV as shown in Fig. 3.28. Using the available

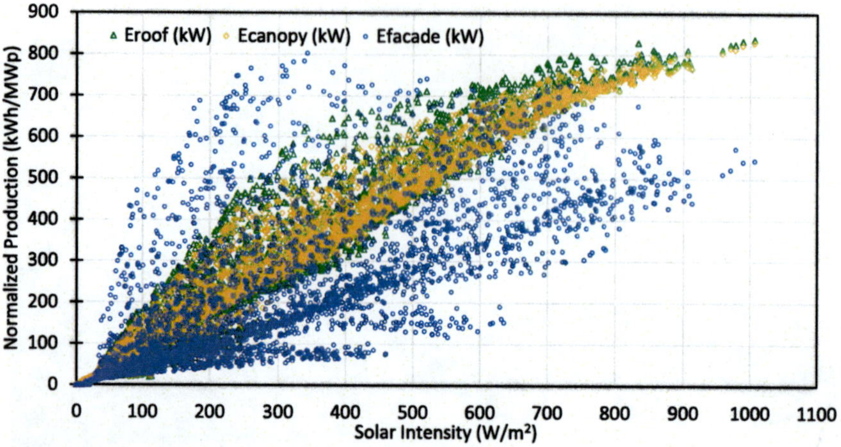

Fig. 3.28 Electricity generation from different building-integrated PVs

surfaces of residential areas can decrease the power output compared to the terrestrial PV plants due to the azimuth and tilt angles of available surfaces. These can be further adjusted in the orientation. Data from [29] is used to show how different residential area usage for solar PVs affects electricity generation in Fig. 3.22.

According to the data from [29] for a typical meteorological year in Frankfurt, Germany, a MW of PV system mounted on building façade (Efacade on Fig. 3.28) generates 829.16 MWh of electricity, a MW of PV system mounted on parking area canopy (Ecanopy on Fig. 3.28) generates 1105.9 MWh of electricity, and a MW of rooftop (Eroof on Fig. 3.28) PV generates 1123.4 MWh of electricity.

3.2.2 Hydro Energy

Hydro energy exploits the potential energy of the elevated water, by utilizing the water movement via hydro turbines. When we look at the countries with high renewable share in their primary energy sourcing, hydropower plays a very critical role. In the Introduction section, successful jurisdiction examples with high renewable share are presented, such as Iceland and New Zealand. The common point for both of these countries is sourcing more than half of the electricity production with hydro energy. When we look at the other countries with a high renewable share, most of these countries have hydropower as the major energy source among renewables [30]. This is not a coincidence, because the high-capacity factor and fewer intermittencies of hydro energy compared to other renewables, make hydro energy a more reliable and effective solution. Hydro energy systems have capacity factors of around 45%. Moreover, intermittencies are more predictable than solar PV or wind, which puts hydro energy in a more advantageous position.

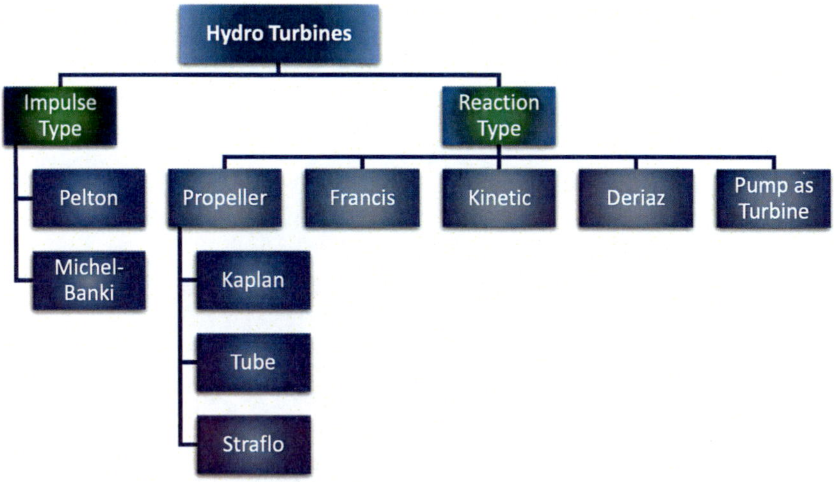

Fig. 3.29 Types of hydro turbines

There are various available hydro turbines, the most commonly used hydro turbines can be mentioned as Pelton, Kaplan, and Francis (the classification of all hydro turbines is given in Fig. 3.29). Different turbine types have different feasibility for different cases; therefore, it is important to select the correct turbine type for specific applications, which can be done by using the chart in Fig. 3.30.

The difference between impulse and reaction turbines is the exploitation difference of hydro energy. Impulse turbines are simply rotated by the water movement where kinetic energy is the driving force. Reaction turbines additionally exploit the pressure difference.

Pelton turbines are impulse type of turbines which utilize only the kinetic energy of water. Among the hydro turbines, Pelton turbines are very suitable for high head applications up until 2000 m. Francis turbines are typically suitable for 3–150 m of heads, up to 600 m of heads in some cases. The number of runner blades for Francis turbines is between 16 and 24. Francis turbines are a reaction type of turbines that also exploit the pressure difference. While energy efficiency can be around 85% for Pelton turbines in typical operations, Francis turbines have around 90% of energy efficiency. On the other hand, Francis turbines have a worse efficiency curve than Pelton turbines since Francis turbines are more sensitive to cavitation, and efficiency drops can be seen when the flow is unstable. Kaplan turbines are suitable options for low water heads, typically between 0.8 and 5 m, up to 100 m in some cases. Kaplan turbines consist of three to eight runner blades, which are adjustable. The rotation speed varies between 600 and 1000 RPMs for Kaplan turbines.

In order to find the electricity generation of hydro energy systems, the useful potential energy needs to be calculated by multiplying the mass flow rate ($\dot{m}$), gravitational force (g), elevation (z) difference after removing the total head losses, and generator efficiency, as follows:

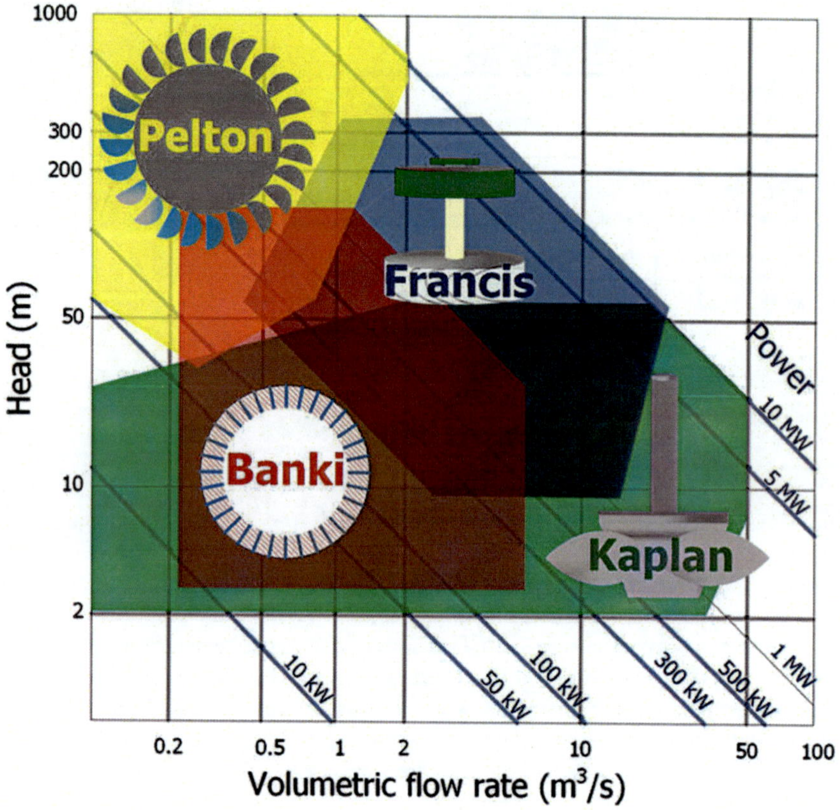

Fig. 3.30 Selection chart of hydro turbines (data from Ref. [31])

$$\dot{W}_{\mathrm{hydro,el}} = \eta_{\mathrm{gen}} \dot{m} g \left(Z_{\mathrm{upper\ reservoir}} - Z_{\mathrm{lower\ reservoir}} - \sum \mathrm{Head\ losses} \right)$$

Case Study 3.7 As mentioned previously, the steady operation characteristics of hydro energy systems make it a potential solution where a steady power supply is needed. Altayib and Dincer [32] developed a hydro-based integrated energy system for multiple useful outputs, as shown in Fig. 3.31. A river was considered as the hydro energy source, which has between 20 and 30 m of head. According to the head and flow rate, they selected the Kaplan turbine to use in the system. The produced electricity is used in methanol synthesis, reverse-osmosis desalination, and power storage processes to generate methanol and fresh water, besides power. They used five Kaplan turbines to exploit the hydro energy. Each Kaplan turbine and the hydro energy system has design parameters as shown in Table 3.4.

They considered that the river's annual average mass flow rate is 426 tonnes/s. Their heads were varying between 20 m and 30 m, which can bring between 85.2

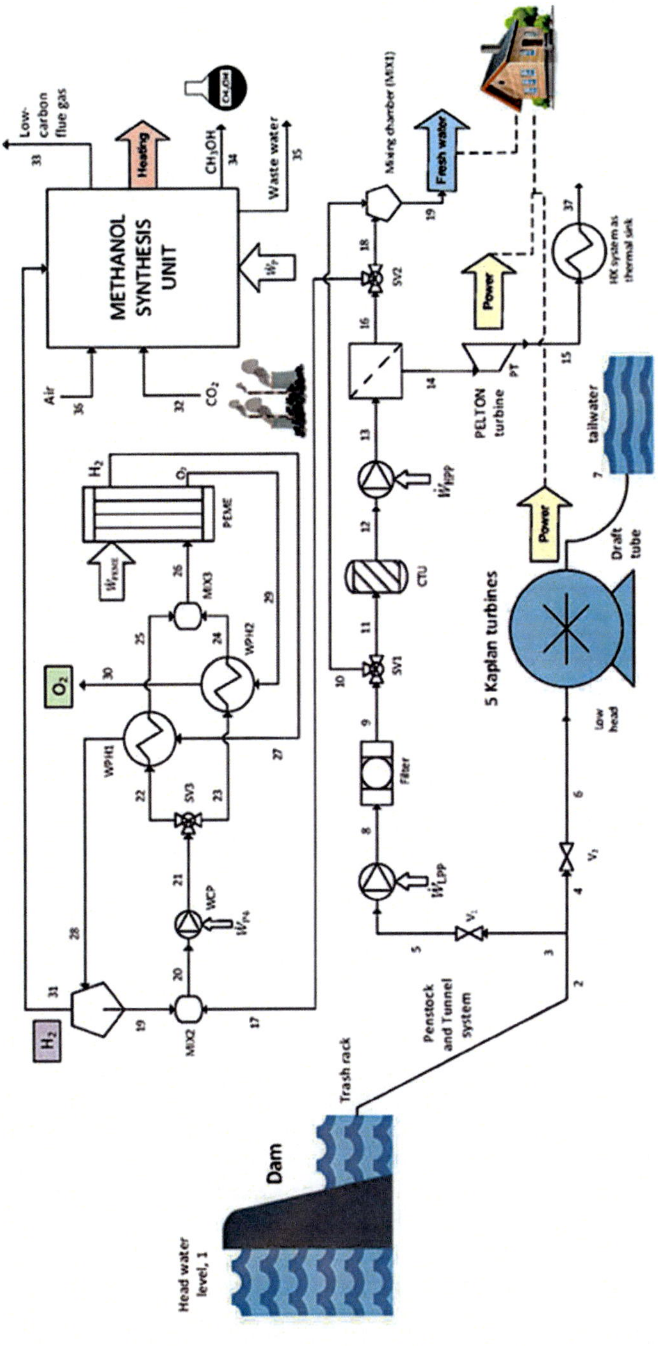

Fig. 3.31 An illustration of hydro energy–based integrated system (adapted from Ref. [32])

Table 3.4 The design parameters for hydro energy system's one hydro turbine

Parameter	Sizing value
Turbine specific speed	$N_S = 500$
Turbine speed	N = 222 RPM
Turbine discharge diameter	$D_d = 3.324$ m
Turbine intake velocity	$\nu_i = 7.16$ m/s
Turbine nameplate power	$P_{KT} = 25$ MW
Turbine intake diameter	$D_i = 4.81$ m/s
Penstock diameter	$D_p = 4.135$ m
Trash rack hydraulic diameter	$D_t = 12$ m
Draft tube velocity	$\nu_d = 6.5$ m/s
Draft tube diameter	$D_d = 3.5$ m
Turbine efficiency	$\eta_T = 92.5\%$
Electric generator efficiency	$\eta_{GEN} = 98.2\%$
Electric transformer efficiency	$\eta_{TRA} = 99.5\%$
Head loss at trash rack	$\Delta z_{TR} = 0.15$ m
Head loss at penstock entrance	$\Delta z_{P,en} = 0.14$ m
Penstock's major head losses	$\Delta z_{P,m} = 0.05$ m
Head loss at penstock exit	$\Delta z_{P,ex} = 0.02$ m
Head loss at tee branch (from state point 2 to 3)	$\Delta z_{T,b} = 2.7$ m
Head loss at tee straight (from state point 2 to 4)	$\Delta z_{T,s} = 0.25$ m
Head loss by valve V1	$\Delta z_{V1} = 0.07$ m
Head loss by valve V2	$\Delta z_{V2} = 0.18$ m
Head loss at the turbine exit	$\Delta z_{KP,ex} = 0.02$ m
Head loss at draft tube	$\Delta z_{DT} = 0.59$ m

Data from Ref. [32]

and 127.8 MW of potential energy input, as shown in Fig. 3.32. Kaplan turbine design efficiency is 92.5%, which can drop to 50% when the load drops to 10% in terms of mass flow rate. On the other hand, if the mass flow rate of the river 10% exceeds the design point, efficiency drops from 92.5% to 91%. A hydro storage system is also designed by considering Ontario, Canada's electricity rates to optimize the benefits.

They integrated a Pelton turbine with the hydro energy storage system. Their calculations showed that 75% of the annual flow of the river is sufficient to meet the baseload of the integrated system, in which a proton exchange membrane electrolyzer and a reverse osmosis unit require 47.85 MW of power. The remaining 25% of excess power is used for storage and also injected into the electrical grid.

Their proposed integrated system can deliver 81.25 MW of electricity to the grid, 86.1 tonnes of methanol per day, 12.3 MW_t heating, 129 tonnes of oxygen per day, and 25,229 tonnes of fresh water in an hour. Their system exploited the potential energy of hydro by generating five different useful outputs. According to their calculations, they presented 61.6% of energy and 58.2% of exergy efficiencies for their overall system.

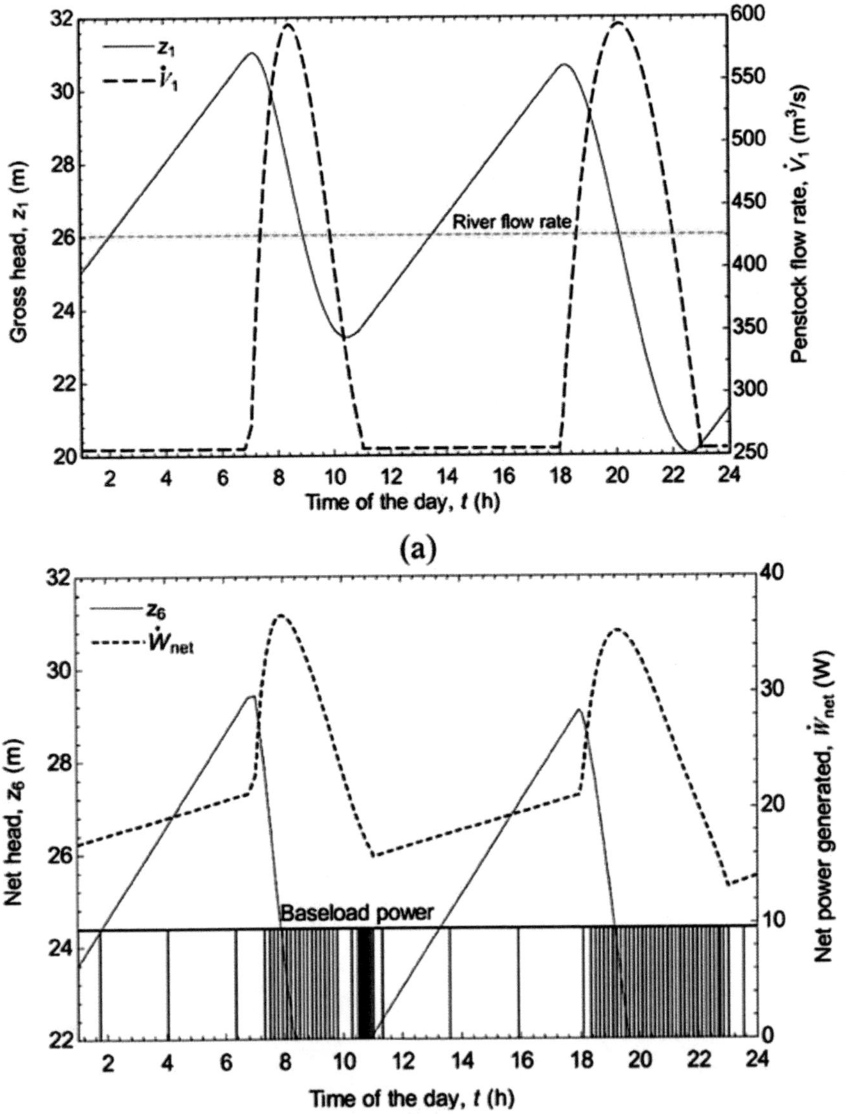

Fig. 3.32 (**a**) Hourly gross heads with penstock flow rate for the system and (**b**) hourly net heads with net power generation for each Kaplan turbine (adapted from Ref. [32])

3.2.3 Wind Energy

During the last decade, wind has been unprecedentedly growing with cost drops and climate action targets from jurisdictions and industries. When we look at the levelized cost of electricity, utility-scale onshore wind is the best option among all types of power generation systems as of 2023. Wind energy possesses a great potential for future energy, and the majority of the wind potential is currently unharnessed and untamed.

When we look at the wind energy history, which is depicted in Fig. 3.33, the first man-made wind harvesting method can be mentioned as ancient sailors, who utilized

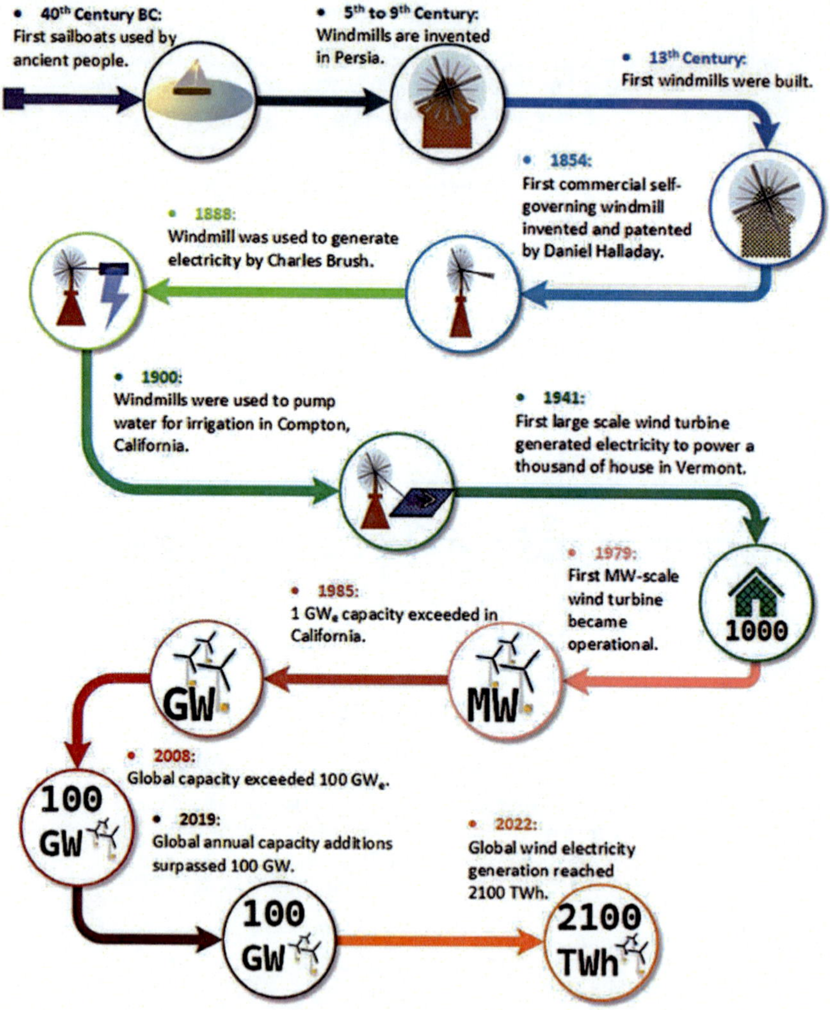

Fig. 3.33 History of wind energy

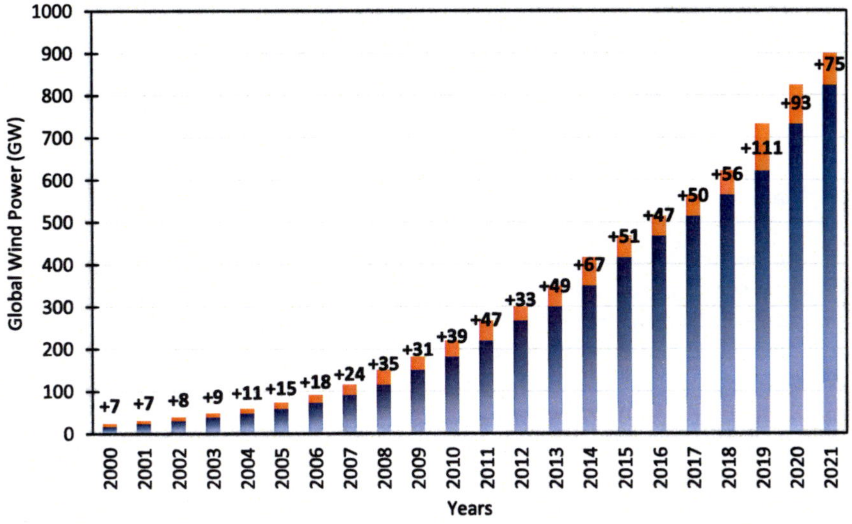

Fig. 3.34 Global wind power installed capacity and annual additions (data from Ref. [35])

winds to mobilize their sailboats in the fourteenth century BC. In the fifth century BC, the first windmills were invented in Persia. The first windmill was built in the thirteenth century, which are used to mill grains. At the end of the nineteenth century, windmills were used for electricity generation purposes for the first time. Irrigation pumps were the first applications that were electrified with windmills. Larger-scale power generation started with a residential application in Vermont in 1941. MW-scale wind turbines became operational in 1979. In 2008, global wind power capacity exceeded 100 GW. After 10 years, annual capacity additions surpassed 100 GW per year. During 2022, wind-based electricity generation reached 2100 TWh. From 2001, global wind power cumulative capacity and capacity additions are shown in Fig. 3.34. Wind-based energy systems are still majorly untamed and unharnessed since the global wind energy potential is 615 PWh, which is enough to 24 times meet the global power consumption [33, 34].

When it comes to the classification of wind turbines, it is important to note that again, renewable energy solutions are not universal but unique from case to case. According to the case, it is important to determine which type of wind turbine is more suitable for specific applications. Among the wind turbine types, triple-bladed horizontal axis wind turbines are the widespread applications, which dominate global wind energy market. There are two main categories that are distinguished by the placement of the axis of rotating shafts. Horizontal wind turbines in general, have a rotor shaft axis which are placed horizontally to the ground surface. Therefore, blades are vertical to the ground surface, which faces toward the wind direction. Vertical wind turbines have a rotor shaft axis which is placed vertically to the ground surface. Horizontal wind turbines are suitable for very large-scale applications, and they can have very large rotor diameters, in some cases, they can be larger than two soccer fields. As of June 2023, the world's largest wind turbine has 16 MW of power generation capacity, 252 m of rotor diameter, and 146 m of hub height built-on the southeastern coast of China [36]. There are other concepts available which have 260 and 280 m of diameters to generate 18 MW of power, however, they are not commercially deployed yet [37, 38]. Due to the limitations, the abovementioned large-scale wind turbines are offshore wind turbine applications. For the onshore applications, one of the largest onshore wind turbines has 7.5 MW of power capacity with 127 m of rotor diameter [39].

Vertical wind turbines have more flexible application areas since the wind direction can be in any direction to drive the vertical wind turbines. However, faster degradation and lower efficiencies compared to the horizontal wind turbines make vertical wind turbines less effective. Although typically horizontal wind turbines are commercially more cost-effective, in special cases, vertical wind turbines can be more effective.

Savonius wind turbines are effective solutions, especially for small-scale applications. They are more advantageous against turbulent wind also. Typically, Savonius turbines are less efficient than Darrieus turbines. Cost is typically lower for Savonius turbines than horizontal wind turbines since Savonius turbines have less complex mechanisms and require less components. Darrieus turbines are also insensitive to wind direction; therefore, there is no yaw mechanism required. Transmission systems, generator, and other parts can be mounted on the ground level, which makes the structure less complicated. For Darrieus turbines, self-start can fail, especially with low starting torque. There are also other designs in the vertical-axis wind turbine category, which attracted many researchers to address the research gaps. However, for typical current commercial cases, horizontal wind turbines possess a highly cost-competitive option. Further classification is shown in Fig. 3.35.

In order to carry out a thermodynamic analysis for a wind turbine, based on the first and second laws of thermodynamics, in other words, energy and exergy aspects, wind energy and forms need to be clarified. Before defining wind energy, the general energy and exergy balance equations can be expressed as follows:

$$\sum_{\mathrm{i}} \dot{\mathrm{m}}_{\mathrm{i}}(h + ke + pe)_{\mathrm{i}} = \sum_{\mathrm{o}} \dot{\mathrm{Q}}_{\mathrm{o}} + \sum_{\mathrm{o}} \dot{\mathrm{W}}_{\mathrm{o}} + \sum_{\mathrm{o}} \dot{\mathrm{m}}_{\mathrm{o}}(h + ke + pe)_{o}$$

$$\sum_{\mathrm{i}} \dot{\mathrm{E}}\mathrm{x}_{\dot{\mathrm{Q}}_{\mathrm{i}}} + \sum_{\mathrm{i}} \dot{\mathrm{W}}_{\mathrm{i}} + \sum_{\mathrm{i}} \dot{\mathrm{m}}_{\mathrm{i}}\mathrm{ex}_{\mathrm{i}} = \sum_{\mathrm{o}} \dot{\mathrm{E}}\mathrm{x}_{\dot{\mathrm{Q}}_{\mathrm{o}}} + \sum_{\mathrm{o}} \dot{\mathrm{W}}_{\mathrm{o}} + \sum_{\mathrm{o}} \dot{\mathrm{m}}_{\mathrm{o}}\mathrm{ex}_{\mathrm{o}} + \dot{\mathrm{E}}\mathrm{x}_{\mathrm{d}}$$

where i and o subscripts denote input and output, and ke and pe denote kinetic energy and potential energy. Specific exergy for the general balance equations can be expressed as follows:

$$ex = (ke + pe + (h - h_0) - T_0(s - s_0))$$

Wind energy can be expressed as the function of kinetic energy which can be expressed as the air at a certain mass flow rate flows through the wind turbine at a certain speed:

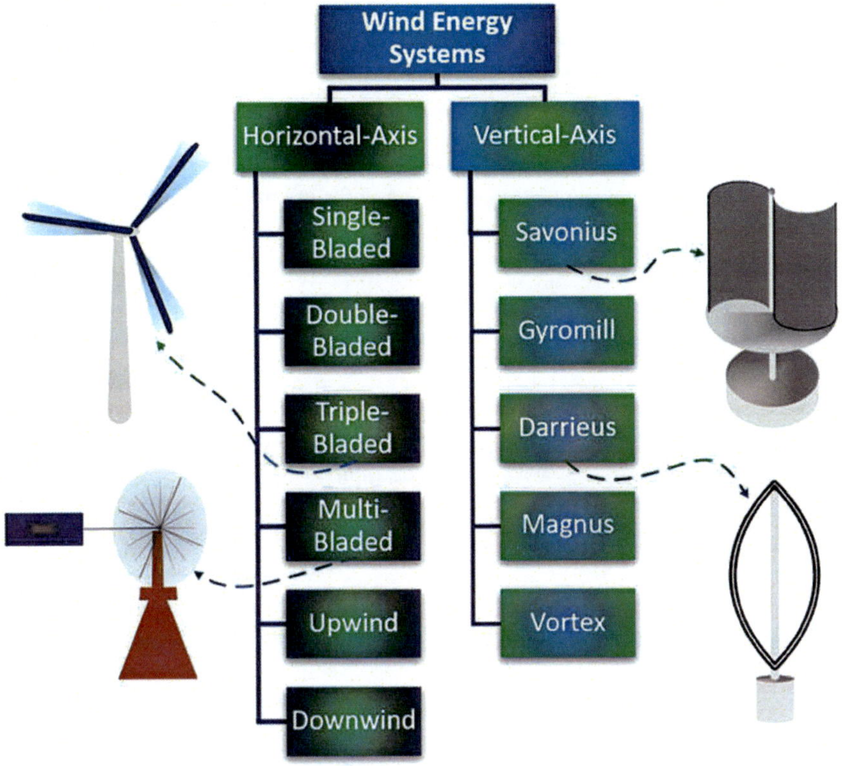

Fig. 3.35 Classification of wind turbines

$$E_{k,\text{wind}} = \frac{1}{2}\dot{m}_{\text{air}} V_{\text{air}}^2$$

where mass flow rate can be also expressed as follows:

$$\dot{m}_{\text{air}} = \rho \pi r^2 V_{\text{air}}$$

The wind power can be expressed as follows:

$$P_{\text{wind}} = \frac{1}{2}\rho \pi r^2 V^3$$

Betz limit is the maximum power coefficient that denotes the exploitable mechanical energy within the kinetic energy of the wind. According to the theory exhibited by Albert Betz, wind turbines cannot exploit more than 59.3% of wind kinetic energy for power generation. The Betz Limit expresses that no wind turbine can generate more power than 59.3% of the kinetic energy of the wind. To account Betz limit, the following expression can be used:

$$\dot{W}_{\mathrm{wt,max}} = C_B P_{\mathrm{wind}}$$

where $\dot{W}_{\mathrm{wt,max}}$ denotes the maximum possible work generation of the wind turbine.

In order to carry out calculation according to the second law of thermodynamics, the change in enthalpy needs to be defined as the change of the flowing matter at two different state points which is shown in Fig. 3.10, which can be expressed as follows (Fig. 3.36):

$$\Delta H_{\mathrm{air}} = \dot{m}_{\mathrm{air}} C_P (T_2 - T_1)$$

where T_1 and T_2 denote the wind chill temperatures at the state points (Fig. 3.37). According to the Joint Action Group for Temperature Indices expression [41], Wind chill temperature can be expressed in °F and mph as follows:

$$T_{wc} = 35.74 + 0.6215\, T_{\mathrm{air}} - 35.75\left(V^{0.16}\right) + 0.4274\, T_{\mathrm{air}}\left(V^{0.16}\right)$$

On the other hand, the entropy difference needs to be defined in order to proceed to the exergy definition. The entropy difference can be written as follows:

$$\Delta S_{\mathrm{air}} = \dot{m}_{\mathrm{air}} T_{\mathrm{ambient}} \left(C_P \ln\left(\frac{T_2}{T_1}\right) - R \ln\left(\frac{P_2}{P_1}\right) - \frac{Q_{\mathrm{loss}}}{T_{\mathrm{at}}} \right)$$

Therefore, exergy of the wind can be calculated as follows:

$$Ex_{\mathrm{wind}} = E_{\mathrm{generated}} + \dot{m}_{\mathrm{air}} \left(C_P (T_2 - T_1) + \left(T_{\mathrm{ambient}} \left(C_P \ln\left(\frac{T_2}{T_1}\right) - R \ln\left(\frac{P_2}{P_1}\right) - \frac{Q_{\mathrm{loss}}}{T_{\mathrm{at}}} \right)\right)\right)$$

where $E_{\mathrm{generated}}$ denotes the generated electricity from the total kinetic energy difference.

When the second law of thermodynamics is considered, it gives more insightful results. A map was created in [40] in order to analyze the wind energy feasibility as shown in Fig. 3.36. Among the parameters to evaluate the feasibility of wind energy, exergy efficiency can be highlighted as the most important parameter that needs to be considered.

Case Study 3.8 A Siemens SWT02.3-101 wind turbine is investigated in a case study with the data from [28]. A wind turbine has 101 m of rotor diameter and 100 m of hub height. When using the meteorological data, height is very crucial for the wind since different wind speeds can be seen at different elevations (Table 3.5).

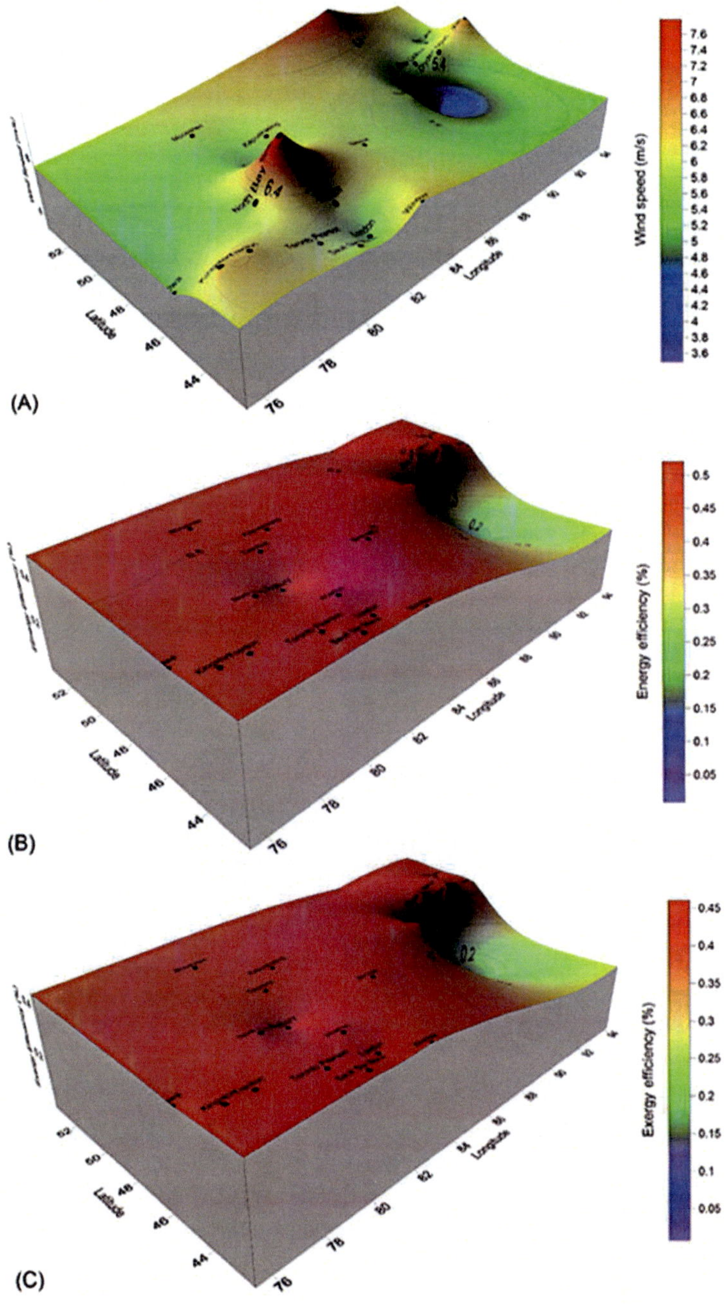

Fig. 3.36 (**a**) Wind speed, (**b**) energy efficiency, (**c**) exergy efficiency maps for Ontario, Canada in July (adapted from Ref. [40])

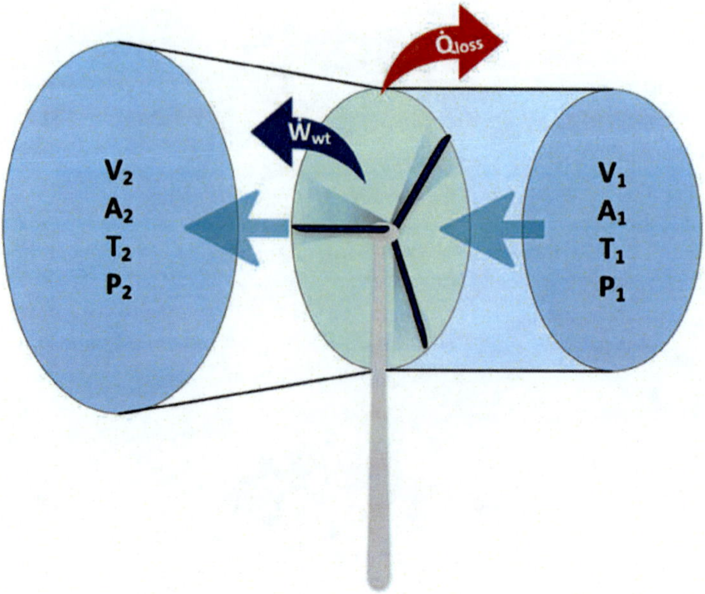

Fig. 3.37 Illustration of wind energy system with state points for inlet and exit

Table 3.5 The technical details of the selected horizontal-axis wind turbine

Parameter	Value
Rated power	2.3 MW
Cut-in speed	3.5 m/s
Rated speed	12.5 m/s
Cut-out speed	25 m/s
Maximum survival speed	55 m/s
Rotor diameter	101 m
Swept area	8000 m^2
Type	Triple-bladed
Tip speed	85 m/s
Power density	287.5 W/m^2
Gear box ratio	1:91
Generator	Asynchronous
Voltage	690 V
Hub height	80 m
Rotor weight	62 tonnes
Nacelle weight	82 tonnes
Tower weight	162 tonnes
Total weight	306 tonnes

Data from Ref. [42]

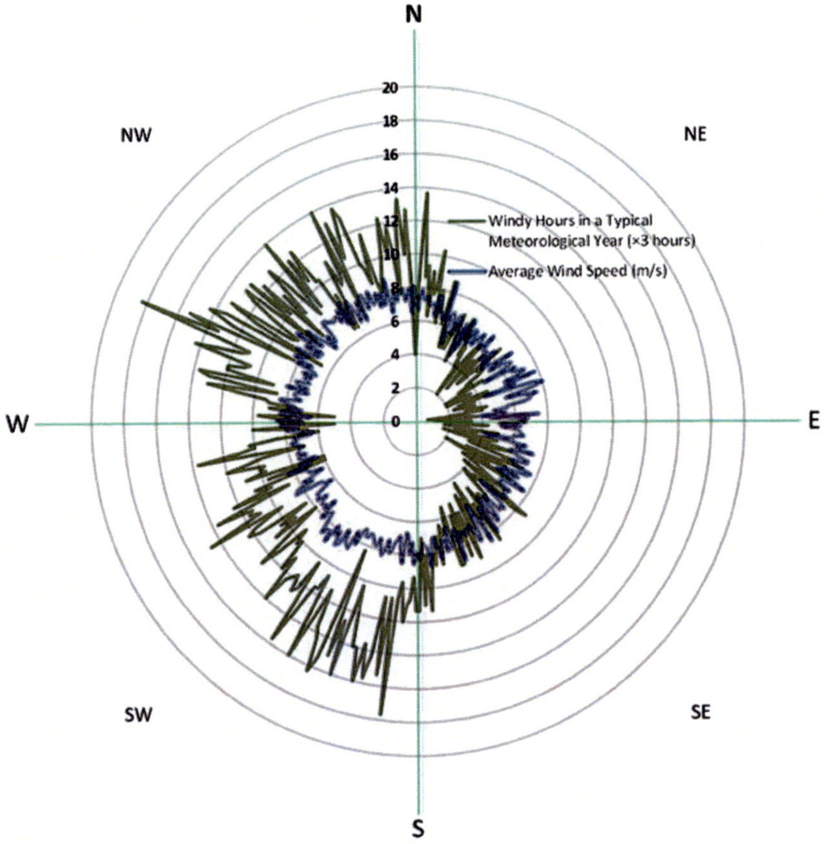

Fig. 3.38 Wind speed and direction during a typical meteorological year in Cochrane, Ontario (data from Ref. [43, 44])

The selected wind turbine is simulated for Cochrane, Ontario, where the wind speed and direction data are shown in Fig. 3.38. In order to calculate the wind power generation, the NREL's SAM software package is used to simulate the wind farm [6].

In order to carry out the calculations, a 23 MW wind farm is designed where each wind turbine has 2.3 MW; therefore, ten wind turbines are used to design the wind farm. The shear coefficient is considered as 0.14. A simple wake model is used with a 0.1 turbulence coefficient and 11.02% constant loss. Availability losses are considered as 3.58% for the turbine, 0.5% for the balance of the plant, and 1.5% for the grid. Electrical losses and parasitic consumption are considered as 2.008% in total. Turbine performance losses are considered as 1.1% for sub-optimal performance, 1.7% for generic power curve adjustment, 0.81% for site-specific power curve adjustment, and 0.4% for high wind hysteresis. Environmental losses are considered 2.4% in total, specifically for 0.21% for icing, 1.8% for degradation,

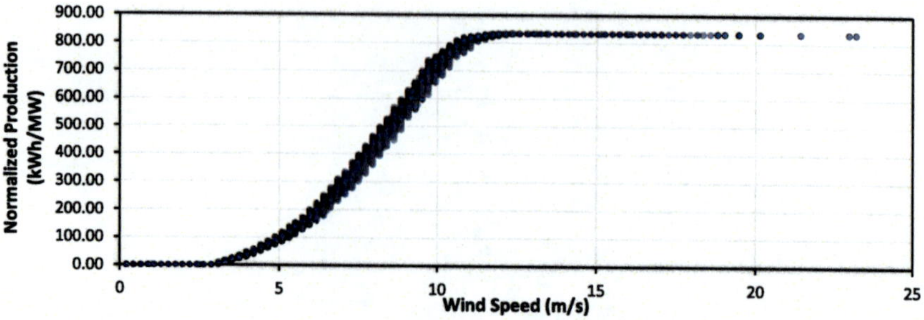

Fig. 3.39 Normalized electricity production of wind farm by wind speed (data from Ref. [28])

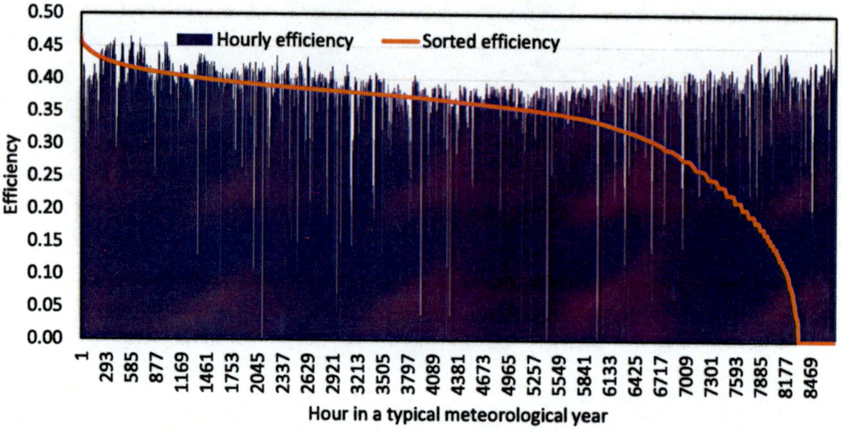

Fig. 3.40 Overall energy efficiency of the wind farm (data from Ref. [28])

and 0.4% for other environmental. The grid, load, environmental, and permit curtailments, as well as operational strategies, driven losses are considered as 2.8%.

Normalized electricity production in the kWh curve for a MW of installed capacity in different wind speeds is shown in Fig. 3.39. According to the curve, best performance can be obtained at wind speeds between 11 and 25 m/s. The proposed wind turbine cannot be driven at wind speeds lower than 3.5 m/s. For security reasons, at more than 25 m/s wind speed, wind turbines cut out and stop (Fig. 3.40).

In order to find the overall energy efficiency of the entire wind farm, net electricity generation needs to be divided into the mechanical power within the wind as follows:

$$\eta_{en,ov} = \frac{\dot{W}_{El,net}}{\frac{1}{2}\rho_{air}\pi R^2 V_{wind}^3}$$

Figure 3.40 shows the overall energy efficiency variations in a typical meteorological year. The maximum energy efficiency of the wind farm is found as 46.44%, and the average energy efficiency of the wind farm is 32.76%. In a typical year, 23 MW of wind farm generated 71.29 GWh of electricity, according to the simulations

3.2.4 Geothermal Energy

Geothermal energy is the heat exploited from the underground. It is a renewable energy solution with a high-capacity factor, which makes geothermal energy a reliable energy source that can replace fossil fuels with less energy storage requirement. When the countries with a high ratio of renewable share are checked, hydro energy and geothermal energy are two of the renewables that have high shares in the energy mix. The main reason for that is both hydro and geothermal energy sources are highly reliable, less fluctuations, less intermittencies, and high-capacity factors, which makes these renewable potential solutions to meet baseload energy demand. When compared to hydro, geothermal energy systems are more reliable since geothermal energy is not sensitive to weather conditions. By virtue of precipitation rates, hydro energy systems have more sensitivity to weather conditions. However, geothermal energy systems require high investment as well as possess investment risks, especially during the initial stages.

There are mainly two different geothermal source types, which are hydrothermal and enhanced geothermal systems. Hydrothermal sources naturally occur and thermofluid comes from the underground and can be pumped and used above the ground. On the other hand, enhanced geothermal systems are newer technology which is used when there is a hot rock but not enough water to use. In these conditions, enhanced geothermal systems pump water and create the reservoir (Fig. 3.41).

Geothermal energy can be used for many purposes, it contains heat that can be used as process heat for any heat-consuming process to generate a variety of useful outputs. Space heating is one of them which does not require a high-level investment, such as power generation systems. Geothermal source temperature is an important criterion to determine which useful outputs and processes are suitable. Space heating is almost possible with any geothermal source, even a few meters underground can be used for space heating purposes. Especially during the wintertime, underground soil temperatures are usually warmer than the ambient air. Therefore, heat pumps can exploit the heat from the underground with an increased coefficient of performance due to the reduced power consumption compared to the air-source heat pumps.

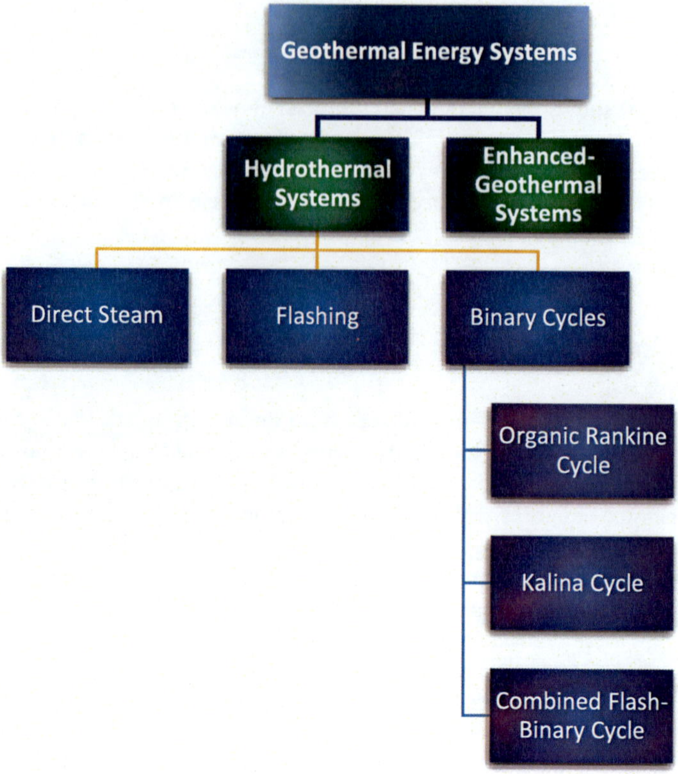

Fig. 3.41 Classification of geothermal energy systems

Another useful output from geothermal energy is power. There are many methods to generate power from geothermal energy. The key to selecting the correct method is the available temperature and geothermal fluid condition. There are three major categories to use geothermal sources for power generation. First is the dry steam power system, where dry steam is the available geothermal source which can be used to drive the steam turbine. The second one is the flash steam power system, where the flashing process occurs by decreasing the pressure of the fluid in order to get steam. According to the temperature and quality of the steam, number of flashing stages varies. Binary steam power systems differ from the first two methods since binary systems use geothermal fluid indirectly with a heat exchanger in order to heat another working fluid within the power generation cycle. Binary cycles can exploit the lower temperatures for power generation since different working fluids can be used, including organic fluids. The organic Rankine cycle and Kalina cycle are two examples of low-temperature geothermal binary cycle power systems. Other than space heating and power generation, application areas and potential services of geothermal-based energy systems have a large variety (Fig. 3.42).

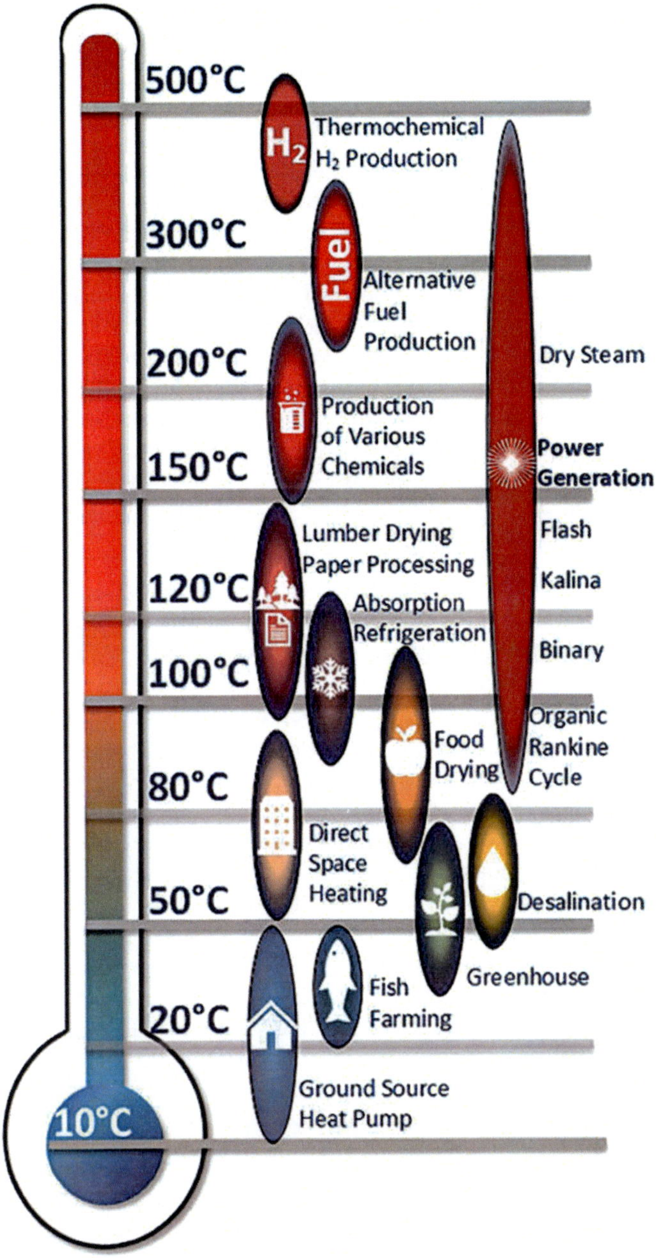

Fig. 3.42 Geothermal energy application areas by temperature

Temperature is the key constraint to select a suitable application. From 10°C, space heating can be done via heat pump systems. When the outside temperature is lower than the ground temperature, the heat pump consumes more energy to upgrade the heat at lower temperatures. 20–50°C geothermal water can be used in fish farms, especially the ones that are located in cold climatic conditions. Greenhouse systems also can benefit from the heat from 25°C temperatures.

Desalination can benefit from low-temperature geothermal energy by the virtue of lower evaporation temperatures at lower pressures. For this reason, low-temperature desalination via low-temperature geothermal energy can be carried out at vacuum pressures. There are two main methods available to desalinate water at low temperatures, namely vacuum distillation and flash evaporation. The vacuum distillation process is to evaporate saline water in an evaporation chamber maintained at vacuum pressure. Evaporated water can be condensed in a different chamber, and fresh water can be obtained from this chamber. Flash evaporation is another low-temperature desalination method, where saline water is transferred to a flash chamber, which is maintained at vacuum pressure. Similar to the vacuum distillation process, evaporated water can be transferred into another chamber to condense and obtain fresh water.

Case Study 3.9 A case study for a binary cycle geothermal power plant is designed and simulated by using NREL's SAM software package [6]. A geothermal resource at 245°C at 1993 m down is considered for the case study. A hydrothermal type of geothermal source is pumped from the production well to the heat exchanger. In the heat exchanger, heat is transferred into the Rankine cycle and superheated steam is generated. Superheated steam is entered into the steam turbine, where it is expanded, and power is generated. The expanded steam, very close to the saturation point, is entered into the air-cooled condenser in order to get condensed and become liquid. Condensed water is entered into the pump where it gets pressurized and restarted to cycle. It is important to note that the power generated from steam turbines should be considered as the gross power, since pumps and other elements consume significant power, especially in geothermal systems since the thermofluid needs to be extracted from the underground. The design parameters are then given in Table 3.6 (Fig. 3.43).

Table 3.6 The design parameters of the case study of a binary cycle geothermal power generation system

Parameter	Value
Pressure changes across the reservoir per 1 tonnes of water	0.06 bar
Reservoir width	500 m
Reservoir height	100 m
Permeability	0.05 Darcy units
Distance from injection to production wells	1500 m
Resource temperature	245°C
Resource depth	1993 m
Production well mass flow rate	110 kg/s
Production pump depth	592.3 m
Production pump size	1289.4 hp

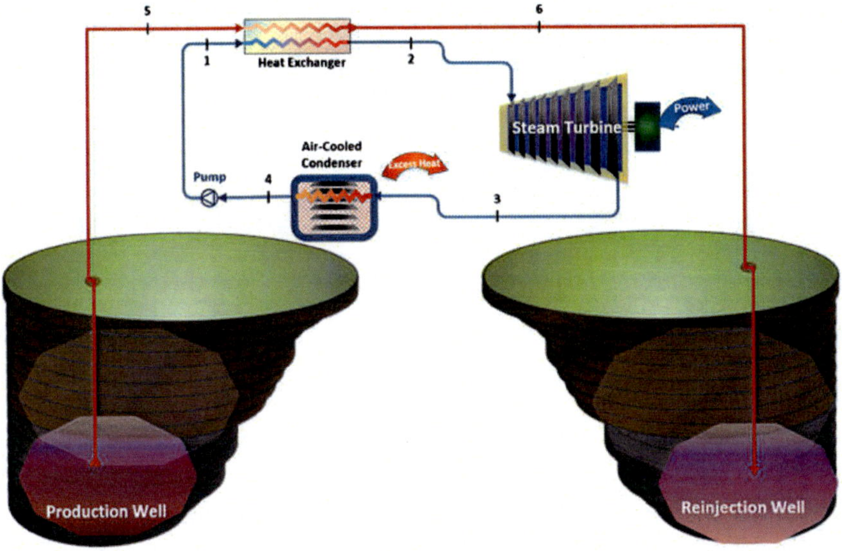

Fig. 3.43 Schematic illustration of a geothermal binary cycle power generation system

The binary cycle system in the case study is investigated by considering the first and second laws of thermodynamics. In order to understand the performance of the system and later compare it with other power generation systems, the energy and exergy efficiencies can be calculated as follows:

$$\eta_{en,ov} = \frac{\dot{W}_{El,net}}{\dot{m}_5 h_5 - \dot{m}_6 h_6}$$

and

$$\eta_{ex,ov} = \frac{\dot{W}_{El,net}}{\dot{m}_5 ex_5 - \dot{m}_6 ex_6}$$

In order to calculate the $\dot{W}_{El,net}$, gross electricity from the generator needs to be used and pump and other power consumptions need to be removed from the gross electricity.

When we look at the obtained results from the simulation and further calculations, from 2056 MW of geothermal heat, 349.6 MW of gross electricity is generated. After removing the pump and other loads, the net electricity production is found as 300 MW. The overall energy and exergy efficiencies for the geothermal binary cycle power generation system are found as 14.59% and 37.79%. For different ambient temperatures, energy, and exergy efficiency variations are presented in Fig. 3.44.

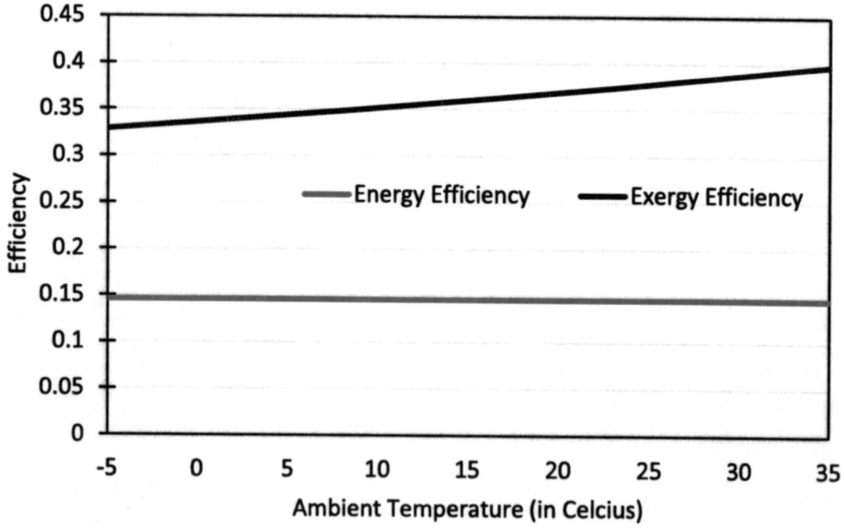

Fig. 3.44 The overall energy and exergy efficiencies of a geothermal binary cycle power generation

3.2.5 Biomass Energy

Biomass energy refers to the contained energy in chemical form within organic or biologic materials such as wood, plants, and wastes. Chemical energy within the biomass can be utilized by physical, thermochemical, and biological conversion processes, as depicted in Fig. 3.45. The global capacity as of 2022 is 148.9 GW_p of power generation [45], which has a US$0.067/kWh of levelized cost of electricity. Moreover, biomass energy systems have high-capacity factors compared to the other renewables which is typically around 68%. When we look at the cost and capacity factors, biomass energy is a promising solution. Regarding the emissions of biomass energy sources need to be considered by balancing it. Photosynthesis process catches the carbon dioxide in order to generate biomass energy source. For this reason, net greenhouse gas emission will be balanced and there will be no extra carbon dioxide emissions related to biomass energy.

There are two main generations of biofuels conventional (first generation), and advanced (second generation). Edible biomass energy sources are utilized, such as corn, sugarcane, sugar beets, and rapeseed, to generate bioethanol or other biofuels, in the conventional biofuel generation process. On the other hand, the advanced biofuel generation process utilizes and treats non-edible biomass energy sources. The reason behind the initiation of the advanced biofuel generation process is to keep edible food products for eating. Pyrolysis, gasification, and anaerobic digestion are among the processes to generate advanced biofuels. There are also third and fourth generations of biofuels which refer to algae-driven biofuel generation (third) and genetically engineered feedstocks and genomically synthesized microorganisms-driven biofuel generation (fourth).

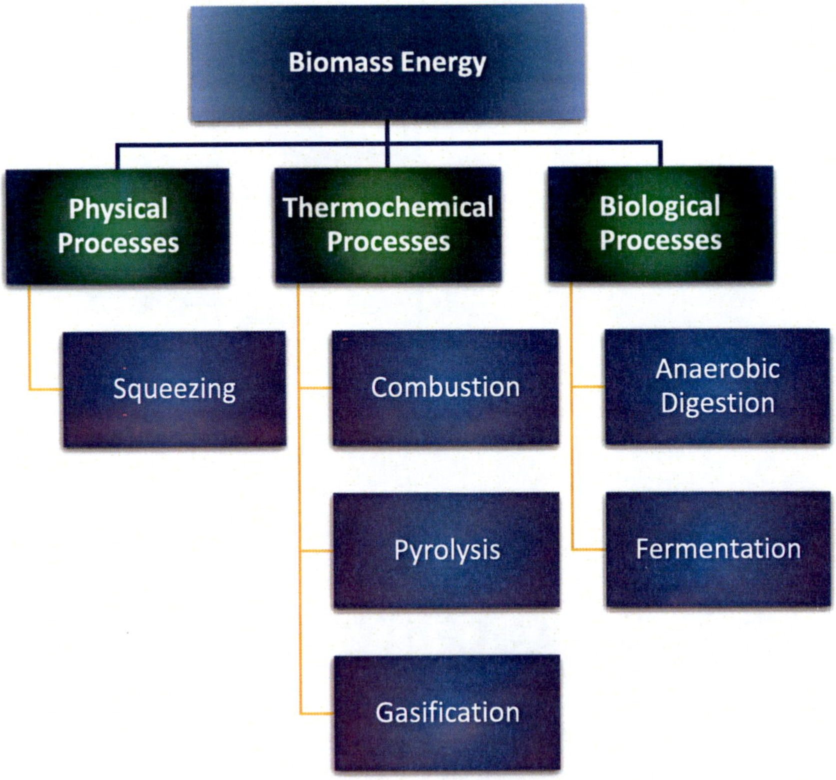

Fig. 3.45 Classification of biomass energy conversion processes

Case Study 3.10 There are also systems that contain both edible and inedible biomass energy sources which can bring synergetic advantages. A case study is carried out by considering the data from [18] for the biomethane production system as shown in Fig. 3.46, where an anaerobic digestion process is considered in the case study where sugarcane and cow manure are mixed along with water. APHA and AWWA-2005 standard methods are considered in order to determine total solids (TS) and total volatile solids (TVS).

Note that the anaerobic digestion is proceeded with the chemical reactions as follows:

$$C_6H_{12}O_6 + 2H_2O \rightarrow 2CH_3COOH + 2CO_2 + 2H_2$$

$$3C_6H_{12}O_6 \rightarrow 4CH_3CH_2COOH + 2CH_3COOH + 2CO_2 + 2H_2$$

$$C_6H_{12}O_6 \rightarrow 4CH_3CH_2CH_2COOH + 2CH_3COOH + 2CO_2 + 2H_2$$

$$CH_3CH_2COOH + 2H_2O \rightarrow CH_3COOH + CO_2 + 3H_2$$

$$CH_3COOH \rightarrow CH_4 + CO_2$$

The data obtained from [18] suggest the design characteristics of sugarcane leaves and cow manure as given in Table 3.7. The process is mainly the micro-organisms

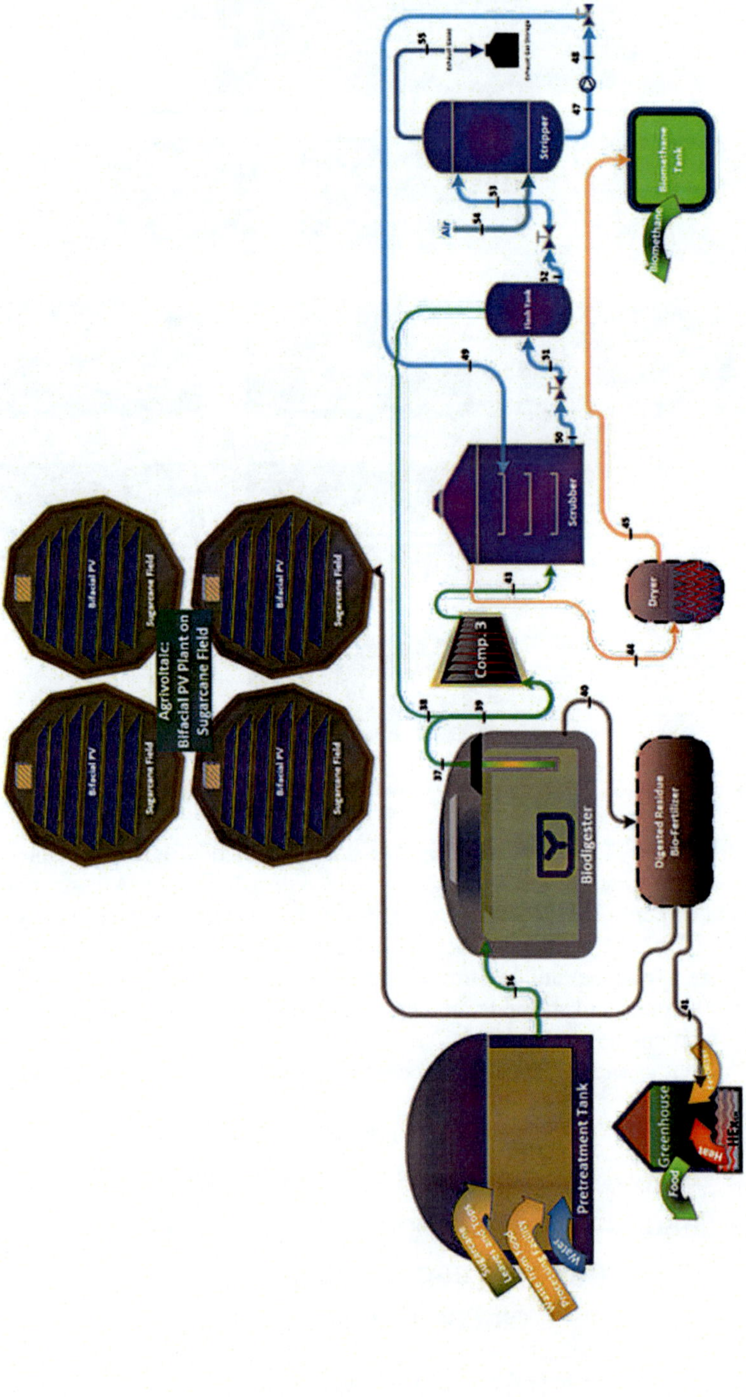

Fig. 3.46 Schematic of an anaerobic digester plant to produce biomethane

Table 3.7 The design Characteristics of sugarcane leaves and tops and cow manure

Parameter	Sugarcane leaves and tops	Cow manure
Moisture (%)	54.07	82.76
Total solids (%)	45.93	17.24
Total volatile solids (% of TS)	94.00	87.21

Adapted from Ref. [18]

Table 3.8 The characteristics of the influent feedstock of pretreated sugarcane leaves and tops manure at various ratios and that of the digestate at the steady state conditions

Pretreated sugarcane leaves and tops and cow manure in 800 mL water	*100:100*
pH of the influent feedstock	8.26
COD (g/L)	24.55
OLR (kg COD/m^3·Day)	0.61
TS (g/L)	20.92
TVS (g/L)	16.07
OLR (kg TVS/m^3·Day)	0.40
TKN (mg/L as N)	565
P (mg/L)	3.62
COD/N	100:2.30
At the steady state conditions	
Average pH	7.74
VFA (mg/L as CH_3COOH)	1443
ALK (mg/L as $CaCO_3$)	4150
VFA/ALK	0.35
Average COD (g/L)	7.65
Average TVS (g/L)	4.416
Amount of total gas (mL/5 day)	7853
Amount of CH_4 (mL/5 day)	3496
Average %CH_4	44.52
% COD degradation efficiency	68.80
% TVS degradation efficiency	72.52
CH_4 yield (L at STP/kg COD degraded)	331
CH_4 yield (L at STP/kg TVS degraded)	480
CH_4 yield (L at STP/kg TS added)	267

Adapted from Ref. [18]

grow by metabolizing the organic material in order to generate methane in an oxygen-free environment. In four phases, the process can be completed, namely hydrolysis, acidogenesis, acetogenesis, and methanogenesis, where different types of microorganisms are used in each different step.

In order to analyze the methane yields, [18] used data taken from the Shimadzu GC-2014 gas chromatograph along with a detector to measure thermal conductivity. In order to find chemical oxygen demand, APHA and AWWA 2005 standard methods were considered. A 100:100 ratio is used where 100 g of sugarcane leaves and 100 g of cow manure were mixed in 800 ml of water. A summary of the results is given in Table 3.8.

3.3 Marine Energy

Marine energy refers to a variety of available energy in the marine environment, such as wave energy, tidal energy, salinity gradient energy, and ocean thermal energy, as categorized in Fig. 3.47. Marine energy systems have the potential to generate 90 PWh of annual electricity, which can triple times meet the current global power demand. Marine energy systems are still in the initial phases that require further developments in order to become more cost-competitive and applicable. Tidal and wave energy systems exploit the mechanical energy and convert it into power. Tidal barrage systems store the water as potential energy and exploit the potential energy to generate power. Tidal current and wave energy systems generate power by exploiting the kinetic energy. Vertical movement of wave energy systems additionally brings potential energy to be exploited by the wave energy systems.

Ocean thermal energy conversion (OTEC) systems exploit the temperature gradient of the ocean, especially when the water is deep, sun's rays cannot travel deeper than 1 km, which makes the water there cooler than the surface. Additionally, it is possible to exploit temperature differences in a reverse way, which is effective especially in cold climates, such as Arctic environments, since the ambient temperatures can be as low as −55°C. Temperature differences between ocean and air can be exploited via specific thermofluids.

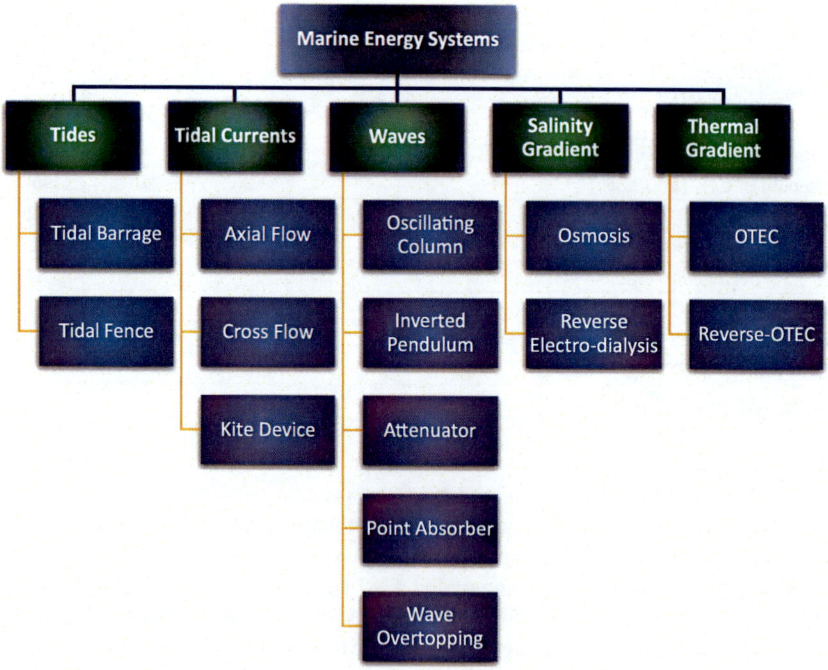

Fig. 3.47 Classification of marine energy systems

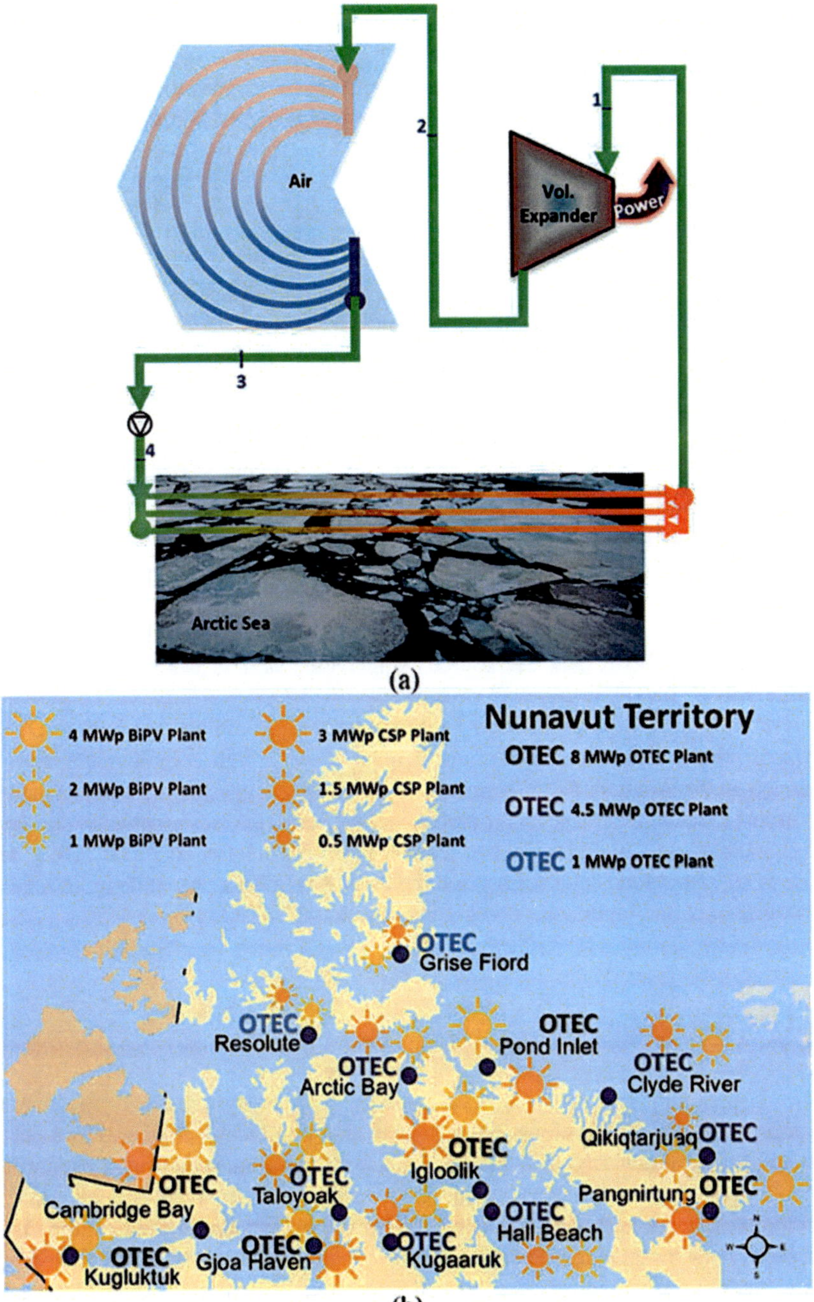

Fig. 3.48 (**a**) An illustration of a reverse type OTEC system and (**b**) the locations identified for polar hamlets and required capacities of reverse-OTEC integrated systems in the Nunavut Territory (modified from Ref. [46])

Table 3.9 The thermophysical properties of the reverse OTEC system

State P.	Substance	Mass flow rate (kg/s)	Temperature (°C)	Pressure (bar)	h (kJ/kg)	s (kj/kg K)	ex (kJ/kg)
1	Ammonia	65.0	−1.0	1.8	1483.0	6.104	73.3
2	Ammonia	65.0	−35.7	0.9	1416.0	6.152	−6.1
3	Ammonia	65.0	−35.0	0.9	41.3	0.381	119.7
4	Ammonia	65.0	−35.0	1.8	41.7	0.382	119.9

Data from Ref. [46]

Case Study 3.11 A case study is carried out based on the data from [46]. A reverse OTEC system can be used by exploiting the low-temperature heat from the Arctic Sea via ammonia. Temperature is decreased to −35°C via air-cooled condenser in order to condense ammonia as shown in Fig. 3.48a. Also, the locations identified for polar hamlets and required capacities of reverse-OTEC integrated systems in the Nunavut Territory are indicated on the map as suitable locations for potential utilization, and therefore, such locations are considered in the subject matter study as originally published earlier [46].

According to the calculations, to produce power by exploiting the temperature difference between the Arctic Sea and the ambient air in the Arctic, 4.62% of energy efficiency and 28.12% of exergy efficiency are calculated when the ambient air is considered as −45°C and the source temperature as 1°C based on the data in Table 3.9.

Furthermore, Temiz and Dincer [46] have recently combined the reverse OTEC system with a solar-based system, which has a very high complementarity. For hamlets in the Arctic circle of Nunavut territory, they suggested different capacities in order to ensure that these communities self-sufficient and sustainable, as shown in Fig. 3.48b. Colder ambient weather usually have less solar energy availability, and vice versa. Therefore, solar energy and the reverse OTEC can generate more reliable power than single-input solar energy or single-input OTEC systems.

3.4 Closing Remarks

The chapter aims to cover renewable energy sources, systems, and applications comprehensively by dwelling on technical and non-technical dimensions, challenges, opportunities as well as historical perspectives. Some countries are identified with a great degree of renewable energy potential and discussed through some of their investments and incentives and the way they mobilized their public. It is, anyway, a clear fact that renewables require unique solutions, where characteristics of sources and required services need to be matched carefully in order to design a potential successful solution to implement. That brings us to the requirement for knowledge in renewable areas. This chapter introduces renewable energy solutions comprehensively, along with unique case studies where some energy and exergy-based analysis and assessment tools and methodologies are discussed. The time resolution of renewable energy system analysis is crucial. Most of the renewables have intermittent availabilities; therefore, assessing renewable solutions for steady-state conditions can only solve half of the problem. In order to investigate the reaction of the system at extreme conditions, it is crucial to analyze systems with high-resolution time data, such as hourly typical meteorological years. When available, case studies considered and presented hourly typical meteorological data in this chapter. Furthermore, simulations are important since software packages consider commercially available products, which makes the analyses more realistic. In case studies, commercially available components and products, along with available high-resolution source data, are used.

In closing, it is further important to combine inputs and outputs in an integrated and smart fashion in order to improve the complementarity and reliability. The current chapter depicts the details of each renewable. The following chapters will include more integrations of renewables in order to improve the efficiencies for better source utilization and better production of useful outputs.

References

1. Ritchie H, Roser M, Rosado P (2022) Energy data. Our world in data
2. United Nations (2015) Iceland's sustainable energy story: A model for the world? | United Nations. UN Chron No. 3 Vol. LII
3. Einstein A (1905) Über einem die Erzeugung und Verwandlung des Lichtes betreffenden heuristischen Gesichtspunkt. Ann Phys 4
4. Sorgulu F, Dincer I (2023) A solar energy driven thermochemical cycle based integrated system for hydrogen production. Energy 269:126834. https://doi.org/10.1016/J.ENERGY.2023.126834
5. JRC Photovoltaic Geographical Information System (PVGIS) – European Commission. https://re.jrc.ec.europa.eu/pvg_tools/en/. Accessed 25 Jul 2023
6. (2020) System Advisor Model Version 2020.2.29 (SAM 2020.2.291). National Renewable Energy Laboratory. Golden
7. National Renewable Energy Laboratory (2023) Linear fresnel projects. In: SolarPACES. https://solarpaces.nrel.gov/by-technology/linear-fresnel. Accessed 27 Jun 2023
8. Kaiser F, Krüger U (2019) Exergy analysis and assessment of performance criteria for compressed air energy storage concepts. Int J Exergy 28:229–254. https://doi.org/10.1504/IJEX.2019.098613
9. Marefati M, Mehrpooya M, Pourfayaz F (2021) Performance analysis of an integrated pumped-hydro and compressed-air energy storage system and solar organic Rankine cycle. J Energy Storage 44:103488. https://doi.org/10.1016/J.EST.2021.103488
10. Chen L, Zhang L, Wang Y et al (2023) Design and performance evaluation of a novel system integrating water-based carbon capture with adiabatic compressed air energy storage. Energy Convers Manag 276:116583. https://doi.org/10.1016/J.ENCONMAN.2022.116583
11. Mohan M, Sharma VK, Kumar EA, Gayathri V (2019) Hydrogen storage in carbon materials – a review. Energy Storage 1:e35. https://doi.org/10.1002/EST2.35
12. Achaibou N, Ferahtia S, Djerioui A et al (2023) Optimal parameters identification strategy of a lead acid battery model for photovoltaic applications. Energy Storage 5:e428. https://doi.org/10.1002/EST2.428

13. Temiz M, Dincer I (2022) Development and assessment of an onshore wind and concentrated solar based power, heat, cooling and hydrogen energy system for remote communities. J Clean Prod 374:134067. https://doi.org/10.1016/J.JCLEPRO.2022.134067
14. Baniasadi E, Ziaei-Rad M, Behvand MA, Javani N (2023) Exergy-economic analysis of a solar-geothermal combined cooling, heating, power and water generation system for a zero-energy building. Int J Hydrog Energy. https://doi.org/10.1016/J.IJHYDENE.2023.01.186
15. Ren Y, Yao X, Liu D et al (2022) Optimal design of hydro-wind-PV multi-energy complementary systems considering smooth power output. Sustain Energy Technol Assess 50:101832. https://doi.org/10.1016/J.SETA.2021.101832
16. Temiz M, Dincer I (2021) Techno-economic analysis of green hydrogen ferries with a floating photovoltaic based marine fueling station. Energy Convers Manag 247:114760. https://doi.org/10.1016/J.ENCONMAN.2021.114760
17. Temiz M, Dincer I (2022) Design and analysis of a floating photovoltaic based energy system with underground energy storage options for remote communities. J Energy Storage 55:105733. https://doi.org/10.1016/j.est.2022.105733
18. Temiz M, Sinbuathong N, Dincer I (2022) Development and assessment of a new agrivoltaic-biogas energy system for sustainable communities. Int J Energy Res. https://doi.org/10.1002/er.8483
19. Safyanu BD, Abdullah MN, Omar Z (2019) Review of power device for solar-powered aircraft applications. J Aerosp Technol Manag 11:e4119. https://doi.org/10.5028/JATM.V11.1077
20. Lencwe MJ, Chowdhury SP, Elgohary HM (2016) Solar photovoltaic integration on locomotive roof top for South African railway industry. In: Proceedings – 2016 51st international universities power engineering conference, UPEC 2016 2017-January:1–5. https://doi.org/10.1109/UPEC.2016.8114087
21. Cheng P, Liu W, Ma J et al (2022) Solar-powered rail transportation in China: potential, scenario, and case. Energy 245:123221. https://doi.org/10.1016/J.ENERGY.2022.123221
22. Sharma P, Harinarayana I (2013) Solar energy generation potential along national highways. Int J Energy Environ Eng 4:1–13. https://doi.org/10.1186/2251-6832-4-16/TABLES/4
23. Di Stefano A, Leotta G, Bizzari F (2017) La Silla PV plant as a utility-scale side-by-side test for innovative modules technologies. In: 33rd European photovoltaic solar energy conference and exhibition EUPVSEC, pp 1978–1982
24. Kopecek R, Libal J (2018) Towards large-scale deployment of bifacial photovoltaics. Nat Energy 3:443–446
25. Hasan A, Dincer I (2020) A new performance assessment methodology of bifacial photovoltaic solar panels for offshore applications. Energy Convers Manag 220:112972. https://doi.org/10.1016/j.enconman.2020.112972
26. Chudinzow D, Haas J, Díaz-Ferrán G et al (2019) Simulating the energy yield of a bifacial photovoltaic power plant. Sol Energy 183:812–822. https://doi.org/10.1016/j.solener.2019.03.071
27. Temiz M, Dincer I (2021) An integrated bifacial photovoltaic and supercritical geothermal driven copper-chlorine thermochemical cycle for multigeneration. Sustain Energy Technol Assess 45:101117
28. Temiz M, Dincer I (2022) Development of solar and wind based hydrogen energy systems for sustainable communities. Energy Convers Manag 269:116090. https://doi.org/10.1016/j.enconman.2022.116090
29. Temiz M, Dincer I (2023) Design and assessment of a solar energy based integrated system with hydrogen production and storage for sustainable buildings. Int J Hydrog Energy 48:15817–15830. https://doi.org/10.1016/J.IJHYDENE.2023.01.082
30. Jacobson MZ (2020) 100% clean, renewable energy and storage for everything. Cambridge University Press, Cambridge. https://doi.org/10.1017/9781108786713
31. Fraenkel P, Parish O, Bolkalders V, et al (1991) Micro-hydro power: a guide for development workers. Micro-hydro Power. https://doi.org/10.3362/9781780442815

32. Altayib K, Dincer I (2022) Development of an integrated hydropower system with hydrogen and methanol production. Energy 240:122780. https://doi.org/10.1016/J.ENERGY.2021.122780
33. Lu X, McElroy MB, Kiviluoma J (2009) Global potential for wind-generated electricity. Proc Natl Acad Sci USA 106:10933–10938. https://doi.org/10.1073/PNAS.0904101106
34. Archer CL, Jacobson MZ (2005) Evaluation of global wind power. J Geophys Res Atmos 110: 1–20. https://doi.org/10.1029/2004JD005462
35. IRENA (2023) Renewable capacity statistics 2023. International Renewable Energy Agency, Abu Dhabi
36. The world's largest wind turbine is now installed in China [update]. https://electrek.co/2023/06/29/the-worlds-largest-wind-turbine-is-now-being-installed-in-china/. Accessed 30 Jul 2023
37. A colossal 18 MW wind turbine is about to debut in China. https://electrek.co/2023/01/07/18-mw-wind-turbine-china/. Accessed 30 Jul 2023
38. Mingyang Goes Beyond 18 MW with New Offshore Wind Turbine | Offshore Wind. https://www.offshorewind.biz/2023/01/13/mingyang-goes-beyond-18-mw-with-new-offshore-wind-turbine/. Accessed 30 Jul 2023
39. Enercon E-126 7.580 – 7,58 MW – Wind turbine. https://en.wind-turbine-models.com/turbines/14-enercon-e-126-7.580. Accessed 30 Jul 2023
40. Dincer I, Rosen MA (2020) Exergy: energy, environment and sustainable development. Elsevier
41. Meteotest AG (2021) Meteonorm 8.0. Bern
42. The World Bank Group (2022) Global Wind Atlas. In: Lyngsby, Denmark. https://globalwindatlas.info/about/TermsOfUse. Accessed 24 Jul 2022
43. Osczevski RJ (2000) Windward cooling: an overlooked factor in the calculation of wind chill. Bull Am Meteorol Soc 81:2975–2978. https://doi.org/10.1175/1520-0477(2000)081<2975:WCAOFI>2.3.CO;2
44. Siemens SWT-2.3 – 101 – 2,30 MW – Wind turbine. https://en.wind-turbine-models.com/turbines/341-siemens-swt-2.3-101. Accessed 31 Jul 2023
45. Ritchie H, Roser M, Rosado P (2022) Energy. OurWorldInData.org
46. Temiz M, Dincer I (2022) A unique ocean and solar based multigenerational system with hydrogen production and thermal energy storage for Arctic communities. Energy 239:122126. https://doi.org/10.1016/j.energy.2021.122126

Chapter 4
Thermal Energy Generation

Nomenclature

A	area (m^2)
DD	decimal degrees
DNI	direct normal irradiance (W/m^2)
EBE	energy balance equation
EnBE	entropy balance equation
ExBE	exergy balance equation
ex	specific exergy (kJ/kg)
$\dot{E}x$	exergy rate (kW)
g	gravitational acceleration (m/s^2)
GHI	global horizontal irradiation (W/m^2)
h	specific enthalpy (kJ/kg)
$\bar{h}$	absolute enthalpy (kJ/kmol)
HP	heat pump
LHV	lower heating value (kJ/kg)
$\dot{m}$	mass flow rate (kg/s)
MBE	mass balance equation
$\dot{N}$	molar flow rate (mol/s)
P	pressure (kPa)
PTC	parabolic trough collector
PV	photovoltaic
Q	heat (kJ)
$\dot{Q}$	heat rate (kW)
S	total entropy (kJ /K)
s	specific entropy (kJ/kg K)
$\bar{s}$	absolute entropy (kJ/kmol K)
$\dot{S}$	entropy rate (kW/K)
T	temperature (°C)

I. Dincer, M. Temiz, *Renewable Energy Options for Power Generation and Desalination*, https://doi.org/10.1007/978-3-031-53437-9_4

w	water, work
$\dot{W}$	power (kW)
X	mass ratio of component

Subscripts

c	complete
co	condenser
comb	combustion
CS	concentrated solar
e	evaporator
f	formation
gtc	gas turbine component
HW	hot water
in	injection, input
LF	linear Fresnel
o	output
P	product
PH	process heat
R	reactant
re-in	re-injection
s	source
sol	solar
SC	space cooling
SH	space heating
u	universal

4.1 Introduction

It is a known fact that a majority of energy conversion processes relies particularly on thermal energy and its utilization. Transportation, power generation, heating, cooling, and many more energy conversion processes utilize the input of thermal energy to generate useful outputs to cover the needs of people and their communities. Since the Industrial Revolution, the dominant method to generate thermal energy has been the combustion of fossil fuels: initially started with coal, later oil, and more recently with natural gas. Utilization of fuels, renewable energy sources, and waste and process heat includes thermal energy generation step in order to proceed to the intended processes to generate useful outputs. Thermal energy can also be directly used for space heating, space cooling, domestic hot water, food drying, greenhouse applications, etc. Technically, it finds a vast amount of opportunities for practical applications.

Figure 4.1a illustrates the general steps of how the primary energy sources, namely fossil fuels, nuclear, and renewables, are utilized to generate the process

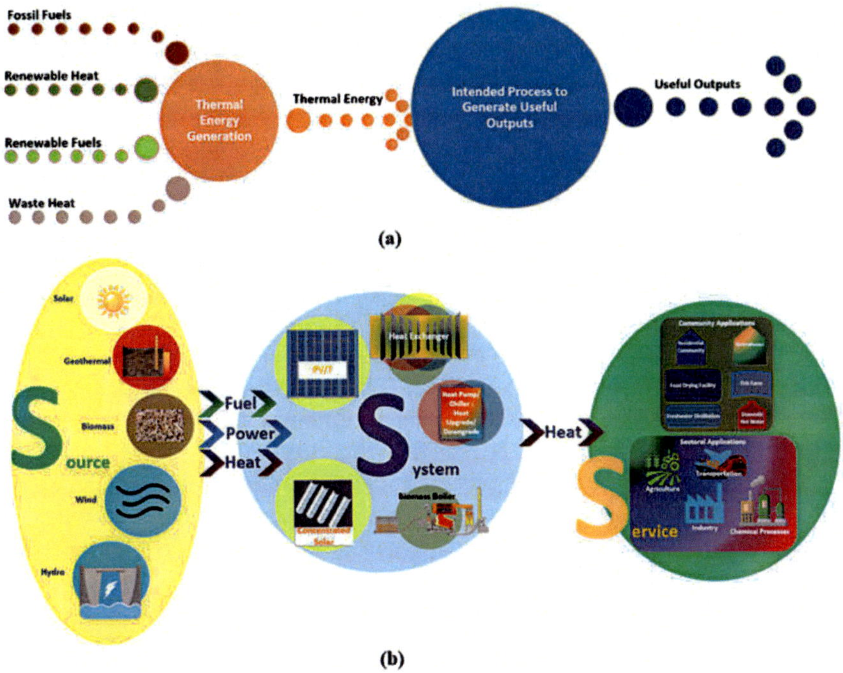

Fig. 4.1 Illustration of (**a**) the pathways to generate useful outputs via thermal energy and (**b**) 3S approach of heat generation from source to service for better sustainability

heat in the form of thermal energy for further processes. From this aspect, chemical energy in fossil fuels and renewable fuels, atomic energy in nuclear fuels, and thermal energy in renewable heat and waste heat can be exploited to generate thermal energy at various temperatures. Temperature is a critical parameter since different processes require different temperature levels in order to be carried out. For example, thermochemical cycles to generate hydrogen typically require temperatures more than 450°C. Rankine cycles have a wider range of temperature requirements, for example, organic Rankine cycles typically operate around 80–120°C temperatures with lower efficiencies, and steam Rankine cycles typically require temperatures more than 300°C at higher efficiencies compared to the organic Rankine cycles. In some cases, thermal energy can be used for cooling processes. The absorption refrigeration cycle utilizes thermal energy, typically around 95°C with a small portion of electricity, in order to remove thermal energy to provide a cooling effect, typically around 5°C temperatures.

Figure 4.1b illustrates the 3S (source-system-service) approach for better sustainability as originally introduced by Dincer [1] in several platforms and publications. The key logic behind this 3S is that a source is always needed for anything, everything, which will then supply the necessary form of energy to a system or systems to generate the useful outputs which are treated as services for people and communities as well as sectors. Furthermore, when we consider any energy spectrum, it can only and effectively be analyzed in three parts as source, system, and service. The renewable thermal energy generation process includes renewable sources, heat generation systems and specific applications for services. Renewable heat is mainly generated by solar, geothermal and biomass sources. However, other renewable sources, such as wind and hydro, can be utilized to generate renewable heat. The process of generating renewable heat from wind, hydro, and other renewables can proceed by utilizing the power from heat pumps or other electric heater in order to generate heat. Photovoltaic (PV) systems are also included in this case, where power is utilized to run electricity-driven heat generators. However, the main sources of renewable heat are solar, geothermal, and biomass. Thermal energy from solar energy can be generated via non-concentrating and concentrating solar thermal systems. Geothermal-based renewable heat can be exploited via direct or indirect systems, where primary loop or secondary loop heat generators can be employed. Biomass-driven systems require combustion or burning processes in order to release their chemical energy as renewable thermal energy.

Among the energy-consuming processes, the power generation sector represents a significant share along with transportation and other industrial processes. When we look at the global power generation process types, the Rankine cycle is the dominant method where thermal energy is extracted from the energy source in order to superheat steam to enable it for expansion. Nuclear, fossil fuel, renewable heat, renewable fuel, and waste heat–driven power generation processes generate heat to drive typically a Rankine cycle or other power generation cycles. On the other hand, solar PV plants, wind farms, and hydropower plants do not require heat generation to generate power. Solar radiation can be exploited via solar photovoltaic (PV) to move electrons from one layer to another to create a potential difference. The kinetic energy of winds can be exploited via wind turbines to generate power. Hydropower plants, on the other hand, exploit the potential energy due to the elevation difference of upper-level and lower-level water reservoirs to generate power. Figure 4.2 shows

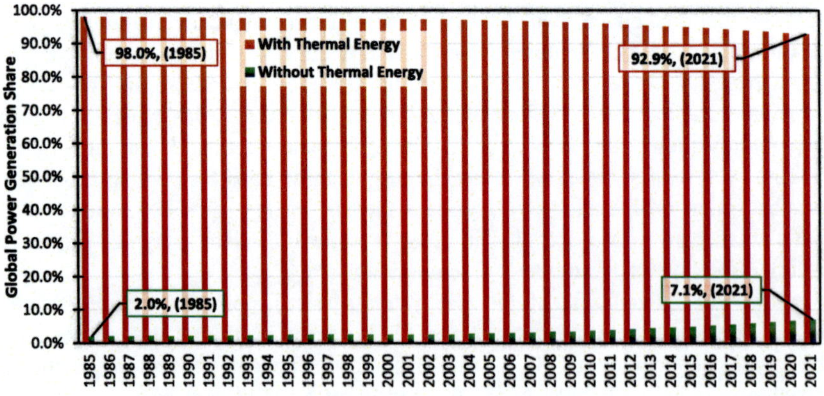

Fig. 4.2 Power generation methods with and without thermal energy generation step (data from Ref. [2])

historical data on power generation methods with and without thermal energy generation. The vast majority of power generation methods include a step where thermal energy generation is carried out. Only a small portion of the power generation methods do not include the thermal energy generation step. This data aligns with the energy consumption data in which fossil fuels have dominance. In order to generate power from fossil fuels, fossil fuels need to be burned in order to generate thermal energy. The growth in renewable power generation systems, especially with hydro-, wind-, and solar-based power generation plants, global power generation methods without thermal energy generation step are increased to 7.1% in 2021, as shown in Fig. 4.2.

There have been ongoing efforts in identifying replacements for fossil fuels and replacing them in many sectors that have been widely used. In this regard, renewable energy options come into the picture where they can offer great opportunities for cleaner solutions. Among the renewable energy, biomass sources are considered in terms of solid matters, which may range from agricultural biomass sources to organic municipal wastes, and biofuels which can easily be combusted to obtain the thermal energy in terms of heat. Thermal energy from this aspect, possesses a critical importance since many energy systems and applications require thermal energy generation step, which is the main reason that the currently available energy infrastructure is built by considering the utilization of thermal energy. Future strategies of jurisdictions do not include fossil fuels as an energy source; however, thermal energy needs will be part of the energy future.

When one looks at the replacement of fossil fuels for thermal energy utilization, low-temperature heat can be generated with many renewable or low-carbon technologies. However, when it comes to the high-temperature heat (more than 200°C), renewable technologies are more limited. Many industrial processes require high-temperature heat [3]; moreover, they emit high carbon dioxide. Nitrogenous fertilizer production requires one of the highest process heat temperatures of 900°C, which is a challenging temperature for the currently available renewable heat generation technologies. In order to reach these temperatures, renewable fuels and high-temperature nuclear reactors are suitable technologies. Concentrated solar thermal collectors also can reach these temperatures, however, the intermittency needs to be mitigated. On the other hand, the petroleum refining industry is one of the major industries which emits high carbon dioxide (124 million metric tons of carbon dioxide equivalent in the United States only). Petroleum refining requires around 630°C process heat. Although it is targeted that petroleum and its products will be no longer required after 2050, the petroleum refining process can be replaced by concentrated solar and nuclear reactors, as well as renewable fuels. Currently available hydrothermal and enhanced types of geothermal systems are not able to surpass temperatures higher than 500°C. However, future geothermal systems, especially with the new developments in enhanced geothermal technologies, may enable them to reach high temperatures. Mining of potash, soda, borate, plastic material, resin manufacturing, alkaline, chlorine, and sodium hydroxide production require process heat temperatures between 180 and 350°C, which emit 36 million metric tons of carbon dioxide equivalent only in the United States.

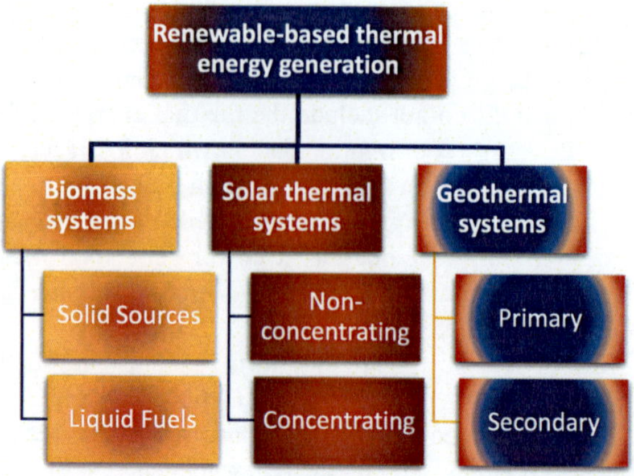

Fig. 4.3 Renewable source–based thermal energy generation methods

In order to replace industrial processes, solar and geothermal are two direct renewable potential heat sources, which can be exploited to generate high-temperature heat. Moreover, indirect renewable heat generation can be done via renewable electricity or renewable fuels. Figure 4.3 presents the renewable thermal energy generation methods, where solar and geothermal-based systems can be considered as direct use of renewable sources; renewable electricity and renewable fuel–based systems can be considered as the in-direct use of renewable sources. Among renewable thermal energy generation methods, high-temperature process heat (more than 700°C) can be generated by using concentrating solar thermal systems, electric heaters, and renewable fuels. For temperatures lower than 500°C, geothermal systems can be included but with a limited source availability. For lower process heat temperatures (lower than 120°C), renewable electricity–based heat pumps and non-concentrating solar thermal systems can be included.

In the current chapter, the key focus is on thermal energy, particularly heat, generated by the use of renewable energy sources, specifically driven by solar, geothermal, and biomass sources. Biomass sources and fuels are also considered for heat generation via combustion. There are also illustrative examples and case studies presented to better illustrate the specific topics with renewables for heat generation. Some analysis methods are thus given for performance assessment and evaluation.

4.2 Biomass Matters and Liquid Fuels

In order to extract the energy content from the organic and biologic materials/matters or liquid fuels, a combustion process needs to be carried out to generate heat from chemical energy. For the combustion process, fuel and an oxidizing agent are needed along with the activation energy. Usually, air is used for its oxygen content as the oxidizing agent in combustion processes. As a result of the process, high-

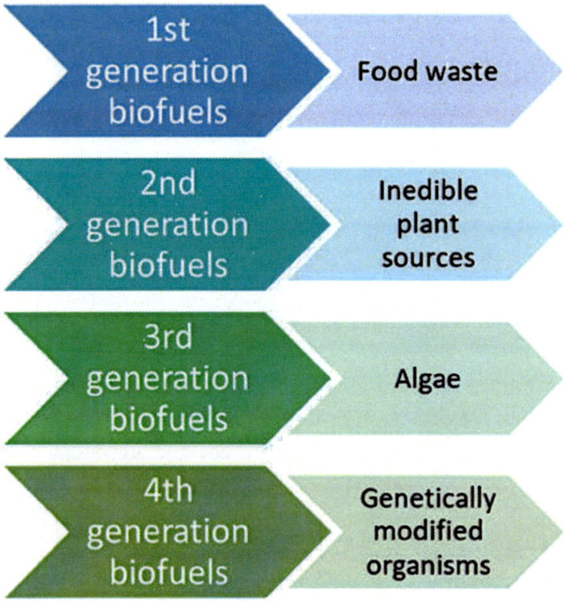

Fig. 4.4 Four generations of biofuels and their sources

temperature heat and combustion gases are extracted. Currently, fossil fuels are the dominant fuels for combustion processes, which are hydrocarbons that naturally form underground. Commercial, residential, industrial, transportation, and other sectors utilize fossil fuels as the major fuel via combustion. Carbon content is critical to determine the emissions; typically, more carbon content causes more carbon dioxide emissions.

It is possible to generate fuels such as biomethanol, bioethanol, or biomethane by using renewable energy and/or waste energy. Wood, wood waste, peat, railroad ties, wood sludges, spent sulphite liquors, food waste, industrial waste, sludge waste, waste alcohol, municipal solid waste, landfill gases, and other wastes can be used to generate biofuels.

In some cases, different sources can be mixed to generate biofuels. When we look at the conversion processes to generate biofuels, fermentation, liquefaction, combustion, mechanical processes, carbonization, gasification, biodigestion, cracking, and esterification.

An anaerobic digestion process is given in Fig. 4.4. Plant waste and/or food waste can be used in the process as the source. In the biodigester component chemical reactions occur as follows:

$$C_6H_{12}O_6 + 2H_2O \rightarrow 2CH_3COOH + 2CO_2 + 2H_2$$

$$3C_6H_{12}O_6 \rightarrow 4CH_3CH_2COOH + 2CH_3COOH + 2CO_2 + 2H_2$$

$$C_6H_{12}O_6 \rightarrow 4CH_3CH_2CH_2COOH + 2CH_3COOH + 2CO_2 + 2H_2$$

The products of the first series of reactions can be digested further via water in order to form carbon dioxide, acetic acid, and water as follows:

$$CH_3CH_2COOH + 2H_2O \rightarrow CH_3COOH + CO_2 + 3H_2$$

Here, acetic acid, as a reactant, can further react in order to produce biomethane and carbon dioxide as follows:

$$CH_3COOH \rightarrow CH_4 + CO_2$$

Furthermore, biomethane and carbon dioxide can be diverted by removing the carbon dioxide in order to purify the methane content in biomethane.

Biodigester generates biomethane methane along with some other by-products, as a mixture of biomethane, carbon dioxide, water, and solid residue. Digested residue can be used as a bio-fertilizer for agricultural purposes. The remaining mixture can be used directly; however, direct use of the mixture can harm the equipment and also can be less efficient. Further treatment processes can help purify methane content in the mixture. After the dryer, biomethane can be generated at high purifications. Figure 4.5 shows the biomethane generation process via anaerobic digestion in a practical setting.

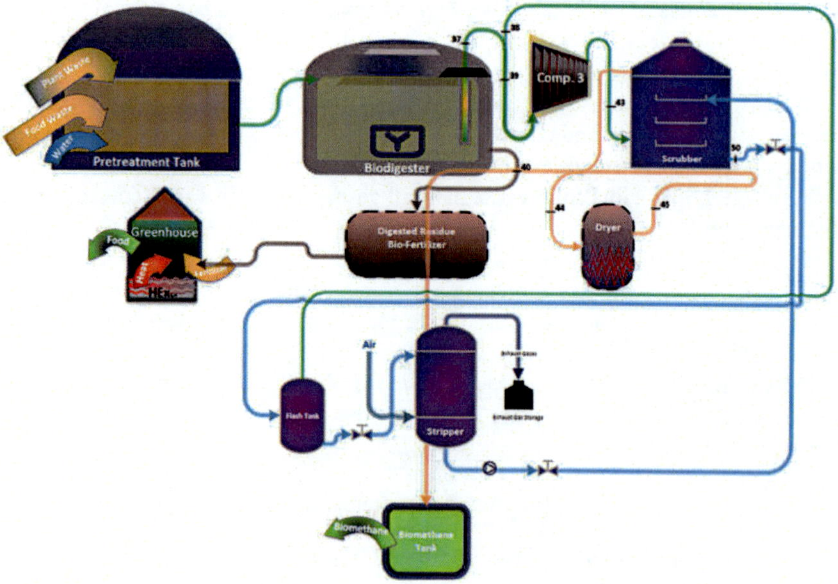

Fig. 4.5 Illustration of biomethane generation process via anaerobic digestion

The combustion process may be categorized based on the occurrence of heat transfer and mass transfer, as well as the air-fuel ratio. According to the heat transfer, there are adiabatic and non-adiabatic combustions. In adiabatic combustion, there is no heat transfer occurs from or to the combustion chamber. In this type of combustion, all heat is collected by the combustion product, which reaches high temperatures. On the other hand, heat is transferred out of the system boundary of the combustion chamber to another medium during the non-adiabatic combustion process.

According to the mass transfer, open and closed types of fuel combustion systems can be mentioned. In the open system, fuel and oxidizing agents enter into the combustion chamber, and products exit from the combustion chamber. In the closed system, no mass transfer occurs from or to the combustion turbine. All substances are kept during the combustion process. Only heat can be transferred.

Further, note that some common types of open, closed, and adiabatic combustion systems are depicted in Fig. 4.6 in which non-adiabatic systems are used in open and closed systems. Apart from the mass and heat transfers, the air-fuel ratio is also used to categorize combustion types. The balance equations for the combustion system can be expressed for mass, energy, entropy, and exergy balances of open, closed, and adiabatic systems.

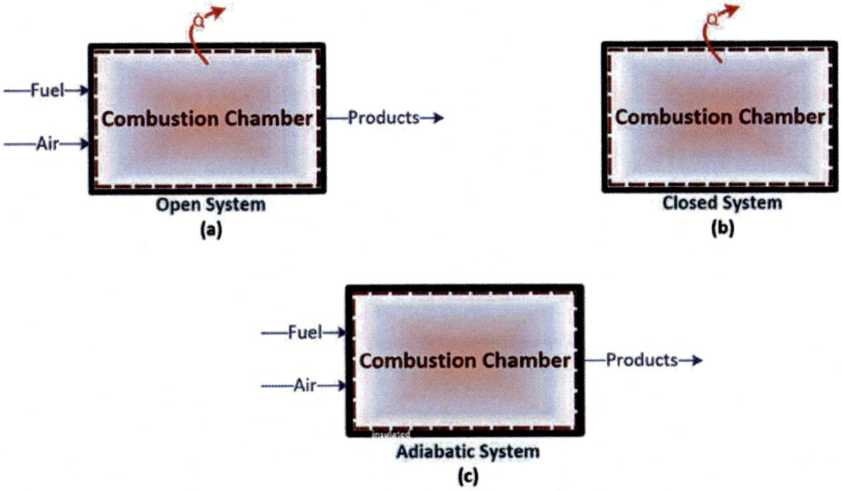

Fig. 4.6 Illustration of three common types of fuel combustion processes for (**a**) open, (**b**) closed, and (**c**) adiabatic systems

The mass balance equations for an open-type combustion system as shown in Fig. 4.6a can be expressed as follows:

$$\sum \dot{n}_R = \sum \dot{n}_P$$

where $\dot{n}$ denotes the mole flow rate of the substances in kmol/s.

The energy balance equation for the open system as given in Fig. 4.6a is this time expressed as follows:

$$H_R = H_P + \dot{Q}$$

where

$$H_R = \sum \dot{n}_R \left(\overline{h_f{}^\circ} + \bar{h} - \overline{h^\circ}\right)_R$$

and

$$H_P = \sum \dot{n}_P \left(\overline{h_f{}^\circ} + \bar{h} - \overline{h^\circ}\right)_P$$

where $\overline{h_f{}^\circ}$ denotes the enthalpy of formation, $\bar{h}$ denotes the enthalpy value obtained at the specific temperature of the substance, and $\overline{h^\circ}$ denotes the enthalpy value at the reference temperature.

The entropy balance equation for the open system as seen in Fig. 4.6a is then written as follows:

$$\dot{S}_R + \dot{S}_{gen} = \dot{S}_P + \frac{\dot{Q}}{T_{comb}}$$

where

$$\dot{S}_R = \sum (\dot{n}\bar{s})_R$$

and

$$\dot{S}_R = \sum (\dot{n}\bar{s})_P$$

where $\bar{s}$ stands for the absolute entropy, and it can be calculated for a specific temperature (T) and pressure (P) as follows:

$$\bar{s}(T,P) = \overline{s^\circ}(T_0,P) - R_u \ln\left(\frac{P}{P_0}\right)$$

where $\overline{s^\circ}$ denotes the reference absolute entropy at the reference temperature and pressure, R_u denotes the universal gas constant, and P_0 refers to the atmospheric pressure, T_0 refers to the reference temperature.

The exergy balance equation for the open system as illustrated in Fig. 4.6a is then expressed as follows:

$$\dot{\mathrm{Ex}}_{\mathrm{R}} = \dot{\mathrm{Ex}}_{\mathrm{P}} + \dot{\mathrm{Ex}}^{\mathrm{Q}} + \dot{\mathrm{Ex}}_{\mathrm{d}}$$

where the exergy destruction ($\dot{\mathrm{Ex}}_{\mathrm{d}}$) is written through the entropy generation and the reference temperature as:

$$\dot{\mathrm{Ex}}_{\mathrm{d}} = T_0 \dot{\mathrm{S}}_{\mathrm{gen}}$$

Moreover, the exergy flows of reactants ($\dot{\mathrm{Ex}}_{\mathrm{R}}$) and products ($\dot{\mathrm{Ex}}_{\mathrm{P}}$) can be expressed as:

$$\dot{\mathrm{Ex}}_{\mathrm{R}} = \sum \dot{\mathrm{n}}_{\mathrm{R}} \left(\overline{h_{\mathrm{f}}}^{\circ} + \overline{h} - \overline{h^{\circ}} - T_0 \overline{s} \right)_{\mathrm{R}}$$

and

$$\dot{\mathrm{Ex}}_{\mathrm{P}} = \sum \dot{\mathrm{n}}_{\mathrm{P}} \left(\overline{h_{\mathrm{f}}}^{\circ} + \overline{h} - \overline{h^{\circ}} - T_0 \overline{s} \right)_{\mathrm{P}}$$

For the closed system as shown in Fig. 4.6b, there is no mass transfer occurring during the combustion process which means that there are no reactants and products crossing the system boundary. The energy balance equation for the closed system is then expressed as follows:

$$U_{\mathrm{R}} = U_{\mathrm{P}} + Q$$

where

$$U_{\mathrm{R}} = \sum n_{\mathrm{R}} \left(\overline{h_{\mathrm{f}}}^{\circ} + \overline{h} - \overline{h^{\circ}} - P\overline{v} \right)_{\mathrm{R}}$$

and

$$U_{\mathrm{P}} = \sum n_{\mathrm{P}} \left(\overline{h_{\mathrm{f}}}^{\circ} + \overline{h} - \overline{h^{\circ}} - P\overline{v} \right)_{\mathrm{P}}$$

The entropy balance equation for the closed system as illustrated in Fig. 4.6b can be expressed as follows:

$$S_{\mathrm{R}} + S_{\mathrm{gen}} = S_{\mathrm{P}} + \frac{Q}{T_{\mathrm{comb}}}$$

where

$$S_{\mathrm{R}} = \sum (n\overline{s})_{\mathrm{R}}$$

and

$$S_R = \sum (n\bar{s})_P$$

where $\bar{s}$ stands for the absolute entropy, and it can be calculated for a specific temperature (T) and pressure (P) as follows:

$$\bar{s}(T, P) = \overline{s^\circ}(T_0, P) - R_u \ln\left(\frac{P}{P_0}\right)$$

The exergy balance equation for the closed system (as shown in Fig. 4.6b) can be written as follows:

$$\mathrm{Ex}_R = \mathrm{Ex}_P + \mathrm{Ex}^Q + \mathrm{Ex}_d$$

where the exergy destruction (Ex_d) can be expressed from entropy generation and the reference temperature as:

$$\mathrm{Ex}_d = T_0 S_{gen}$$

Moreover, the exergy of reactants (Ex_R) and products (Ex_P) can be expressed as:

$$\mathrm{Ex}_R = \sum n_R \left(\overline{h_f}^\circ + \bar{h} - \overline{h^\circ} - P\bar{v} - T_0\bar{s}\right)_R$$

and

$$\mathrm{Ex}_p = \sum n_P \left(\overline{h_f}^\circ + \bar{h} - \overline{h^\circ} - P\bar{v} - T_0\bar{s}\right)_P$$

For the adiabatic system (as clearly seen in Fig. 4.6c), the mass balance equation can be expressed as follows:

$$\sum \dot{n}_R = \sum \dot{n}_P$$

The energy balance equation for the adiabatic system (as seen in Fig. 4.6c) can be expressed as follows:

$$H_R = H_P$$

where

$$H_R = \sum \dot{n}_R \left(\overline{h_f}^\circ + \bar{h} - \overline{h^\circ}\right)_R$$

and

$$H_{\mathrm{P}} = \sum \dot{n}_{\mathrm{P}} \left(\overline{h_{\mathrm{f}}{}^{\circ}} + \overline{h} - \overline{h^{\circ}} \right)_{\mathrm{P}}$$

The entropy balance equation for the adiabatic system (as seen in Fig. 4.6c) is then obtained as follows:

$$\dot{S}_{\mathrm{R}} + \dot{S}_{\mathrm{gen}} = \dot{S}_{\mathrm{P}}$$

where

$$\dot{S}_{\mathrm{R}} = \sum (\dot{n}\overline{s})_{\mathrm{R}}$$

and

$$\dot{S}_{\mathrm{R}} = \sum (\dot{n}\overline{s})_{\mathrm{P}}$$

where $\overline{s}$ stands for the absolute entropy, and it can be calculated for a specific temperature (T) and pressure (P) as follows:

$$\overline{s}(T,P) = \overline{s^{\circ}}(T_0,P) - R_{\mathrm{u}} \ln\left(\frac{P}{P_0}\right)$$

The exergy balance equation for the adiabatic system (as shown in Fig. 4.6c) is then written as follows:

$$\dot{\mathrm{E}}\mathrm{x}_{\mathrm{R}} = \dot{\mathrm{E}}\mathrm{x}_{\mathrm{P}} + \dot{\mathrm{E}}\mathrm{x}_{\mathrm{d}}$$

where the exergy destruction ($\dot{\mathrm{E}}\mathrm{x}_{\mathrm{d}}$) can be expressed from entropy generation and the reference temperature as:

$$\dot{\mathrm{E}}\mathrm{x}_{\mathrm{d}} = T_0 \dot{S}_{\mathrm{gen}}$$

Moreover, the exergy flows of reactants ($\dot{\mathrm{E}}\mathrm{x}_{\mathrm{R}}$) and products ($\dot{\mathrm{E}}\mathrm{x}_{\mathrm{P}}$) is then expressed as:

$$\dot{\mathrm{E}}\mathrm{x}_{\mathrm{R}} = \sum \dot{n}_{\mathrm{R}} \left(\overline{h_{\mathrm{f}}{}^{\circ}} + \overline{h} - \overline{h^{\circ}} - T_0 \overline{s} \right)_{\mathrm{R}}$$

and

$$\dot{\mathrm{E}}\mathrm{x}_{\mathrm{P}} = \sum \dot{n}_{\mathrm{P}} \left(\overline{h_{\mathrm{f}}{}^{\circ}} + \overline{h} - \overline{h^{\circ}} - T_0 \overline{s} \right)_{\mathrm{P}}$$

The heating values and air-fuel ratio (AFR) are important parameters when it comes to the combustion or burning process. There are two types of heating values, one is called as higher heating value (HHV) or gross calorific value and the other is a lower heating value (LHV) or net calorific value. Higher and lower heating values are distinct from each other due to the heat of vaporization of the water. The difference between the higher and lower heating values can be calculated as follows:

$$\mathrm{HHV}\left(\frac{\mathrm{kJ}}{\mathrm{kg}}\right)=\mathrm{LHV}\left(\frac{\mathrm{kJ}}{\mathrm{kg}}\right)+x_{\mathrm{H_2O}}h_{\mathrm{fg,H_2O}}$$

where the HHV at reference conditions is:

$$\mathrm{HHV}=\sum n_{\mathrm{P}}\left(\overline{h_{\mathrm{f}}}^{\circ}\right)_{\mathrm{P}}-\sum n_{\mathrm{R}}\left(\overline{h_{\mathrm{f}}}^{\circ}\right)_{\mathrm{R}}$$

Here, HHV includes the enthalpy of water/steam mixture, on the other hand, LHV does not include the enthalpy of water/steam mixture.

When we look at the AFR, it can be defined as the ratio of air input to fuel input to the combustion, in terms of mass, which can be written as follows:

$$\mathrm{AFR}=\frac{m_{\mathrm{air}}}{m_{\mathrm{fuel}}}$$

According to the air-fuel ratio, there are complete and incomplete combustions can be mentioned. Complete combustion refers to the process where all fuel input completely burns with enough amount of the oxidizing agent and forms combustion gases. For complete combustion, the amount of air required is called stoichiometric air. If the air amount is higher than the stoichiometric air, then it is called as excess air. An incomplete combustion, on the other hand, refers to the combustion process where there is less amount of air then the stoichiometric air.

It is important to understand the combustion process, since it is the major process for energy consumption, and it needs to be replaced by sustainable processes in order to reach the climate remediation goals. From this aspect, all the infrastructure for combustion processes needs to be replaced by the infrastructure for sustainable processes. This sustainable energy transition requires unprecedented efforts. The humanity has been witnessing major transitions in information and communication technologies (ICT), and biotechnology. However, neither of these transitions require large-scale physical changes. Sustainable energy transition can be considered distinct from ICT and biotechnology transitions due to the large-scale physical infrastructure changes. Numerous researchers have been studying options for a transitionary period, in order to mitigate greenhouse gas emissions. Yapicioglu and Dincer [4] investigated the combustion process with renewable fossil fuel blends. This approach enables the use of the same infrastructure with significantly less emissions.

Illustrative Example 4.1 An illustrative example is investigated by using the data from [4]. Ammonia fuel is blended with hydrogen, bioethanol, and biomethanol.

When we look at the complete combustion of the above-mentioned fuels, chemical formulas are given below.

For the combustion of ammonia, chemical equation can be expressed as follows:

$$NH_3 + a_c(O_2 + 3.76\,N_2) \rightarrow 1.5H_2O + 3.76a_cN_2$$

where a_c is 0.75 for the complete combustion ammonia.

For the combustion of hydrogen, the chemical equation can be expressed as follows:

$$H_2 + a_c(O_2 + 3.76\,N_2) \rightarrow H_2O + 3.76a_cN_2$$

where a_c is 0.5 for the complete combustion of hydrogen.

It is important to highlight that, hydrogen and ammonia combustions do not emit greenhouse gases. This shows that blending will help to mitigate carbon dioxide emissions from the combustion process, since the hydrogen or/and ammonia portion of the fuel blend will not emit carbon dioxide while generating heat.

For the combustion of bioethanol, chemical equation can be expressed as follows:

$$C_2H_5OH + a_c(O_2 + 3.76\,N_2) \rightarrow 2CO_2 + 3H_2O + 3.76a_cN_2$$

where a_{th} is 3 for the complete combustion of bioethanol.

For the combustion of biomethanol, chemical equation can be expressed as follows:

$$CH_3OH + a_c(O_2 + 3.76\,N_2) \rightarrow CO_2 + 2H_2O + 3.76a_cN_2$$

where a_{th} is 1.5 for a complete combustion of biomethanol.

When one looks at the combustion characteristics of these fuels, lower heating value and air/fuel ratio by weight are two of the important parameters to check. Figure 4.7 presents the lower heating value and air/fuel ratio by weight of various fuels. Hydrogen has the highest lower heating value as 120 MJ/kg. Among the compared fuels, ammonia has the lowest lower heating value as 18.8 MJ/kg. Moreover, hydrogen has the highest air/fuel ratio by weight as 34.32. On the other hand, ammonia has a 6.05 air/fuel ratio by weight. According to the blending ratio of fuels, those parameters need to be considered in order to arrange the air intake and the blended fuel intake. The lower heating value of the blended fuel is the parameter which needs to be calculated in order to determine the mass flow rate of the blended fuel due to the heat generation capacity of the combustion chamber and the power generation capacity of the connected electromagnetic generator.

Note that the ammonia blending process is carried out by blending the ammonia with air at air intake, and then air-ammonia mixture is mixed with fuel at the spark igniting generator engine [4]. The illustration of the ammonia-blended combustion process is depicted in Fig. 4.8.

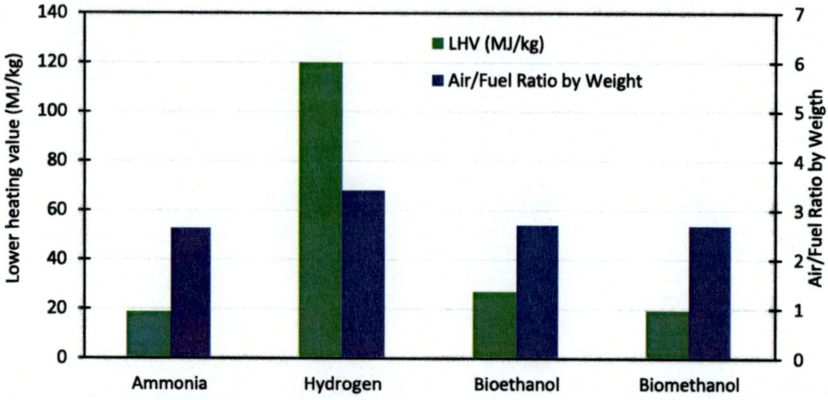

Fig. 4.7 LHV and AFR of different fuels

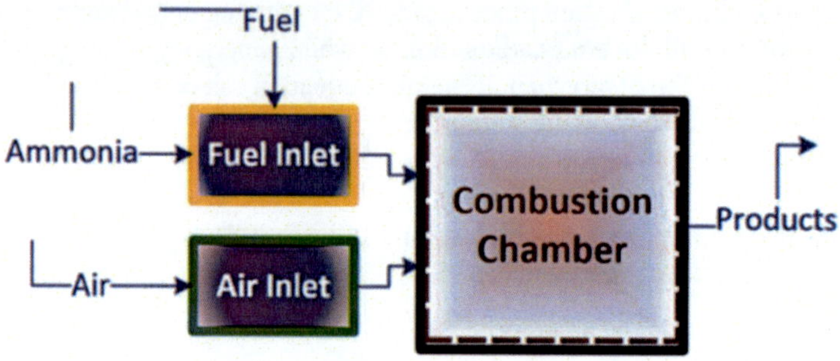

Fig. 4.8 Illustration of ammonia-blended combustion system

The ammonia-blended combustion was carried out by using a minimum of 20% ammonia to 80% fuel and 80% ammonia to 20% fuel blends. Ammonia-blended hydrogen–driven power generation process has energy efficiencies between 23.8% and 24.7%, and exergy efficiencies between 28.48% and 29.57%; energy and exergy efficiencies for ammonia-blended bioethanol are between 27.04% and 27.13%, and between 32.9% and 34.1%; energy and exergy efficiencies for ammonia-blended biomethanol are between 25.81% and 25.84%, and between 32.4% and 33.02%. Figure 4.9 shows the energy efficiencies of the different ammonia fuel blends for the fuel-to-electricity generation process via combustion and electromagnetic electricity generation.

The results further show that ammonia-bioethanol mixture has the highest energy and exergy efficiencies among ammonia fuel blends. When we look at the emissions of different ammonia fuel blends, hydrogen-ammonia mixture does not contain any carbon; therefore, no greenhouse gasses are emitted. On the other hand, especially at the 80% ammonia to 20% fuel mixture case, more than 80% of the CO_2 emissions are eliminated.

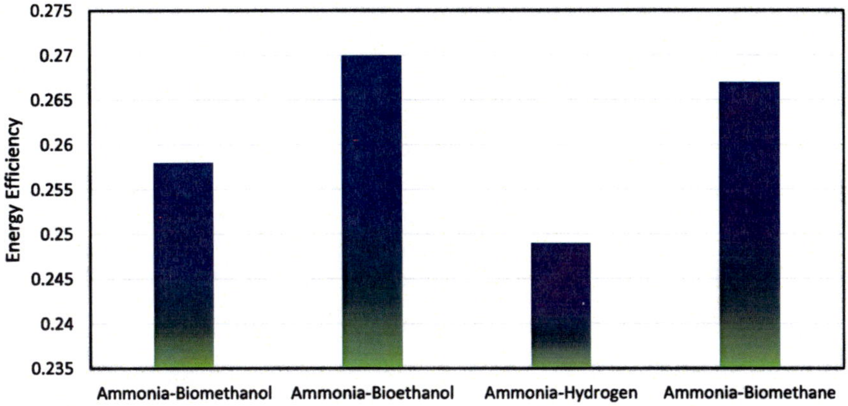

Fig. 4.9 Energy efficiency of fuel to power generation process for various ammonia-blended fuels

4.3 Solar Thermal Systems

Thermonuclear fusion reaction happens in the sun, which releases energy from the fusion process of hydrogen to form helium-4 via stellar nucleosynthesis. The released energy reaches the earth via radiation, which can be harvested via solar energy systems. Solar energy is already harvested by natural systems: plants can utilize the radiation in the photosynthesis process to generate glucose. Human-made systems can be used to harvest solar energy. Solar energy systems usually harvest solar energy by collecting the radiation heat and transferring it to another medium, except photovoltaics.

Typically, water or thermal oil is used as the thermal fluid in solar thermal energy systems. According to the application area, the temperature requirement of the process heat from solar thermal energy systems varies. Figure 4.10 depicts the temperature range of heat generated by concentrating and non-concentrating solar systems, and the potential application areas.

For residential heating, greenhouse heating, fish farming, and other space heating applications, the temperature requirement of the process heat can be as low as 25°C, even lower temperatures are useful with additional heat pumps. Desalination process can be carried out typically more than 50°C process heat temperatures. The food drying process usually requires more than 70°C of process heat. Space cooling can also be done by utilizing the solar thermal energy, in which absorption refrigeration cycles can be carried out by utilizing the process heat at more than 90°C temperatures.

Power generation can be done by utilizing a wide range of process heat temperatures, typically more than 250°C; however, lower temperatures can be exploited also via cycles such as organic Rankine cycles. Higher temperatures can be utilized in many other processes, such as hydrogen generation or alternative fuel production. All above-mentioned processes require thermal energy generation step, which can be

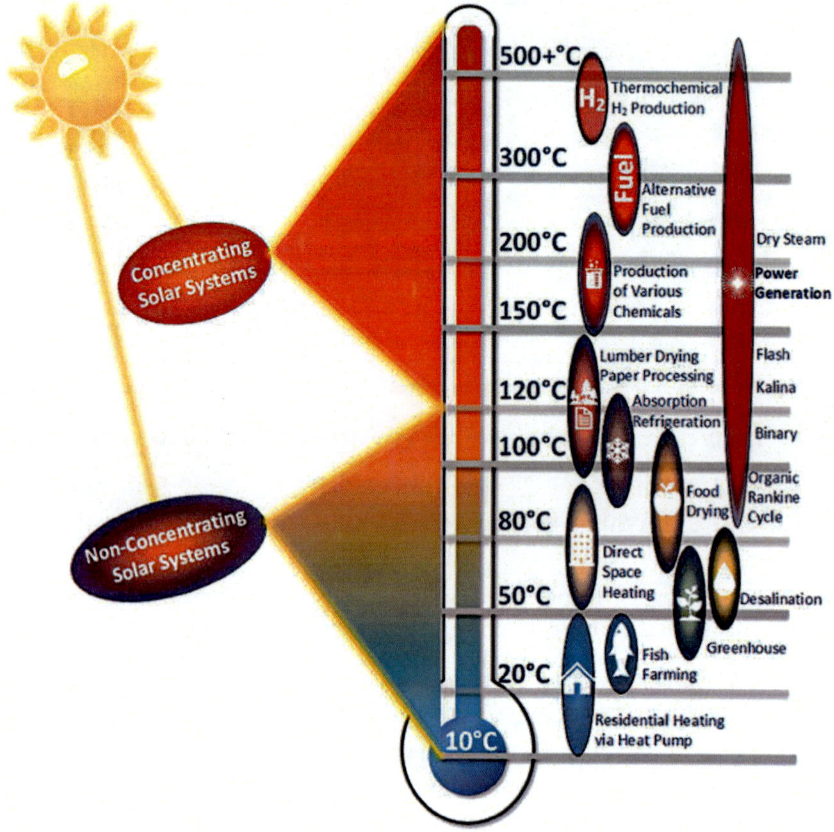

Fig. 4.10 Illustration of temperatures of heat generated from concentrating and non-concentrating solar systems and their potential application areas

generated by transferring heat from sun's rays to the working fluid. A wider perspective can consider the reaction in sun and can claim that the thermal energy generation via solar thermal systems starts with thermonuclear fusion in the sun. However, this is not a practical approach in the open literature, and we usually consider solar radiation as the source of energy.

4.3.1 Non-concentrating Solar Thermal Systems

Non-concentrating solar thermal collectors directly utilize solar energy without multiplying the solar radiation, where the absorber area and interception area are equal. This brings simplicity compared to the concentrating solar thermal collectors. Evacuated tubes and flat plates can be mentioned as the most common type of

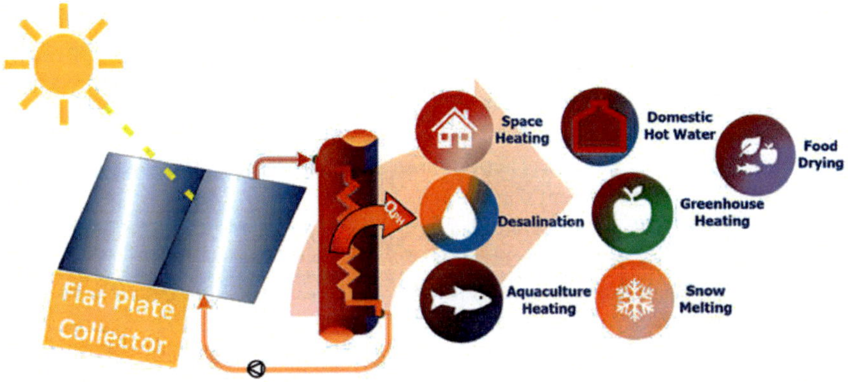

Fig. 4.11 Illustration of flat plate collector–type non-concentrating solar thermal system

non-concentrating solar thermal collector systems. Application areas can be seen in Fig. 4.11; space heating and domestic hot water are the typical applications with non-concentrating solar systems. Heat generation cannot reach high temperatures, and application areas are limited. However, there are integrated systems with flat plate collectors in which multiple useful outputs can be generated, such as electricity, fresh water, hydrogen, and cooling effect.

Figure 4.11 depicts a flat plate–type solar collector. These are very cost-effective systems which exploit solar irradiance in order to heat a working thermofluid, which feeds a heat exchanger. Heat from the heat exchanger can be used for a variety of applications. Temperature can reach up to 200°C for the process heat generated in the heat exchanger. Process heat can be further employed for residential or commercial space heating, water desalination, aquaculture heating such as fish farms, domestic hot water, greenhouse heating, snow melting, food drying, etc. A low-cost system can meet very important commodities for communities. This is a key to sustainable development, where a low-cost solution can meet many needs of communities by protecting the environment and not consuming nonrenewable sources.

When we compare the flat plate and vacuum tube collectors, flat plates can be mentioned as the simplest form to collect solar radiation. The material used to absorb solar radiance is usually copper or aluminum due to the high conductivity of copper and the low cost of aluminum. The thermal energy generation is directly related to the area of the flat plate collector. Heat generation can reach up to 200°C, but typically lower temperatures are used with flat plate collectors. Vacuum tube collectors, on the other hand, are more advanced versions of flat plates. Vacuum tubes can generate heat up until 300°C, usually lower temperatures are used in typical applications. The vacuum plays an important role in minimizing convective heat losses. A chemical heat transfer liquid is used inside the evacuated tube which changes phase and collects solar radiance by changing phase from liquid to gas and transfers the heat to the water inside the tank by changing phase again from gas to liquid.

Case Study 4.1 A case study is investigated by considering a study on a flat plate collector–based hydrogen generation system [5]. Solar energy is collected via a flat plate collector and salt gradient solar pond system. An organic Rankine cycle system is considered in order to utilize solar energy to generate electricity. An electrolysis system is considered to use electricity to generate hydrogen. In order to minimize intermittencies, a solar pond system is used for heat storage purposes. Their proposed system is shown in Fig. 4.12.

The flat plate collectors heated water from 60°C to 91.61°C at 200 kPa. In the organic Rankine cycle, ammonia is utilized. In the evaporator, 91.61°C of water transferred its heat to ammonia and superheated it to 80°C at 2000 kPa. Superheated ammonia is expanded to 45°C and 1750 kPa. After the condensation of ammonia in the condenser, ammonia is pressurized to 2000 kPa. Thermophysical properties at state points are presented in Table 4.1.

Their proposed design stores energy via the storage zone of the solar pond which has 300 m^2 of surface area. Inlet and outlet temperatures of the exchanger-1 were

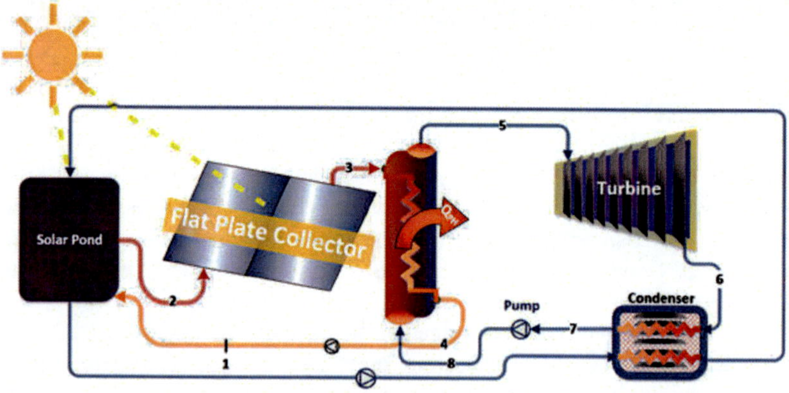

Fig. 4.12 Illustration of flat plate– and solar pond–based hydrogen generation system

Table 4.1 The thermophysical properties of state points

State point	Substance	Temperature (°C)	Pressure (kPa)	Mass flow rate (kg/s)	Specific enthalpy (kJ/kg)	Specific entropy (kJ/kg°C)	Specific exergy (kJ/kg)
0	Water	30	101.3	0	125.8	0.44	0
1	Water	40	200	0.5	167.7	0.57	0.79
2	Water	60	200	0.5	251.3	0.83	5.98
3	Water	91.61	200	0.5	383.8	1.21	23.36
5	Ammonia	80	2000	0.08	1625	5.45	19.31
6	Ammonia	45	1750	0.08	1493	5.12	50.75
7	Ammonia	30	1750	0.08	342	1.49	101.7
8	Ammonia	31	2000	0.08	346.9	1.5	101.2

Data from Ref. [5]

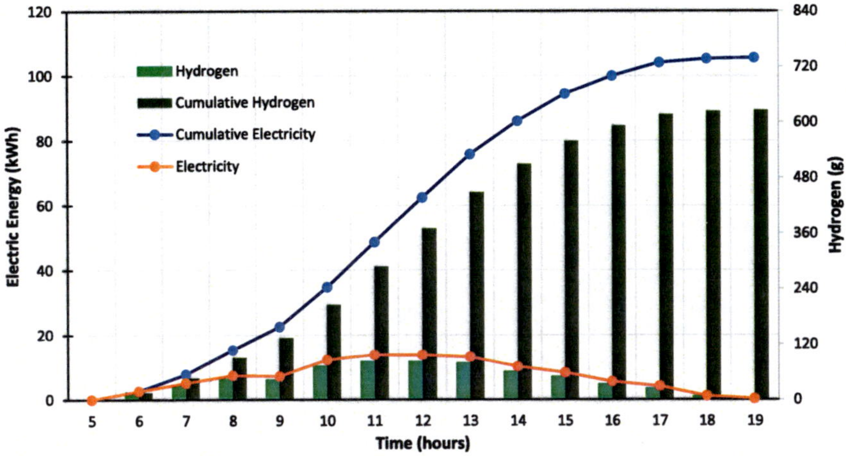

Fig. 4.13 Electricity and hydrogen generation from flat plate and solar pond–based hydrogen generation system (data from [5])

considered to be 40 and 60°C. They used meteorological data for Adana, Türkiye, from 2010 to 2013. During the month of July, 62341.72 kWh of solar energy and 58163.1 kWh of solar exergy reached the solar pond; on the other hand, 18557.5 kWh of solar energy and 17391.1 kWh of solar exergy reached during the month of January. Furthermore, 77927 kWh of solar energy and 72703.9 kWh of solar exergy reached the flat plate collector during the month of July.

Generated electricity is used for the water electrolysis process in order to disassociate water to generate hydrogen and oxygen. Although there are heat generation and power generation steps, their proposed system can be considered as a single-input and single-output system since all heat and power generated from the system was used to generate hydrogen. For this reason, the hydrogen generation curve aligns with power generation and heat generation curves. The heat generation curve is a flattened version of the solar radiation curve due to the solar pond heat storage. Figure 4.13 shows power and hydrogen generation from the system. When we look at the power generation on July 15, a 105.6 kWh of electricity is generated from 5 am until 7 pm. The electrolysis system consumes all the electricity generated from the organic Rankine cycle. On the same day, 2.25 kg of hydrogen was generated from the system.

4.3.2 Concentrating Solar Thermal Systems

As shown in Fig. 4.14, concentrating solar thermal systems can generate higher temperature process heat for further applications. Concentrating solar thermal systems include mirrors and a collector where mirror(s) are used to concentrate sun's

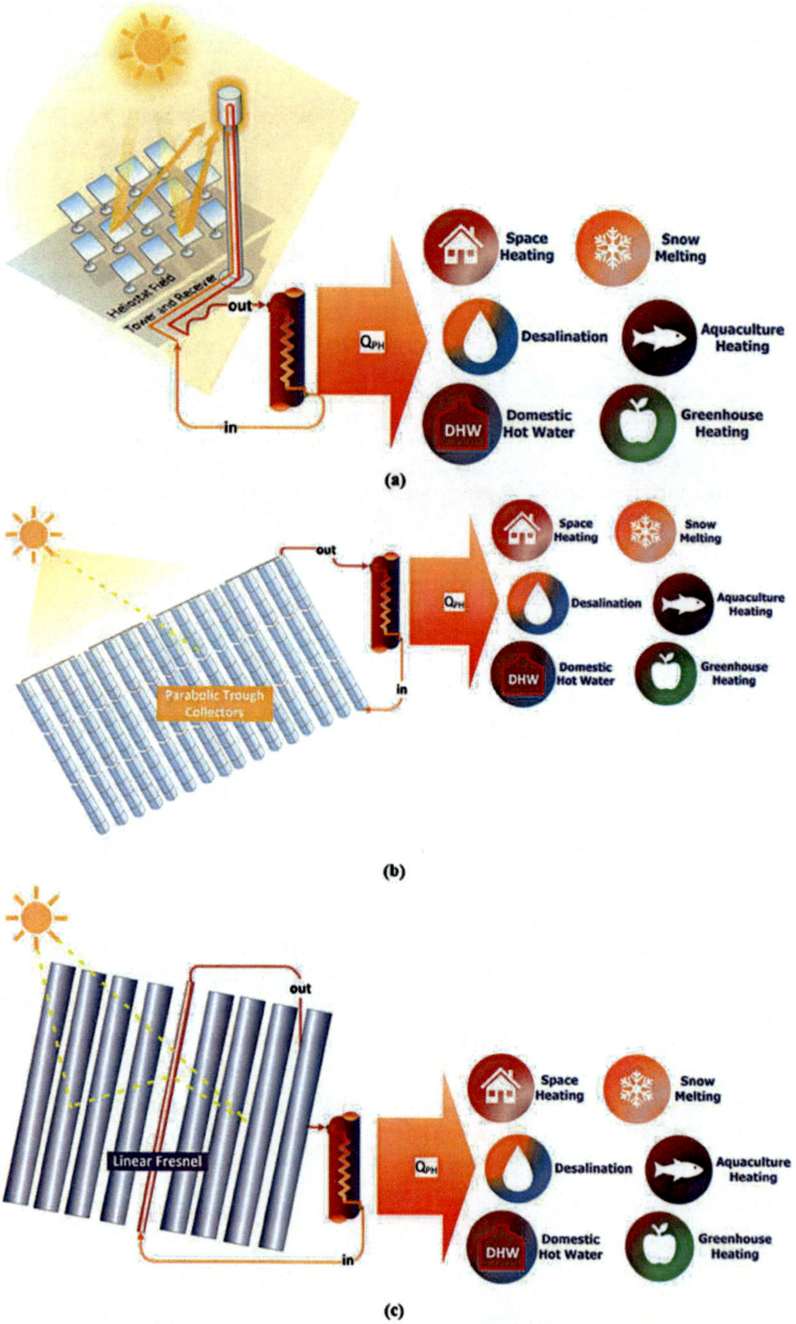

Fig. 4.14 Illustration of (**a**) solar tower, (**b**) parabolic trough, and (**c**) linear Fresnel–type concentrating solar systems for thermal energy generation

rays on a focal line or focal surface in order to reach higher temperatures. A variety of power generation cycles are suitable for concentrating solar thermal systems [6]. In addition to that, thermochemical hydrogen production [7], alternative fuel production [8], and various chemical production processes are suitable for concentrating solar thermal systems.

Solar tower, parabolic trough, linear Fresnel and dish-Stirling collectors are the types of concentrating solar systems. Among these technologies, solar tower and parabolic trough can be highlighted as commercially proven technologies. Other technologies are still in progress to become commercially mature technologies. In solar tower–type concentrating solar systems, heliostat mirrors focus the sunlight on a receiver, which is placed on top of a tower. Parabolic trough–type concentrating solar systems consist of parabolically curved mirrors, which are highly reflective polished metals which focus sunlight on the receiver tube, which lays on the focus of curved mirrors. Linear Fresnel–type concentrating solar systems consist of flat mirrors which focus the sunlight to receiver tubes that are placed on top of the mirror group.

Thermofluids absorb the heat from the receiver and carry the heat for specific applications or to the thermal energy storage tanks. Typically, thermal energy storages are used with concentrating solar plants. Usually, two tanks are used to store thermal energy, which can hold the thermofluid inside. Thermofluids carry the heat from the collector to the thermal energy storage tanks or to the field. Regardless of the application, the first step of any concentrating solar-based system is thermal energy generation. Thermal energy generation happens by utilizing the concentrated solar irradiance from the receiver via thermofluid. By using the first and second laws of thermodynamics, the thermal energy generation step of concentrating solar systems can be calculated by multiplying the specific direct normal irradiance per unit collector area and the total aperture area of the collector of the concentrating solar system as follows:

$$\dot{q}_{\mathrm{DNI,CS}} \times A_{\mathrm{aperture,CS}} = \dot{Q}_{\mathrm{DNI,CS}}$$

where $A_{\mathrm{aperture,CS}}$ denotes the total aperture area of the concentrating solar collector, $\dot{q}_{\mathrm{DNI,CS}}$ denotes the direct normal irradiance reaching the unit surface area of the concentrating solar collector for the selected location, and $\dot{Q}_{\mathrm{DNI,CS}}$ denotes the direct normal solar irradiance reaching the concentrating solar system.

In order to calculate the exergy version of thermal energy input, the second-law analysis needs to be carried out. The second-law analysis requires the reference and source temperatures. For solar energy systems, the temperature of sun can be considered as the source temperature, which is around 5780 K [9]. For the reference temperature, typical calculations can consider 25°C as reference temperature. Thermal exergy input to the system can be calculated as follows:

$$\dot{E}x_{Q_{DNI,CS}} = \dot{Q}_{DNI,CS} \times \left(1 - \left(\frac{T_0}{T_s}\right)\right)$$

Once the thermal energy from solar radiation is collected, thermofluid can carry the heat for further applications. A significant portion of solar radiation reaching the collector is lost due to many reasons. Source and useful output of the concentrating solar system can be considered as solar radiation reaching the aperture area of the collector and available process heat generated by the system. In this scenario, energy and exergy efficiencies of concentrating solar system can be calculated as follows:

$$\eta_{en} = \frac{\dot{Q}_{PH,CS}}{\dot{Q}_{DNI,CS}}$$

and

$$\eta_{ex} = \frac{\dot{E}x_{Q_{PH,CS}}}{\dot{E}x_{Q_{DNI,CS}}}$$

where $\dot{Q}_{PH,CS}$ denotes the available process heat generated by the concentrating solar system.

Case Study 4.2 A case study is comparatively assessed for three different concentrating solar collector technologies. Solar tower, linear Fresnel, and parabolic trough types of solar collectors are considered in order to evaluate their heat generation performance. Figure 4.14 depicts the concentrating solar systems for heat generation.

Each system is designed to have the same solar radiation input. According to their system characteristics, heat generation will be different. The objective of the current case study is comparatively assessing different types of concentrating solar systems due to their heat generation.

Meteorological data from the national solar radiation database in the System Advisor Model [14] is used for Daggett, California, United States location, for three of the concentrating solar systems. Location-related parameters are given in Table 4.2. According to that, 2127.95 kWh of global horizontal irradiance reaches the unit surface area. The ambient temperature changes between −3°C and 44°C in a typical year, where the average temperature is 17°C.

In a solar tower concentrating solar thermal system, a tower is designed at 193.5 meters of height with a receiver area with a length and width of 21.6 and 17.6 meters. Stainless steel AISI316 with 0.88 coating emittance and 0.94 coating absorbance is considered. According to the total aperture reflective area, parabolic trough and linear Fresnel systems are designed. The parabolic trough concentrating solar thermal system design parameters are given in Table 4.3.

Some simulations are carried out by using the System Advisor Model software package. According to the simulation results in a typical year, each solar tower,

Table 4.2 The details of the location for case study 4.2

Parameter	Detail
Location	Daggett, CA, United States
Latitude	34.85 DD
Longitude	−116.78 DD
Elevation	561 m
Meteorological data	National Solar Radiation Database
Weather station ID	91,486
Meteorological data type	Typical meteorological year
Annual global horizontal irradiance	2127.95 kWh/m^2
Annual direct normal irradiance	2799.55 kWh/m^2
Annual diffuse horizontal irradiance	456.25 kWh/m^2

Table 4.3 The parabolic trough–type concentrated solar thermal system design parameters considered

Parameter	Value
Heat absorption capacity	820.5 MW$_t$
Collector exit temperature	391°C
Collector inlet temperature	293°C
Number of collector loops	238
Heat transfer fluid	Therminol VP-1
Row spacing	15 m
Stow angle	170°
Deploy angle	10°
Washes per year	63
Total aperture reflective area	1,208,730.0 m^2
Solar field area	772 acres
Total land area	1081 acres
Tracking power	238 kW
Pump efficiency	0.85
Piping thermal loss coefficient	0.45 W/m^2 K
Collector	SkyFuel sky trough
Collector reflective aperture area	656 m^2
Aperture width	6 m
Length of collector assembly	115 m
Number of modules per assembly	10
Receiver	Schott PTR80
Absorber tube inner diameter	0.076 m
Absorber tube outer diameter	0.08 m

linear Fresnel, and parabolic trough–type concentrating solar system collected 3384 GWh of solar radiance via 1,208,730 m^2 of total aperture reflective area. On the other hand, 3210 GWh of the exergy version of solar radiance is collected by each type of concentrating solar system, in a typical year. However, each system generated different thermal energy and exergy at different temperatures. Solar tower–type concentrating solar thermal systems generated 1531.5 GWh of process heat in a typical year. Maximum process heat temperature reached 573.98°C, in a typical year. Linear Fresnel–type concentrating solar thermal system generated 1021 GWh of process heat in a typical year, where the maximum process heat temperature is obtained as 536.04°C. Parabolic trough–type concentrating solar thermal system generated 1482.8 GWh of process heat in a typical year. Maximum process heat temperature is found as 436.86°C for the parabolic trough collector. Among the collector types, solar tower–type concentrating solar systems generated the highest heat in a typical year, whereas linear Fresnel–type concentrating solar systems performed as the worst in terms of heat generation. The parabolic trough–type concentrating solar system is slightly lower than the solar tower–type.

When we look at the exergy versions of collected solar radiance and generated heat, 3210 GWh of exergy of solar radiance is collected by each type of concentrating solar system. Useful thermal exergy outputs of concentrating solar systems are found as 992.06 GWh for solar tower, 812.22 GWh for parabolic trough, and 602.9 GWh for linear Fresnel. Figure 4.15 shows the energy and exergy versions of useful heat generation on different days in a typical year.

The energy and exergy efficiencies are calculated by dividing the energy and exergy versions generated by process heat to the total direct normal irradiances. Among the concentrating solar systems, solar tower type has the highest overall energy and exergy efficiencies as 45.25% and 30.9%. For the parabolic trough concentrating solar system, the overall energy and exergy efficiencies are found as 43.81% and 25.3%. Overall energy and exergy efficiencies for the linear Fresnel–type concentrated solar system is found as 30.18% and 18.78%, as shown in Fig. 4.16.

Case Study 4.3 A case study is investigated by using data and information from [10]. They combined two renewable heat sources as biomass and solar, on order to generate multiple useful outputs. They utilized heat via two Rankine cycles, two gas turbine cycles, and an absorption refrigeration cycle in order to generate power, heat, domestic hot water, and cooling effect for a sustainable city. When we look at the solar thermal system, a concentrating-type solar system is considered with solar tower with heliostat-type collectors. On the biomass side, a biomass combustion system is designed due to the dry olive pits as biomass source. Figure 4.17 depicts the overall system layout.

In order to determine exergy destructions of each component, exergy balance equations are used. The exergy destruction for the gas turbine 1 component is:

$$\dot{m}_6 ex_6 = \dot{m}_7 ex_7 + \dot{W}_{gtc1} + \dot{Ex}_{dgtc1}$$

Exergy destruction for Rankine turbine 1 is:

$$\dot{m}_{26} ex_{26} = \dot{m}_{27} ex_{27} + \dot{W}_{t1} + \dot{Ex}_{dt1}$$

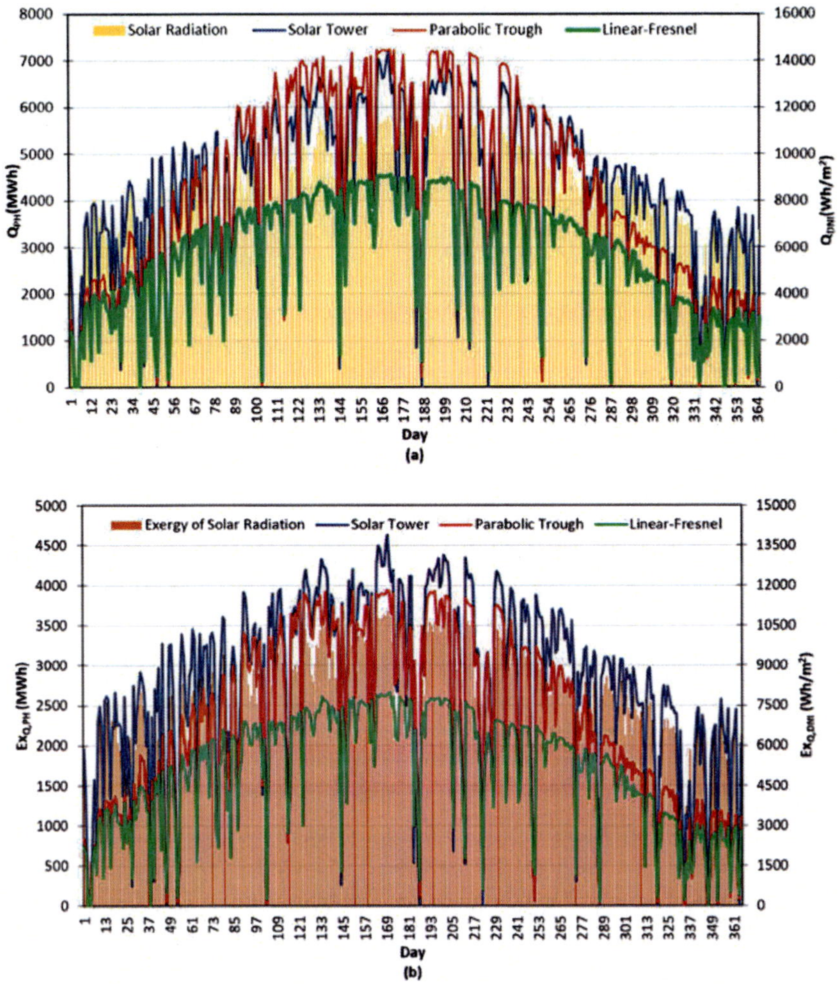

Fig. 4.15 (**a**) Energy and (**b**) exergy of process heat generation from solar tower, parabolic trough, and linear Fresnel concentrating solar systems by energy and exergy contents of direct normal irradiance

Exergy destruction for condenser 1 can be written as follows:

$$\dot{m}_{27}ex_{27} + \dot{m}_{11}ex_{11} = \dot{m}_{28}ex_{28} + \dot{Q}_{co1}\left(1 - \frac{T_0}{T_s}\right) + \dot{E}x_{dc1} + \dot{m}_{12}ex_{12}$$

Exergy destruction for the evaporator can be written as follows:

$$\dot{m}_{37}ex_{37} + \dot{Q}_e\left(\frac{T_e}{T_0} - 1\right) = \dot{m}_{38}ex_{38} + \dot{E}x_{de}$$

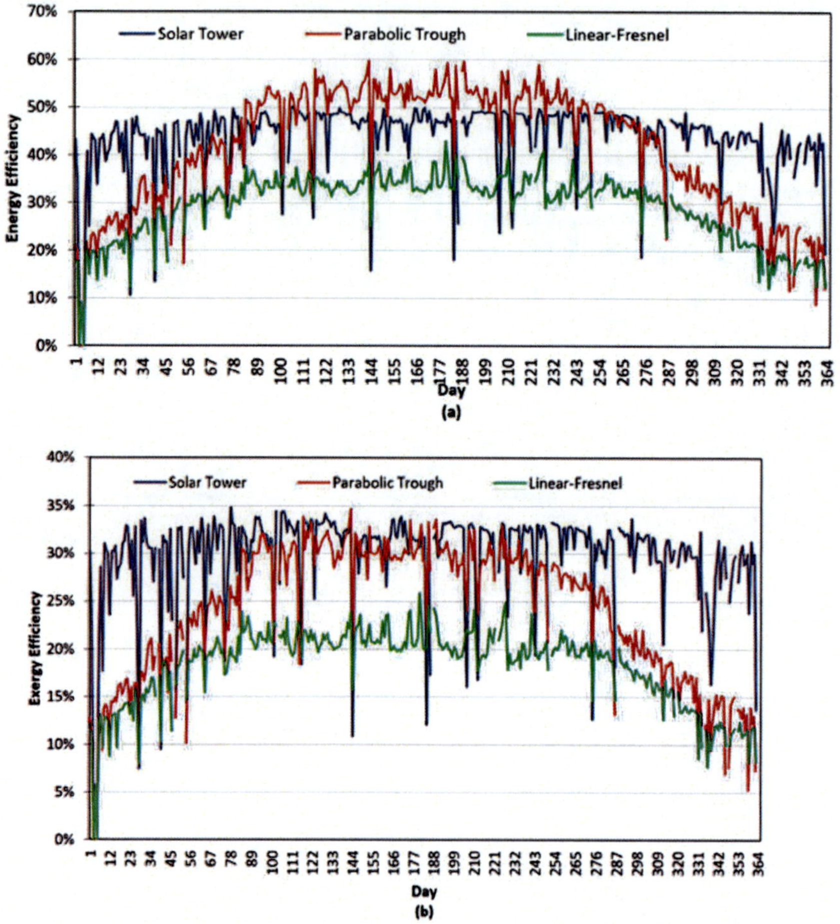

Fig. 4.16 (a) Energy and (b) exergy efficiencies of solar tower, parabolic trough, and linear Fresnel concentrating solar thermal systems

Exergy destruction for the absorber can be written as follows:

$$\dot{m}_{44}ex_{44} + \dot{m}_{38}ex_{38} = \dot{m}_{39}ex_{39} + \dot{Q}_a\left(1 - \frac{T_0}{T_s}\right) + \dot{Ex}_{da}$$

The overall integrated system's performance can be calculated by considering the useful outputs from gas turbines, steam turbines, refrigeration cycle, hot water, space heating systems, and inputs of solar and biomass subsystems.

$$\eta_{en} = \frac{(\dot{W}_{gtc1} + \dot{W}_{t1} + \dot{W}_{gtc2} + \dot{Q}_{SC} + \dot{W}_{t2} + \dot{Q}_{DHW} + \dot{Q}_{DHW})}{\dot{m}_4 LHV_{biomass} + \dot{Q}_{Solar}}$$

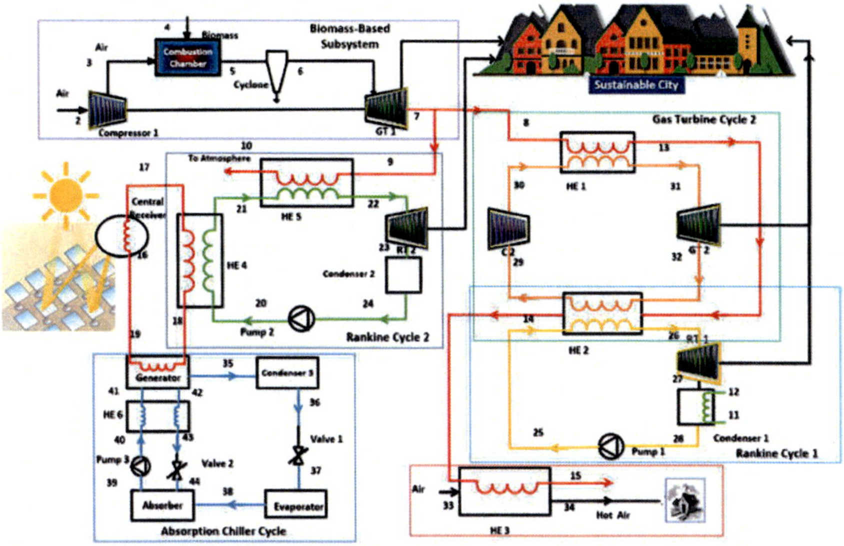

Fig. 4.17 Illustration of system layout for case study 4.3 (modified from Ref. [10])

and

$$\eta_{ex} = \frac{(\dot{W}_{net,gtc1} + \dot{W}_{net,t1} + \dot{W}_{net,gtc2} + \dot{E}x_{Q_{SC}} + \dot{W}_{net,t2} + \dot{E}x_{Q_{SH}} + \dot{E}x_{Q_{HW}})}{\dot{m}_4 ex_{biomass} + \dot{E}x_{sol}}$$

Table 4.4 shows thermophysical properties of substances at each state point. Note that the reference temperature and pressure are considered 298 K and 100 kPa

The exergy destructions indicate the importance of practically what is lost and how a system can further be improved. This way it helps understand which component destroys how much exergy. This brings more insightful results if the system is further modified for improvement. When we look at the exergy destruction results for components of the integrated system, the combustion chamber shows the highest exergy destruction rate at 85.9 MW during steady-state operation. Other components have significantly lower exergy destructions according to the results. Figure 4.18 shows the rates of exergy destruction for different components of the integrated system.

When we look at the system performance, energy, and exergy efficiencies are two of the important parameters to compare the system with other systems. For the integrated system, the overall energy and exergy efficiencies are calculated as 66.5% and 39.7%. The results are promising and competitive when conventional systems or single input single output systems are considered.

Table 4.4 The input and calculated process data for the multigeneration system

State Point	Substance	Pressure (kPa)	Mass flow rate (kg/s)	Temperature (K)	Specific enthalpy (kJ/kg)	Specific exergy (kJ/kg)
Ref.	Water	100	–	298	104.8	0
Ref.	Air	100	–	298	298.4	0
1	–	100	–	4500	–	–
2	Air	100	40	298	298.4	0
3	Air	500	40	502.7	506.1	187.5
4	Biomass	100	4.2	298	18,588	24,747
5	Combustion gases	500	44.2	1600	2335	526.1
6	Combustion gases	500	44.2	1600	2335	526.1
7	Exhaust gases	100	44.2	999.4	1325	−4.323
8	Exhaust gases	100	30.94	999.4	1325	−4.323
9	Exhaust gases	100	13.26	999.4	1325	−4.323
10	Flue gases	100	13.26	301.4	385.7	−391.9
11	Water	100	35	298	104.2	0
12	Water	100	35	353	334.3	18.94
13	Exhaust gases	100	30.94	860.9	1142	−129
14	Exhaust gases	100	30.94	556.6	738.1	−360.2
15	Flue gases	100	30.94	402.1	542.9	−469.8
16	Oil	100	10	363	125.3	10.31
17	Oil	100	10	493	379.2	86.8
18	Oil	100	10	393	179.8	22.01
19	Oil	100	10	363	125.3	10.31
20	Water	4000	5	349.5	322.7	20.58
21	Water	4000	5	443.3	721.5	118.4
22	Water	4000	5	673	3213	1201
23	Water	40	5	349	2404	321.4
24	Water	40	5	349	317.6	16.37
25	Water	8000	6	367.7	402	37.38
26	Water	8000	6	673	3138	1246
27	Water	80	6	366.7	2324	390.7
28	Water	80	6	366.7	391.7	28.78
29	Air	100	10	300	300.4	0.006713
30	Air	800	10	586.4	593	267
31	Air	800	10	1100	1161	629
32	Air	100	10	681.2	693.4	142.3
33	Air	100	100	298	298.4	0
34	Air	100	100	358	358.8	5.336
35	Water	7.4	0.1778	354.51	2652	2344
36	Water	7.4	0.1778	313.05	167.7	57.71
37	Water	0.68	0.1778	274.475	167.7	56.69

(continued)

Table 4.4 (continued)

State Point	Substance	Pressure (kPa)	Mass flow rate (kg/s)	Temperature (K)	Specific enthalpy (kJ/kg)	Specific exergy (kJ/kg)
38	Water	0.68	0.1778	274.475	2503	2180
39	Water/LiBr	0.68	3	310	103.9	30.63
40	Water/LiBr	7.4	3	312.07	107.9	34.36
41	Water/LiBr	7.4	3	357.39	196.9	116.7
42	Water/LiBr	7.4	2.822	363	223.2	142.9
43	Water/LiBr	7.4	2.822	312.07	128.6	55.32
44	Water/LiBr	0.68	2.822	316.96	128.6	54.61

Data from Ref. [10]

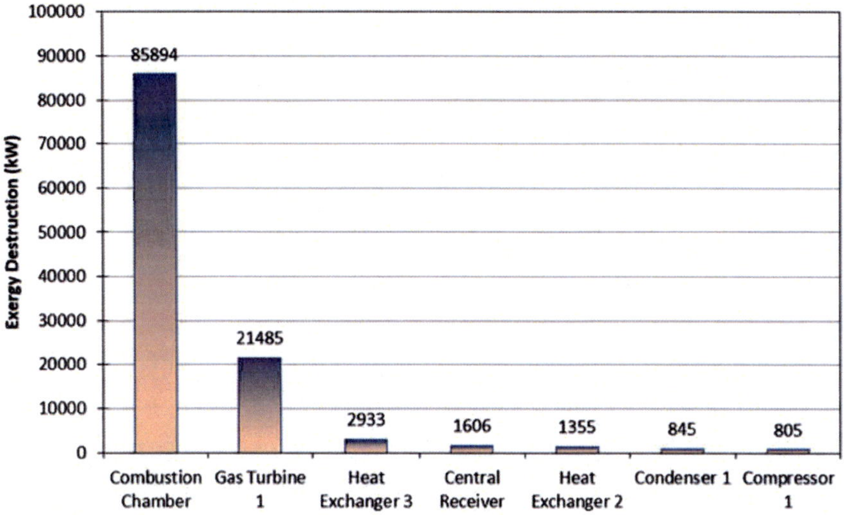

Fig. 4.18 The rates of exergy destruction for different components of the system (data from Ref. [10])

4.4 Geothermal-Based Thermal Energy Generation

In contrast to popular renewables wind and solar, geothermal energy is a reliable energy source that can be utilized with high-capacity factors. Geothermal energy can be used for a wide variety of applications. Although geothermal energy systems require high capital costs, low-temperature applications can be applied with lower capital costs.

Geothermal energy can be available in different forms in different types of reservoirs. Geothermal reservoirs can be categorized according to the reservoir temperature, as presented in Table 4.5.

Table 4.5 The classification of some geothermal reservoirs

Classification	Reservoir T	Fluid phase in reservoir	Fluid state at utilization
Very low temperature	<100°C	Liquid	Liquid
Low temperature	100°C to 190°C	Liquid	Steam-water mixture
Moderate temperature	190°C to 230°C	Liquid	Steam-water mixture
High temperature	230°C to 300°C	Liquid-dominated two phase	Saturated steam
Very high temperature	300°C+	Liquid-dominated two phase	Superheated steam

Various categories may be given for the classification of geothermal resources as follows:

- Hot water from oil/gas fields
- Magma bodies
- Conductive sedimentary
- Hydrothermal
- Geo-pressure
- Radiogenic
- Enhanced geothermal

When we look at the first category, abandoned oil and gas fields can be exploited for low-temperature geothermal energy. Deep reservoirs can be warmer than the surface, which can be easily utilized for geothermal applications.

Another category can be mentioned as molten rocks or magma bodies. Underground below active volcanoes typically consist of molten rack magma. Temperature can increase even more than 650°C. The pressurized water can be pumped to the reservoir and circulated back for geothermal energy usage.

Sedimentary rocks have a characteristic of high porosity and permeability. For this reason, deep wells with sedimentary rocks can be used for high-temperature geothermal water with high flow capacities.

In hydrothermal systems, hot water naturally exists at the natural reservoir. In some cases, steam can be used directly in the system to produce a useful output, which can be also called as primary system. If the existing water is not suitable due to contaminations, etc., then a heat exchanger can be used to transfer heat from the geothermal water to another thermofluid for the process.

Geo-pressure-type geothermal systems consists of high water at very high pressures up until 600 bar. Due to the high pressure, geothermal water has methane and other dissolved solids. For this reason, a secondary loop is typically used with geo-pressure-type geothermal systems.

In order to calculate both energy and exergy efficiencies of geothermal-based systems with re-injection, energy efficiency can be calculated as follows:

$$\eta_{en} = \frac{\dot{Q}_{PH}}{\dot{m}_{in}h_{in} - \dot{m}_{re-in}h_{re-in}}$$

where mass flow rate of injection water is denoted by $\dot{m}_{in}$, enthalpy of injection water is denoted by h_{in}, mass flow rate of re-injection water is denoted by $\dot{m}_{re-in}$, and enthalpy of re-injection water is denoted by h_{re-in}.

If there is no re-injection, the energy efficiency is defined as follows:

$$\eta_{en} = \frac{\dot{Q}_{PH}}{\dot{m}_{in}h_{in} - \dot{m}_{in}h_0}$$

In terms of exergy, the exergy efficiency of geothermal-based systems with re-injection is defined as follows:

$$\eta_{ex} = \frac{\dot{E}x_{Q_{PH}}}{\dot{m}_{in}ex_{in} - \dot{m}_{re-in}ex_{re-in}}$$

If there is no re-injection, the exergy efficiency can be calculated as follows:

$$\eta_{ex} = \frac{\dot{E}x_{Q_{PH}}}{\dot{m}_{in}ex_{in}}$$

In the energy efficiency with no reinjection, the reference enthalpy is extracted at the denominator, however, the exergy efficiency with no reinjection does not have any extraction at the denominator. This occurs due to the calculation of exergy, where the reference enthalpy and reference entropy is already extracted in order to calculate exergy, which can be seen as follows:

$$\dot{m}_{in}ex_{in} = \dot{m}_{in}(h_{in} - h_0 - T_0(s_{in} - s_0))$$

which proves that exergy of injected water is a value which already calculated by extracting the reference values.

Considering geothermal-based thermal energy applications, space heating, and cooling are among some popular applications in practice. Absorption refrigeration chillers can be used to utilize geothermal energy to generate cooling. Goswami cycle also can be considered to generate both power and cooling. On the other hand, geothermal energy can be used for desalination purposes to generate fresh water. Figure 4.19 shows how geothermal energy is generated for many application areas.

Two main categories can be mentioned for geothermal energy systems: primary loop and secondary loop. In primary loop systems, geothermal energy can be directly used to generate useful outputs. From a renewable heat generation perspective, a geothermal-driven system can generate space heating, fresh water, and domestic hot water as well as heating for dry food, plant, and aquaculture applications. Figure 4.20a depicts how geothermal energy can be directly utilized for the system

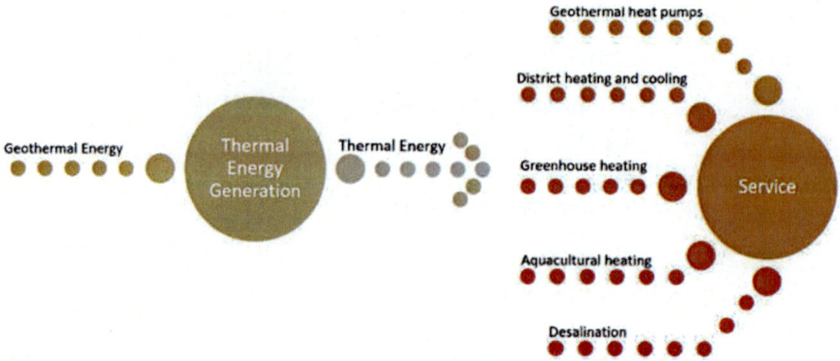

Fig. 4.19 Illustration of geothermal-based heat applications

to generate useful outputs. The second utilization method is depicted in Fig. 4.20b. In order to design primary and secondary loop systems, contamination is an important parameter. Some geothermal sources consist of harmful contents such as sulfur or other heavy metals. In those cases, using a secondary loop will prevent the transfer of harmful content from the geothermal water to the utilization area.

Farsi and Dincer [11] carried out a recent study where they utilized geothermal energy for multigeneration. An integrated system was designed to utilize hot underground steam for power generation via the Rankine cycle and organic Rankine cycle. The remaining energy was used to generate cooling via the ammonia-water absorption refrigeration cycle. Two different desalination systems are used multi-effect distillation desalination and direct contract membrane distillation. According to their result, their proposed system was capable of generating 165.6 tonnes of fresh water per day, which is enough for 1100 people. The exergy efficiency of geothermal-driven direct contact membrane desalination system was found to be 12.3% at 50°C of bulk feed temperature. It is important to highlight the importance of process heat temperature. If lower-temperature geothermal heat was used to generate 25°C of bulk feed temperature, the exergy efficiency of the direct contact membrane desalination system was 7.1%. The exergy efficiency of their combined system was found as 24%, which includes the power, fresh water, and cooling generation as useful output and underground hot water as the input.

Case Study 4.4 A case study is investigated by using data and information from Oshawa, Ontario, Canada. A residential complex with 109 residents wants to switch to a zero carbon heating option. They currently utilize natural gas–driven boilers. Their natural gas supplier reports that 10 years of average natural gas consumption of the residential complex is 151,031 m^3 per year, which is equal to 1242.08 MWh of annual heating with 80% efficiency of the natural gas–driven boiler.

The management of the residential complex received two offers from a heat pump company. A geothermal source is considered in one of the cases where a vertical ground source heat pump is employed to generate residential heat. Figure 4.21

Fig. 4.20 Illustration of (a) primary and (b) secondary loop geothermal energy systems

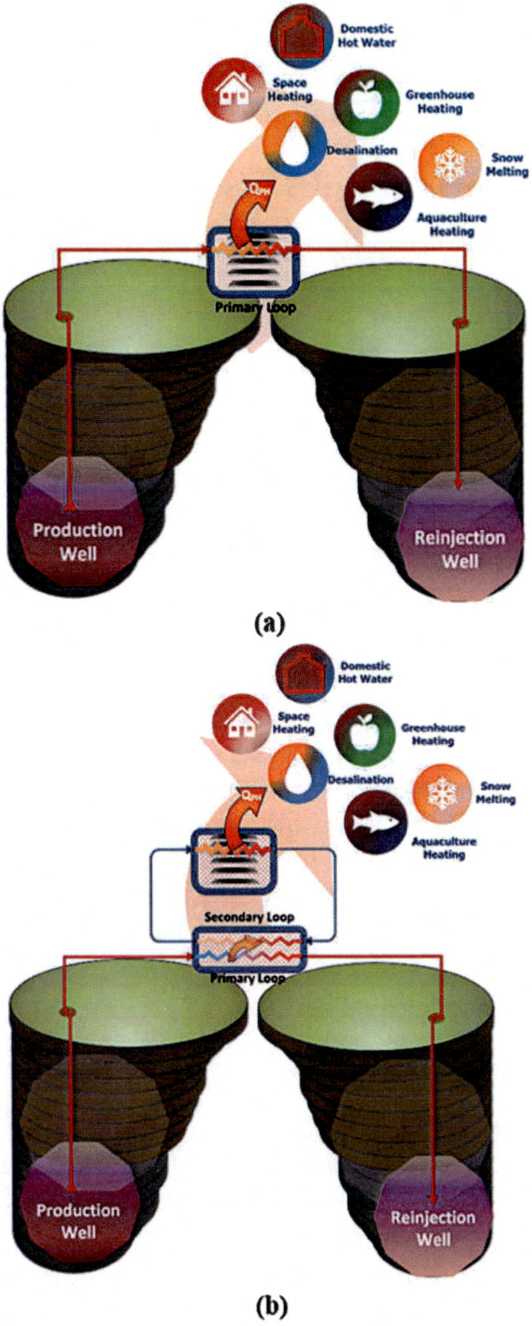

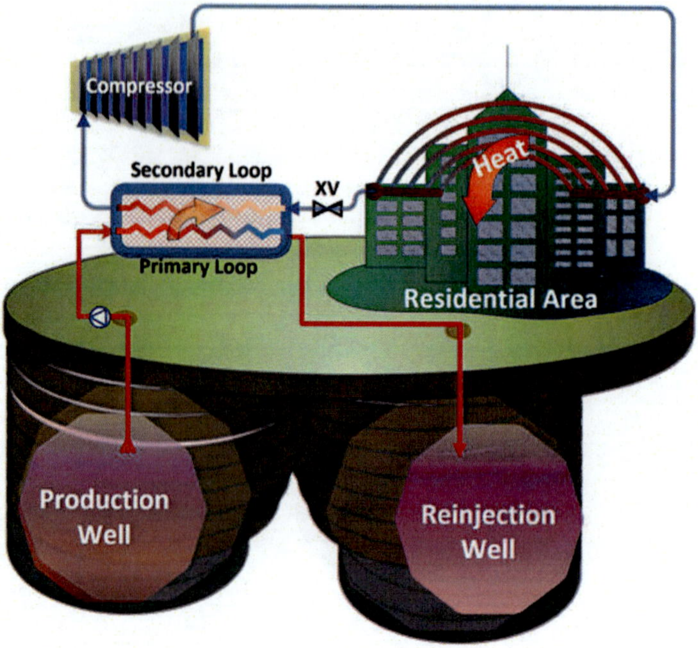

Fig. 4.21 Illustration of ground source heat pump option for residential heating

depicts the illustration of the ground source heat pump for the residential area. Another option is considered an air source heat pump which can upgrade the outside temperature to the required temperatures.

According to the results ground source heat pump consumes 206.6 MWh of electricity in a typical year, in order to meet 1242 MWh of heating load of 109 people in a residential complex in Oshawa, Ontario, Canada. On the other hand, the air source heat pump option consumes 477.72 MWh of electricity in a typical year for the same load in the same climate, as shown in Fig. 4.22.

Considering the coefficient of performance (COP) values for air source and ground source heat pumps for case study 4.4, the air source heat pump performs at COP of 2.6 on average in a typical year. The vertical ground source heat pump has a COP of 6.01 on average in a typical year as shown in Fig. 4.23.

The annual consumption of the currently available natural gas–driven boiler was 151,031.0 m^3, which is the source of carbon dioxide emissions. It can be replaced by 477.7 MWh of electricity consumption in the air source heat pump case or 206.6 MWh of electricity in the ground source heat pump case.

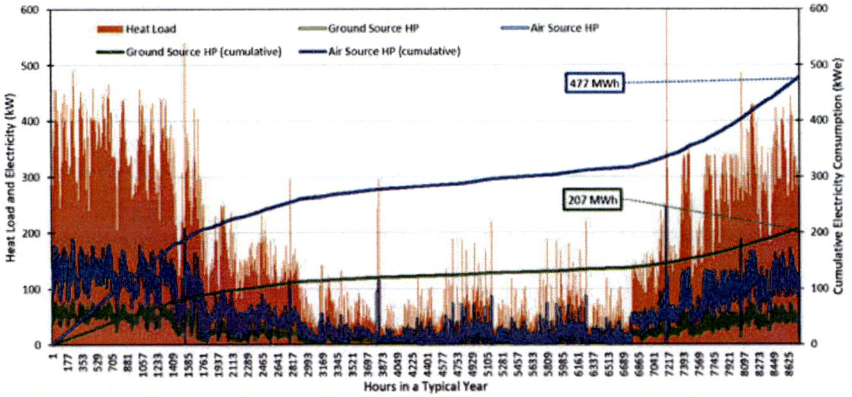

Fig. 4.22 Heat load and heat pump consumptions for air source heat pump and ground source heat pump cases

Fig. 4.23 Coefficient of performance values of air source and ground source heat pumps for case study 4.4

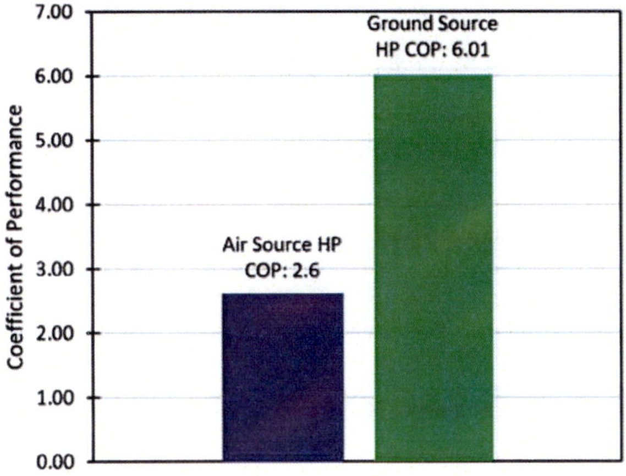

4.5 Closing Remarks

The current chapter aims to cover the thermal energy generation process mainly from renewable energy sources. Currently, the combustion or burning of fossil fuels is the primary driver of any energy-related process from space heating to power generation, transportation to industrial processes. In order to mitigate carbon emissions, it is crucial to remove combustion processes. That brings to the requirement of knowledge in the thermal energy generation area. This chapter introduces thermal energy generation comprehensively, starting from combustion to renewables.

In closing, it is important to note that thermal energy generation is the first step of many applications. Therefore, further sections will provide more details on how to use thermal energy to generate useful outputs, including fresh water. Moreover, it is important to integrate renewables with multiple cycles and subsystems in order to improve the utilization of thermal energy, which are discussed in the following chapters.

References

1. Dincer I (2021) Thermodynamics: a smart approach, 1st edn. Wiley, Chichester
2. Ritchie H, Roser M, Rosado P (2022) Energy data. Our world in data. Retrieved from: https://ourworldindata.org/energy (online resource)
3. McMillan CA, Boardman R, McKellar M, et al (2016) Generation and Use of Thermal Energy in the U.S. Industrial Sector and Opportunities to Reduce its Carbon Emissions. Idaho Falls, ID (United States)
4. Yapicioglu A, Dincer I (2018) Performance assessment of hydrogen and ammonia combustion with various fuels for power generators. Int J Hydrog Energy 43:21037–21048. https://doi.org/10.1016/J.IJHYDENE.2018.08.198
5. Erden M, Karakilcik M, Dincer I (2017) Performance investigation of hydrogen production by the flat-plate collectors assisted by a solar pond. Int J Hydrog Energy 42:2522–2529. https://doi.org/10.1016/J.IJHYDENE.2016.04.116
6. Achkari O, El Fadar A (2020) Latest developments on TES and CSP technologies – energy and environmental issues, applications and research trends. Appl Therm Eng 167:114806. https://doi.org/10.1016/J.APPLTHERMALENG.2019.114806
7. Temiz M, Dincer I (2021) Concentrated solar driven thermochemical hydrogen production plant with thermal energy storage and geothermal systems. Energy 219:119554. https://doi.org/10.1016/J.ENERGY.2020.119554
8. Kani NC, Gauthier JA, Prajapati A et al (2021) Solar-driven electrochemical synthesis of ammonia using nitrate with 11% solar-to-fuel efficiency at ambient conditions. Energy Environ Sci 14:6349–6359. https://doi.org/10.1039/D1EE01879E
9. Miroshnichenko L (2023) Main characteristics of the sun. In: Solar-terrestrial relations, pp 7–20. https://doi.org/10.1007/978-3-031-22548-2_2
10. Khalid F, Dincer I, Rosen MA (2015) Energy and exergy analyses of a solar-biomass integrated cycle for multigeneration. Sol Energy 112:290–299. https://doi.org/10.1016/J.SOLENER.2014.11.027
11. Farsi A, Dincer I (2019) Development and evaluation of an integrated MED/membrane desalination system. Desalination 463:55–68. https://doi.org/10.1016/J.DESAL.2019.02.015

Chapter 5
Power Generation

Nomenclature

A	area (m^2)
AC	alternative current
B	Betz limit
DNI	direct normal irradiance
EBE	energy balance equation
EnBE	entropy balance equation
ExBE	exergy balance equation
ex	specific exergy (kJ/kg)
$\dot{E}x$	exergy rate (kW)
FF	fill factor
g	gravitational acceleration (m s^{-2})
GHI	global horizontal irradiation
h	specific enthalpy (kJ/kg)
I	current
LHV	lower heating value (kJ/kg)
$\dot{m}$	mass flow rate (kg/s)
MBE	mass balance equation
η	efficiency
N	north
$\dot{N}$	molar flow rate (mol/s)
NREL	National renewable energy laboratory
OTEC	ocean thermal energy conversion
P	pressure (kPa)
PV	photovoltaic
Q	heat (kJ)
$\dot{Q}$	heat rate (kW)
LCOE	levelized cost of electricity

I. Dincer, M. Temiz, *Renewable Energy Options for Power Generation and Desalination*, https://doi.org/10.1007/978-3-031-53437-9_5

S	total entropy (kJ /K)
s	specific entropy (kJ/kg K)
$\dot{S}$	entropy rate (kW/K)
T	temperature (°C)
V	voltage (V)
w	water, work
$\dot{W}$	power (kW)
X	mass ratio of component
SAM	system advisor model

Subscripts

con	condenser
El	electricity
LF	linear Fresnel
m	maximum
oc	open circuit
p	pump
s	source
sc	short circuit
SC	solar collector
t	turbine

5.1 Introduction

Historically, power generation using energy sources, primarily fossil fuels, has been an essential goal for human society, especially after the industrial revolution. Using natural sources from the nature enabled human beings to carry out processes which could not normally be done by using their human power. Natural sources were harnessed to be able to move things for managing things for daily activities to enhance life quality and move things, such as boats, to provide mobility for people. From very early times (going back to the fortieth century BC) to the years of 1888, there were various historical landmarks were achieved by various people and nations through various discoveries and developments, aiming to harness energy from nature for various activities, as illustrated in Fig. 5.1.

The earliest attempts to utilize energy can be mentioned as the control of fire, which began roughly a million years ago due to the microscopic traces of wood ash. Typically, wood is used as a source to generate controlled fire for various purposes, including cooking. However, the utilization of energy sources to generate motion begins before mills. Sailboats were used to exploit wind power to generate motion, in ancient Egypt, where a piece of cloth was hung and pulled on a simple rigging system of the boat, from around 4000 BC to 3500 BC.

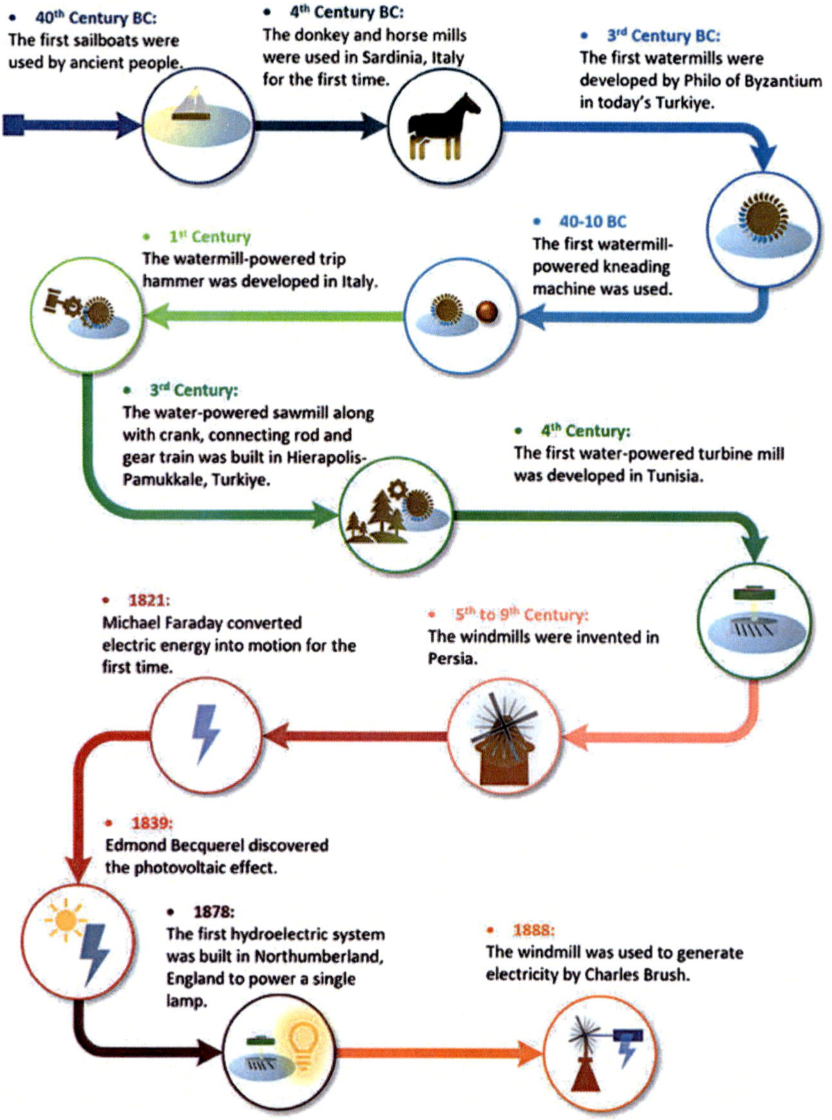

Fig. 5.1 A historical illustration of power generation inventions and developments

In the power generation area, the invention of mills can be mentioned as a major milestone. The first milling machine is invented in ancient Greece in the fifth century BC, which is the Olynthus Mill. The first mills were used for grinding grains, agriculture, mining, and construction work. However, the first milling machines were driven by humans as the power source. The first nonhuman milling machine is introduced in Sardinia Island, Italy, in the fourth century BC. When we look at the inventions of nonhuman and non-animal-driven milling machines, in the third century BC, Philo of Byzantium invented the first watermill in today's Türkiye. Until the fourth century, watermills were used for a variety of purposes from construction to mining, agriculture to kneading. In the first century, watermills were used for military purposes to manufacture swords with the water-powered forging mill, in Italy. A crank, connecting rod, and gear train mechanism is introduced with a water-powered sawmill in Hierapolis-Pamukkale, Türkiye, around the third century. The first turbine-type mill was invented and used in Tunisia around the fourth century. Around fifth to ninth centuries, the first windmills were built in Persia, which is the first time mills were used with wind power.

Electricity brought another aspect to the power generation. On September 3, 1821, Faraday demonstrated the first electromagnetic experiment by converting electrical energy into rotary motion, which is a very important milestone in the power generation area. The conversion of electrical energy into motion enables other energy sources to turn into motion as well as a reverse conversion of the motion to electrical energy. Solar is one of the sources which can be used to generate power. In 1839, Edmond Becquerel discovered the photovoltaic effect which leads to photovoltaic energy conversion with the contributions of other scientists including Albert Einstein and his Nobel Prize winner explanation of the photoelectric effect.

Utilization of renewables to generate electricity begins with hydropower and wind power systems. In 1878, the first hydroelectric project was demonstrated in Cragside, Northumberland, England, by William Armstrong. The first system for the conversion of wind energy into electricity was built by Charles Brush, which is called The Giant Brush Windmill with 17 m of rotor diameter and 144 rotor blades from cedar wood material. The Giant Brush Windmill was successfully operated for 20 years in Cleveland, Ohio, the United States, where 12 kW of electricity was generated to power Charles Brush's mansion.

Today, a wide variety of energy sources can be converted into power by using energy systems with the contributions of inventions in the area of energy. When we look at renewable power generation, biomass, geothermal, solar, wind, hydro, and marine-based power generation systems can be mentioned, where renewable energy sources are utilized with either thermal or non-thermal systems in order to generate power as the useful output, in a sustainable fashion. This particular chapter primarily focuses on covering all renewable energy sources utilized particularly for power generation to cover the needs of people and communities. There are also many illustrative examples presented to highlight the importance of every renewable energy source.

5.2 Renewable Energy Sources for Power

Today modern society particularly relies on electric power with exponentially increasing demand. From lighting to mobility, industry to information and communication sectors, every sector and area demands electric power. Electricity can be mentioned as the most versatile energy transfer medium that can be transmitted without major losses. Electricity can be converted into many forms of useful outputs. From electricity, thermal energy can be generated by using electric heaters or heat pumps. Water disassociation can be carried out by using an electrolyzer in order to generate hydrogen and oxygen from water. Electrochemical processes can be applied in order to generate other chemicals by using electricity. There are also many other useful outputs that can be generated by using electricity, including fresh water. Electricity-driven desalination methods can be mentioned as mechanical vapor compression, reverse osmosis, electrodialysis, and forward osmosis.

Digitalization contributed to the exponential growth of electricity demand in countries. Moreover, key events of the modern era, such as the COVID-19 pandemic, brought a global scale digital transformation, which led to a new industrialization. When we look at the largest global companies, Apple, Microsoft, Amazon, Alphabet-Google, and Meta form the five of the top six companies in terms of market capitalization. Moreover, crypto technologies, data utilization, data storage, and blockchain technologies are expected to grow exponentially. It is expected that information and communication technology will consume a quarter of global energy consumption by 2030. Currently, a total of 10% of global electricity consumption is for information and communication technologies.

Currently, nonrenewables are the major energy sources for electricity generation as shown in Fig. 5.2. Almost 20,000 TWh of electricity is generated by using nonrenewable sources such as oil, gas, coal, and nuclear, which is 70% of global electricity generation. When we look deeper into 2022 energy data [1] for nonrenewables, one-third of electricity generation comes from coal sources, which is equal to 10,190.7 TWh of electricity. Natural gas is the second global electricity

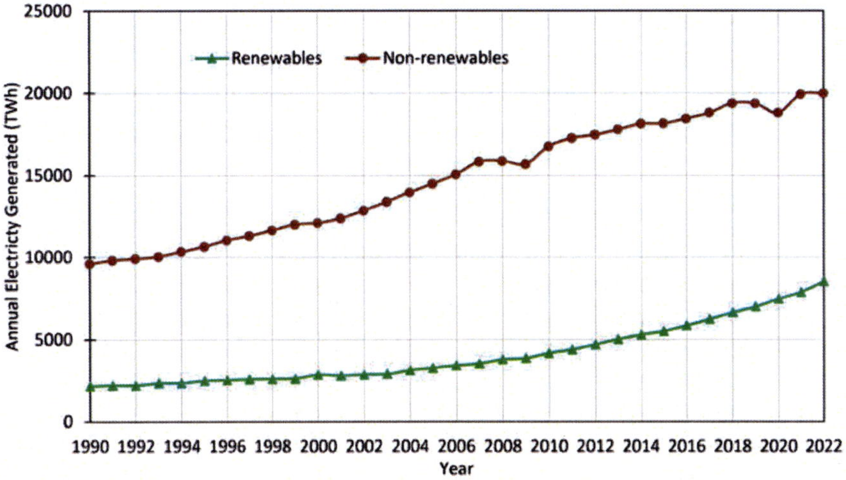

Fig. 5.2 Global annual electricity generation from renewable and nonrenewable sources (data from Ref. [1])

Fig. 5.3 Illustration of Source-System-Service (3S) approach for renewable power generation

source after coal, where 6309.5 TWh of electricity is generated globally by utilizing natural gas. Nuclear and oil are other nonrenewable sources of electricity generation, which are utilized to generate 2610.1 TWh of electricity via nuclear and 884.9 TWh of electricity via oil globally as of 2022.

After the realization of global warming due to the increasing greenhouse gases in the atmosphere, jurisdictions agreed on sets of actions in order to mitigate global warming by eliminating greenhouse gases. Among greenhouse gases, carbon dioxide is the major contributor to the global warming. Therefore, jurisdictions decided to switch to carbon-free societies by accelerating renewable source utilization in order to replace fossil fuel utilization. As a sustainable process, power generation from renewables with a 3S approach [2] can be seen in Fig. 5.3.

Renewable power generation methods are then classified in Fig. 5.4. Direct conversion methods include power generation methods without a thermal energy conversion step. Wind, hydro, and marine sources consist energy in the forms of kinetic and potential energy. Kinetic and potential energies are the form of energy where 100% of energy is associated with motion or position. Wind energy is

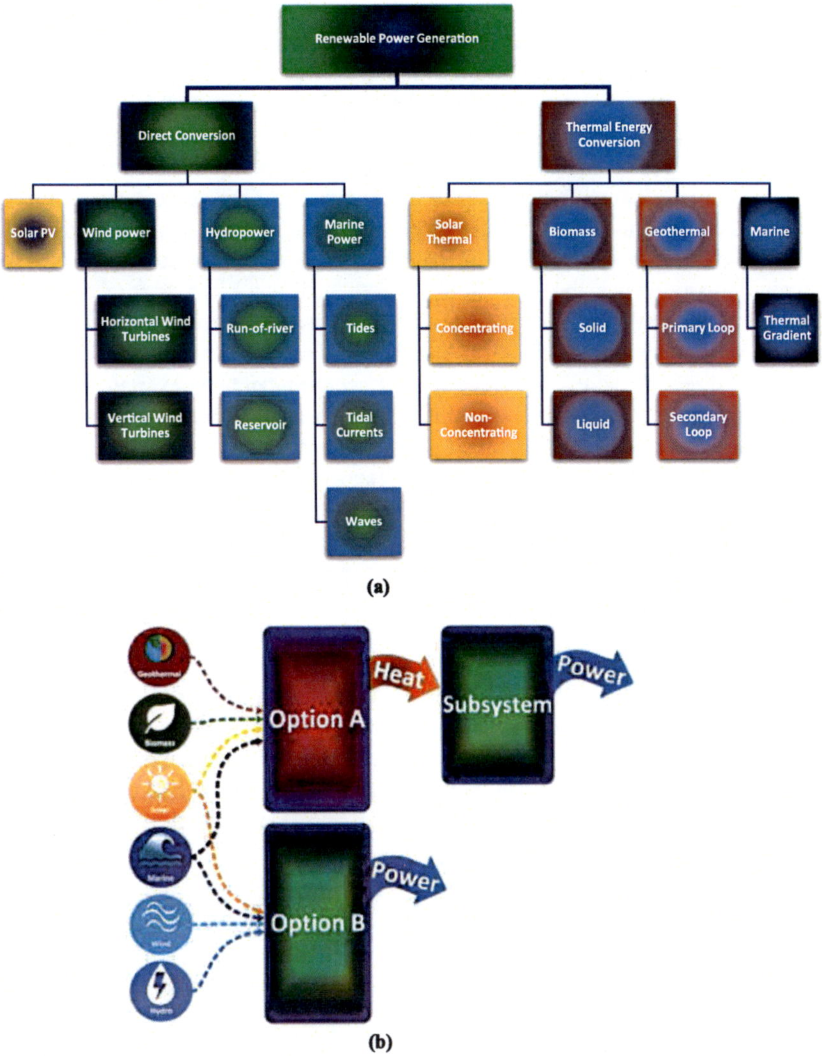

Fig. 5.4 (**a**) Classification of renewable power generation and (**b**) renewable to power conversion

available in the form of kinetic energy, which is associated with the motion of the wind. Hydro energy can be related to position, however, during the hydropower generation process, first potential energy is converted into motion energy and then the motion energy is turned into power via turbines and electricity via electromagnetic generators.

On the other hand, solar and marine energy sources can be utilized via direct or thermal energy conversion methods. Solar PV systems directly turn solar radiation into direct current (DC) electricity. Marine power consists of potential or kinetic energies associated with position or motion. Solar thermal and ocean thermal systems require a thermal energy conversion step in order to utilize solar or marine sources to generate power. Thermal energy conversion–involved renewable power generation methods include solar thermal, biomass, geothermal, and marine thermal gradient systems as shown in Fig. 5.4a. The majority of renewable power generation methods with thermal energy conversion include a power generation cycle, typically a Rankine cycle. For lower temperatures, the Kalina cycle, organic Rankine cycle, and ammonia trilateral cycle can be considered. Higher temperatures can be exploited by using the steam Rankine or organic Rankine cycles. Figure 5.4b depicts the conversion process of different renewable sources either directly or via the thermal energy conversion step. Briefly, geothermal and biomass sources need to be used via at least two steps where heat is collected and utilized at the first step, and power is generated from the heat collected. Wind and Hydro sources can be exploited directly where kinetic or potential energies can be converted into power with the correct components. According to the type of system, solar and marine energy can be harvested either directly or via thermal energy conversion.

When we look deeper into renewable energy utilization for global electricity generation in 2022, 30% of global electricity generation is driven by renewable sources, which is equal to 8532.6 TWh of electricity. Among renewables, hydro represents 50% of global renewable electricity generation with 4326.8 TWh of electricity. Wind and solar come after hydro with 2139.2 TWh of electricity generation from wind and 1286.3 TWh of electricity from solar. Biomass contributed to 2% of global electricity generation in 2022, with 677.6 TWh of electricity, respectively.

Note that recent global efforts have become more visible in order to implement renewable source utilization to replace fossil fuel utilization. During the last 10 years (between 2012 and 2022), renewable electricity generation has almost doubled, as shown in Fig. 5.5. From the late twentieth century until today, hydro has been the

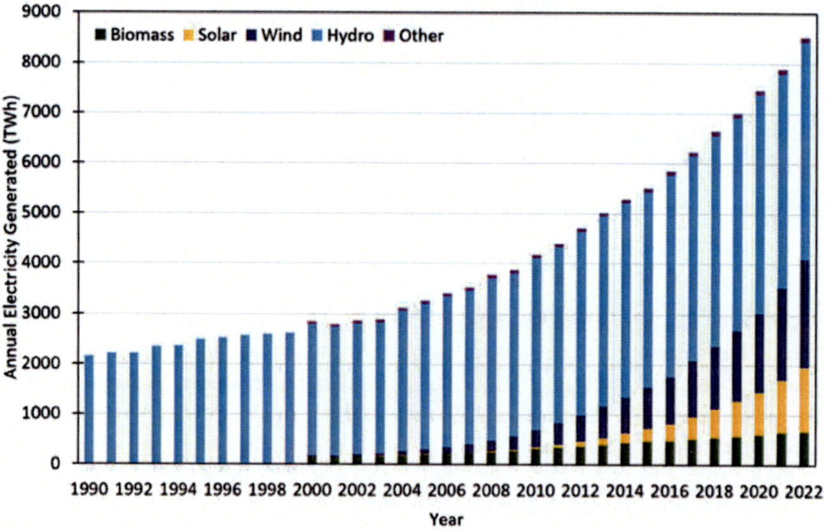

Fig. 5.5 Global annual electricity generation by using renewable sources (data from Ref. [1])

major renewable source for electricity generation. However, global hydro-based electricity generation only doubled since 1990, while total renewable electricity generation has tripled. Among renewables, solar and wind are the fastest-growing renewables, especially for the last 20 years. After 2003 wind-based electricity generation started and accelerated, starting at 63 TWh in 2003 and reaching 2139.2 TWh only in 20 years. On the other hand, solar-based electricity generation was 95 TWh in 2012, and only in 10 years it reached 1286.3 TWh.

There is a strong correlation between the acceleration of renewable source utilization and the levelized cost of electricity. When we look at the tremendous drop in solar and wind-based electricity generation costs as shown in Fig. 5.5, during the last 10 years of available data (between 2011 and 2021), it matches with the tremendous growth of wind and solar-based electricity generation, as shown in Fig. 5.6. In 2010, among popular renewables, solar energy utilization for electricity generation was the most expensive method, at $0.42/kWh. Hydropower systems were the major, as well as the cheapest renewable electricity generation method with $0.039/kWh, in 2010. When it comes to 2021, a kWh of solar-based electricity costs $0.048, which is only 10% of the cost in 2010. On the other hand, onshore wind became the cheapest method among renewables, at $0.033/kWh.

When we look at global electricity generation, we observe that some of the countries successfully utilize renewable sources. Iceland, Norway, Brazil, Canada, Denmark, and New Zealand are some of the examples which use renewables as the major energy source for electricity generation as shown in Fig. 5.7. Among these countries, Brazil is one of the largest countries in terms of population and electricity demand. The majority of Brazilian grid is powered by hydropower systems. Fossil

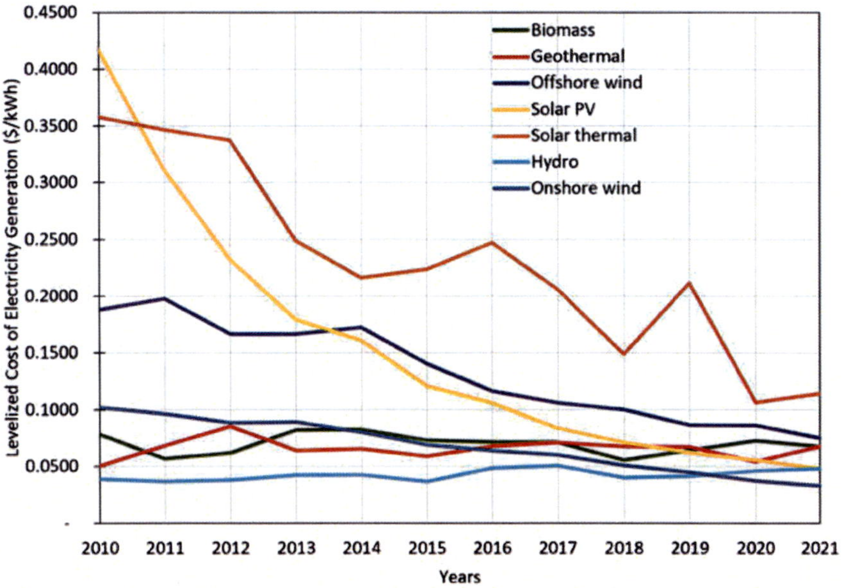

Fig. 5.6 Global levelized cost of electricity values of renewable sources (data from Ref. [3])

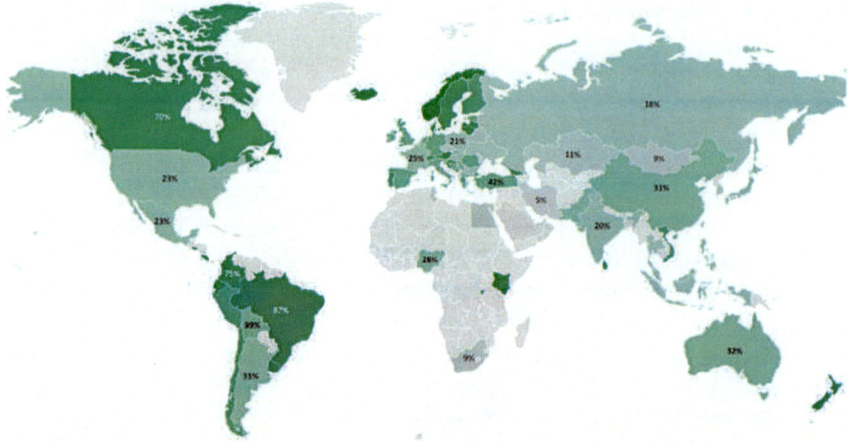

Fig. 5.7 Renewable electricity generation shares of countries (data from Ref. [1])

fuels, biomass, nuclear, wind, and solar come after hydropower, where 87% of the electricity generation is supplied by using renewables.

For many countries, achieving 100% renewable electricity is an important step to achieving net-zero goals. Albania, Iceland, and Paraguay already achieved a 100% renewable electricity grid. Norway can be the next country to become 100% renewable, which is using 99% renewable electricity, currently. The majority of electricity in Iceland is generated via hydropower systems. The remaining part is generated by using geothermal and other renewables in Iceland. Geothermal energy is also used for heating purposes in Iceland, which helped to achieve 85% renewable source utilization among primary energy sources, including transportation and other sectors. New Zealand is another example of the high utilization of renewables. Canada, as the second largest country in terms of land size, achieved 70% of the renewable electricity generation share. In total, Canada generated 638 TWh of electricity, and 445 TWh is generated from renewables where 392.5 TWh of electricity is generated via hydropower systems, 38.13 TWh of electricity is generated via wind power systems, 5.56 TWh of electricity is generated via solar power systems, and 9.06 TWh of electricity is generated by using biomass-driven power generation systems. It is important to note that, nuclear represents 13% of Canada-wide power generation processes; for that reason, only 17% of power generation is driven by fossil fuel sources in Canada. The United States, as the biggest economy in the world in terms of gross domestic product, targets 2035 to make its power sector 100% carbon pollution-free, as President Joe Biden stated in 2023 [4].

It is possible for all countries to become 100% carbon pollution-free when available sources are considered. In order to achieve carbon pollution-free power generation globally, a trendline analysis is carried out as shown in Fig. 5.8. Short-term data was used in order to keep the highest expectations. Renewable shares of global power generation are considered between 2010 and 2022 according to [1]. According to the coefficient of determination, in other words, the value of R_2,

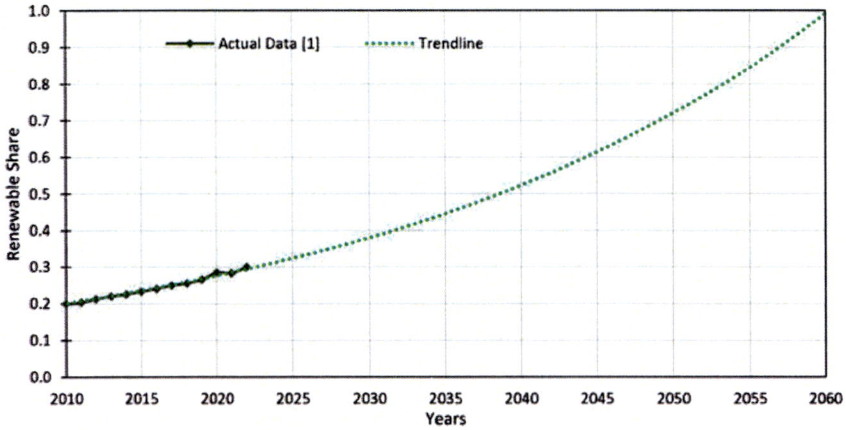

Fig. 5.8 A trendline analysis on future renewable power generation shares

the prediction shows an excellent relationship with a value of 0.989% of the relationship between dependent and independent variables.

5.3 Solar-Based Power Generation

Solar energy can be mentioned as the main driver of all other renewables on the Earth. Solar radiation from the sun drives wind, hydro, marine, geothermal, as well as biomass, indirectly. For biomass, the biosphere uses solar energy to proceed with the photosynthesis process, which can be used later as biomass to generate heat. Air and water movements occur as a result of the heating effect of solar radiation. Figure 5.9 depicts solar energy as the main driver on the Earth.

When we look at the solar-based power generation systems, two main types can be mentioned as solar photovoltaic systems and solar thermal–based power generation systems. The main difference between solar photovoltaic and solar thermal systems lies behind the conversion of solar energy. Solar photovoltaic systems utilize solar radiation for direct conversion into DC electricity. On the other hand, solar thermal–based power generation systems transfer solar radiation into a thermofluid and then proceed to the power generation cycle. While solar photovoltaic systems can turn solar radiation into electric power in one step, the solar thermal–based system requires an integrated system such as a steam Rankine cycle to generate electric power from solar radiation. According to the process heat temperature generated by solar thermal–based system, the power generation system can be determined. For lower temperatures, organic Rankine, Kalina, and trilateral ammonia cycles can be considered, which are typically considered with non-concentrating solar thermal systems. Concentrating solar systems usually generate higher temperatures, more than 300°C, which can be utilized with better energy and exergy efficiencies compared to the lower temperature solar-thermal systems.

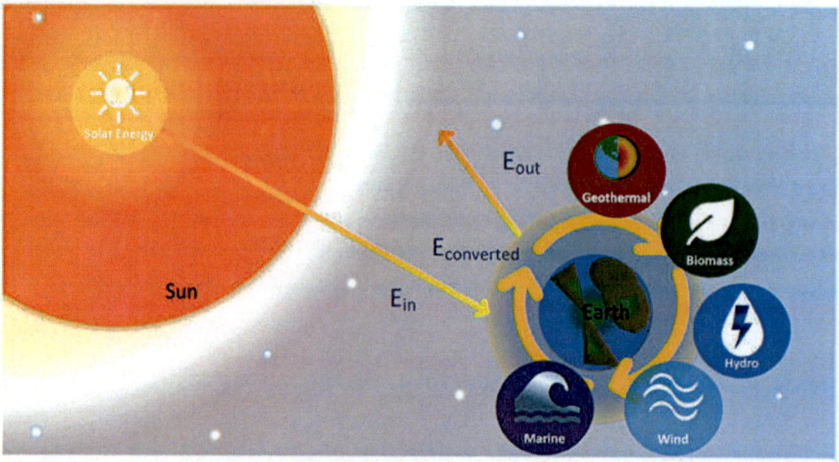

Fig. 5.9 Illustration of solar energy as the main driver of renewables

The current section aims to present power generation methods by using solar photovoltaic and solar thermal–based systems. It is important to note that, solar energy is an intermittent energy source; therefore, a compensation is required, such as integration with other renewable sources or energy storage systems, for continuous energy supply.

5.3.1 Solar Photovoltaic Systems

After the invention of photoelectric effect [5], first solar PV cell is demonstrated by Bell Labs in 1954. In the power generation sector, the accelerated growth of solar PVs started after 2010 with the effect of dropping costs. Today, solar PV has become one of the major renewable technologies to produce electricity.

During the daily movement of sun, the amount of solar radiation on PV cells changes, which leads to current and voltage changes. In order to calculate the maximum electric power that can be generated from a solar PV module, open circuit (OC) voltage and short circuit (SC) current need to be multiplied with fill factor (FF), as follows:

$$\dot{W}_{\mathbf{el},m} = \mathrm{FF} \times V_{\mathrm{oc}} \times I_{\mathrm{sc}}$$

where the maximum electric power is equals to maximum voltage and maximum current, as follows:

$$\dot{W}_{\mathbf{el},m} = V_{\mathrm{m}}\, I_{\mathrm{m}}$$

and the fill factor is then expressed as follows:

$$FF = \frac{V_m I_m}{V_{oc} I_{sc}}$$

where the values of FF vary between 0.5 and 0.82.

In order to calculate the total solar energy collected by a solar PV plant, first of all, area of each PV module (a_{PV}) needs to be multiplied by number of PV module (N_{PV}), as follows:

$$A_{PV} = N_{PV} a_{PV}$$

where A_{PV} gives the total area of PV modules in the plant. The total area of PV modules in the plant needs to be multiplied by global horizontal irradiation per unit area ($\dot{q}_{GHI}$), as follows:

$$\dot{Q}_{GHI,PV} = \dot{q}_{GHI} \times A_{PV}$$

On the other hand, for the exergy analysis of the solar energy, the Parrot's model can be used, which can be expressed as follows:

$$\dot{Ex}_{\dot{Q}_{GHI,FPV}} = \dot{Q}_{GHI,FPV} \left(1 - \frac{4}{3} \frac{T_o}{T_{sun}} (1 - \cos\delta)^{\frac{1}{4}} + \frac{1}{3} \left(\frac{T_o}{T_{sun}}\right)^4\right)$$

where the half angle of the disc of sun is denoted with δ.

To calculate the energy and exergy efficiencies of solar PV systems, electricity generated by PV modules can be considered as the useful output and the energy and exergy versions of solar energy reaching the area of PV modules can be taken as the total energy and exergy inputs. From this aspect, energy and exergy efficiencies of solar PV systems are as follows:

$$\eta_{en,ov} = \frac{\dot{W}_{El,PV}}{\dot{Q}_{GHI,PV}}$$

and

$$\eta_{ex,ov} = \frac{\dot{W}_{El,PV}}{\dot{Q}_{GHI,PV}}$$

Illustrative Example 5.1 An illustrative example is carried out by considering [6], in order to investigate the performance of PV applications on different latitudes in the Northern Hemisphere. In total, six different locations are analyzed as follows:

- Polaris, Nunavut, Canada at 75°N, 96.8°W
- Kangirsuk, Quebec, Canada at 60°N, 70°W

- Richardson, New Brunswick, Canada at 45°N, 64.8°W
- Ponte Vedra Beach, Florida, the United States of America at 30°N, 81.4°W
- Aztlan, Chiapas, Mexico at 15°N, 92.7°W
- Coaque, Ecuador at 0.1°N, 80.1°W

The specific analyses are carried out in order to find performances at different latitudes. It is important to note that the latitudes correlate with the sun's angle. However, in terms of total solar energy reaching the location, meteorological data and global horizontal irradiation parameters are the major sources. There are also parameters available with minor effects, such as temperature coefficient. The temperature coefficient of PV modules affects the electricity generation performance according to the operating temperature. Higher temperatures typically lead to lower electricity generation. From this aspect, colder ambient temperatures are particularly beneficial for the operation of solar PV modules.

In total, six different PV plants are designed with the same parameters. Each PV plant is designed at 75 MWp of capacity. Each MWp of PV capacity is considered as a section, where each section consists of 2499 units of LG 400 N2T-A5 bifacial PV modules which are connected 17 in series and 147 in strings. For each MWp section, 100 units of SUN200-100KTL-M1 inverters are considered in the simulations. While designing the PV plant, 7 m of pitch distance and 3 m of height are considered. For the structure, single axis horizontal tracker systems on the north-east axis are designed in order to simulate the PV plants.

While analyzing the PV plants, the Parrot's model is considered for the exergy analysis, Perez transposition mode is considered in order to model the direct and global irradiation components in the simulation (Fig. 5.10).

When we look at the result, in Coaque, Ecuador, located on 0° latitude, 1926.8 MWh of alternative current (AC) electricity is generated from each 1 MW of PV plant section, where 1590.5 kWh global horizontal irradiation is reached the unit area of PV modules in a year. 75 MWp of PV plant generated 144.5 GWh of electricity in a typical year in Coaque, Ecuador location on 0° latitude. In Aztlan, Mexico location on 15°N latitude, 2332 MWh of alternative current (AC) electricity is generated from each 1 MW of PV plant section, where 1896.4 kWh global horizontal irradiation is reached the unit area of PV modules in a year. 75 MWp of PV plant generated 174.9 GWh of electricity in a typical year in Aztlan, Mexico location on 15°N latitude. In Ponte Vedra, Florida location on 30°N latitude, 2132.4 MWh of alternative current (AC) electricity is generated from each 1 MW of PV plant section, where 1664.7 kWh global horizontal irradiation is reached the unit area of PV modules in a year. 75 MWp of PV plant generated 159.9 GWh of electricity in a typical year in Ponte Vedra, Florida location on 30°N latitude (Fig. 5.11).

In Richardson, New Brunswick location on 45°N latitude, 1732.6 MWh of alternative current (AC) electricity is generated from each 1 MW of PV plant section, where 1259.2 kWh of global horizontal irradiation is reached the unit area of PV modules in a year. 75 MWp of PV plant generated 129.95 GWh of electricity in a typical year in Richardson, New Brunswick location on 45°N latitude. In Kangirsuk,

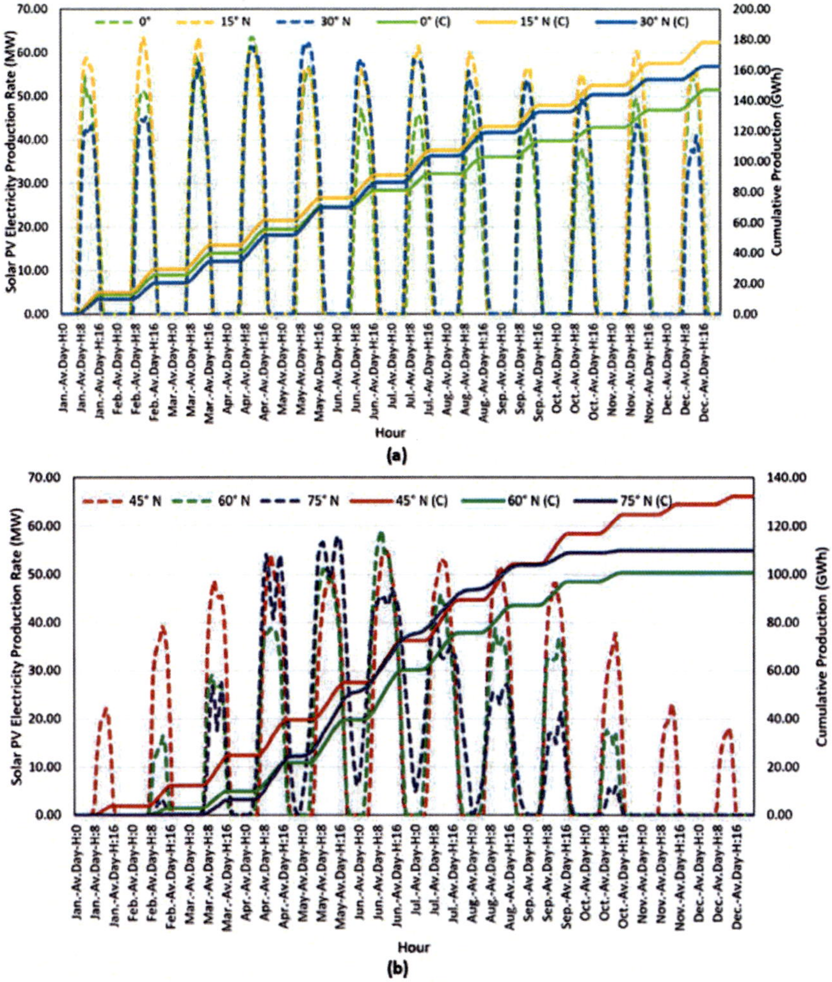

Fig. 5.10 Electricity production rate and cumulative production during the year for (**a**) 0°, 15°N, 30°N and (**b**) 45°, 60°N, 75°N latitude cases (data from Ref. [6])

Quebec location on 60°N latitude, 1321.6 MWh of alternative current (AC) electricity is generated from each 1 MW of PV plant section, where 899.9 kWh global horizontal irradiation is reached the unit area of PV modules in a year. 75 MWp of PV plant generated 99.12 GWh of electricity in a typical year in Kangirsuk, Quebec location on 60°N latitude. In Polaris, Nunavut location on 75°N latitude, 1445.7 MWh of alternative current (AC) electricity is generated from each 1 MW of PV plant section, where 829.8 kWh global horizontal irradiation is reached the unit area of PV modules in a year. 75 MWp of PV plant generated 108.43 GWh of electricity in a typical year in Polaris, Nunavut location on 75°N latitude. A summary of the results is given in Fig. 5.11.

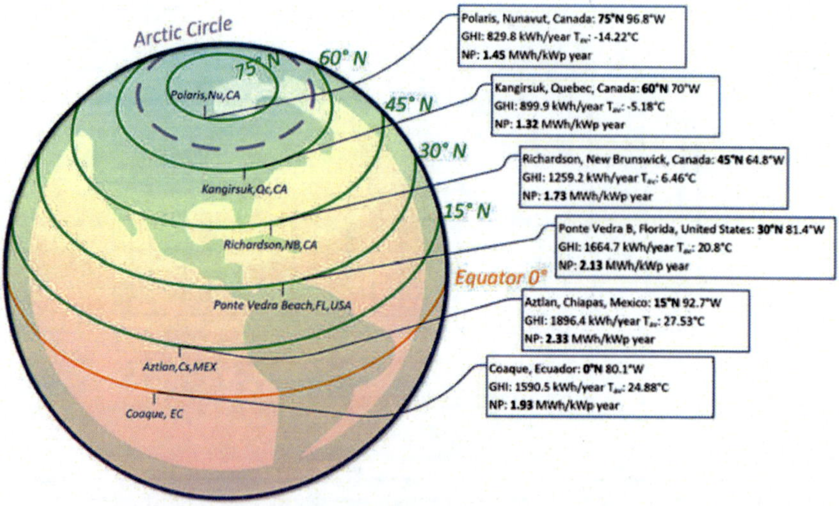

Fig. 5.11 Global horizontal irradiation (GHI) and normalized power (NP) production of 1 MWp of PV system on different latitudes (data from Ref. [6])

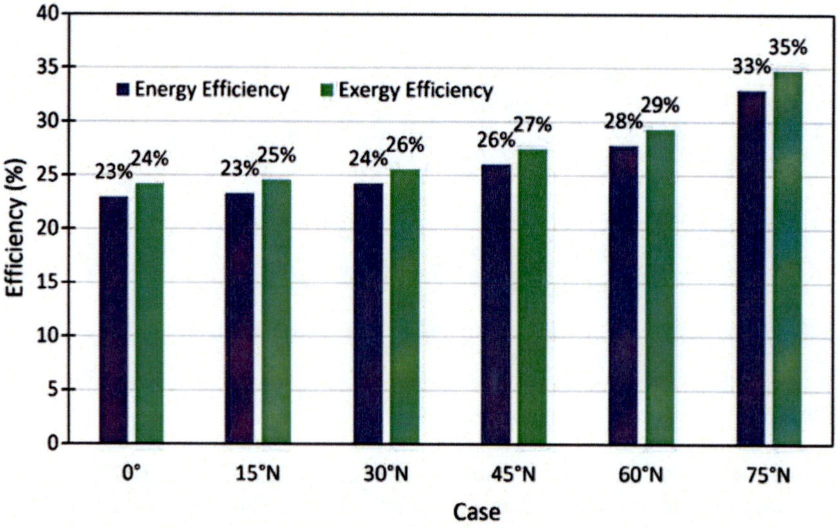

Fig. 5.12 Energy and exergy efficiencies of a bifacial PV plant (data from Ref. [6])

When it comes to the energy and exergy efficiencies of the simulated systems on different latitudes. Bifacial PV plant in Polaris, Nunavut location on 75°N latitude has significantly higher energy and exergy efficiencies as 33% and 35%. The main reason behind this result is due to the low temperatures in the area, which leads higher electricity generations from the system. On the other hand, in terms of electricity generation from the same capacity of PV plant, the simulated system on 15°N latitude generated highest electricity. The energy and exergy efficiencies obtained for all cases are given in Fig. 5.12.

5.3.2 Solar Thermal–Based Power Generation

Solar thermal–based power generation systems collect solar energy and transfer it into a heat transfer fluid, in other words, thermofluid. Later on, thermofluid can be used to drive power generation systems such as Rankine cycle. There are two main categories available in solar thermal–based power generation systems, non-concentrating and concentrating systems.

Solar thermal–based power generation systems are the potential solutions to replace fossil fuel–based energy systems. High temperatures can be achieved via solar thermal systems, especially with concentrating solar systems. Among renewables, direct high-temperature heat can be achieved via solar and geothermal systems. However, solar thermal systems are currently more able to achieve higher temperatures such as 500°C with the current state of the art. The importance of high-temperature process heat can be mentioned as enabling processes. Some of the processes require high-temperature heat. Among renewables, solar thermal systems are promising alternatives to drive high-temperature processes, such as thermochemical hydrogen production cycles. Moreover, glass or steel industries and other hard-to-electrify industrial processes require high-temperature heat. From this aspect, solar thermal systems possess high importance.

Moreover, solar thermal–based power generation systems are suitable for multigeneration. As a result of the power generation cycle, a heat around 40–90°C can be released from the condenser. Rather than wasting the heat released from the condenser, it can be utilized by integrating different subsystems. Desalination and district heating and cooling systems can be mentioned to those subsystems which can utilize the waste heat from the condenser.

(a) *Non-concentrating Solar Thermal–Based Power Generation Systems*
Currently, non-concentrating solar thermal systems are used for lower value processes, since the process heat generated from non-concentrating systems is usually at low temperatures, which typically has lower exergy content compared to higher temperature process heat. As a result, non-concentrating solar thermal systems are more suitable for processes such as space heating, space cooling, or desalination. In order to generate power, typically, lower temperature cycles, such as organic Rankine, Kalina, and ammonia trilateral cycles, can be considered, since those are typically utilized to employ the low-temperature heat.

If one looks at the types of non-concentrating solar thermal–based power generation systems, the collector types can be mentioned. There are mainly two types of non-concentrating collectors: evacuated tube and flat plate collectors. Evacuated tube collectors are more advanced systems since the thermofluid inside the collector is in vacuum condition, which makes it easier to evaporate. Usually, evacuated tube collectors are more expensive option due to the cost of more advanced components, compared to the flat plate collectors. Flat plate collectors are, on the other hand, recognized as a low-cost option which consists of a flat glass and heat absorbing area at the backplate along with multiple tubes. Solar radiation is collected at the heat absorbing area, and the heat transfer fluid is heated.

Illustrative Example 5.2 This illustrative example is carried out for non-concentrating solar thermal systems to generate power in Daggett, California, the United States selected as a specific location. The evacuated tube collector–type system in Fig. 5.13 is considered and compared with the flat plate collector in Fig. 5.14. In order to find energy and exergy efficiencies of the system, balance equations need to be found for each component.

For the pump component in each system, the mass balance equation (MBE), energy balance equation (EBE), entropy balance equation (EnBE), and exergy balance equation (ExBE) can be expressed as:

MBE: $\dot{m}_1 = \dot{m}_2$
EBE: $\dot{m}_1 h_1 + \dot{W}_p = \dot{m}_2 h_2$
EnBE: $\dot{m}_1 s_1 + \dot{S}_{gen,p} = \dot{m}_2 s_2$
ExBE: $\dot{m}_1 ex_1 + \dot{W}_p = \dot{m}_2 ex_2 + \dot{E}x_{d,p}$

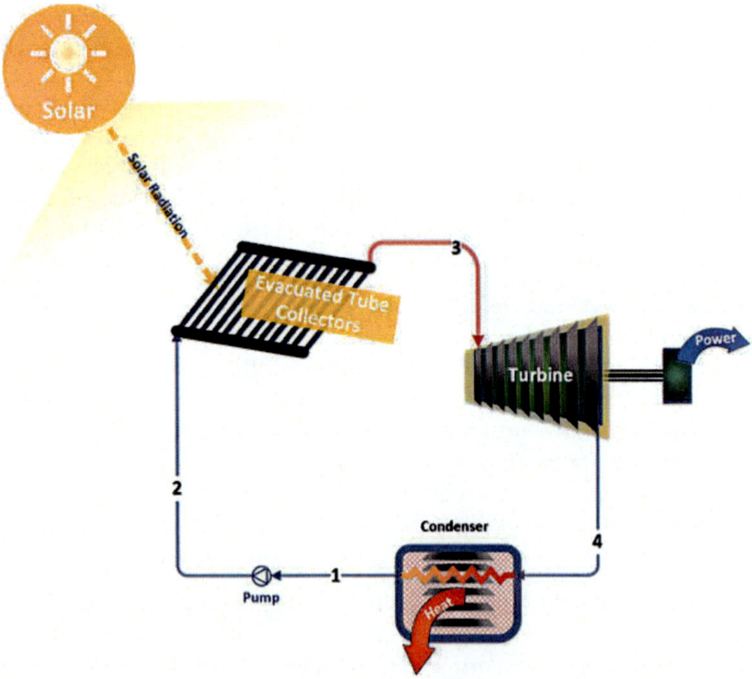

Fig. 5.13 Illustration of evacuated tube collector–driven power generation system

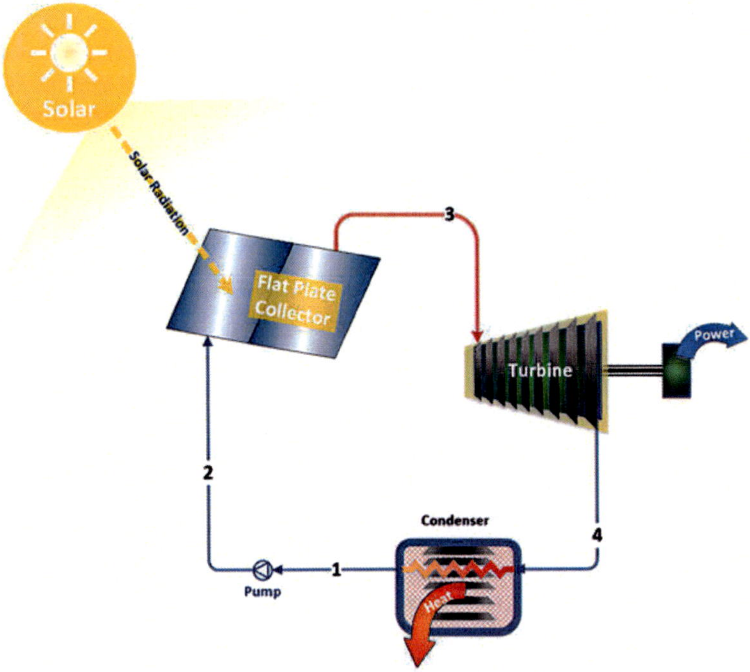

Fig. 5.14 Illustration of flat plate collector–driven power generation system

The balance equations for the turbine in each system can be written as:

MBE: $\dot{m}_3 = \dot{m}_4$
EBE: $\dot{m}_3 h_3 = \dot{W}_t + \dot{m}_4 h_4$
EnBE: $\dot{m}_3 s_3 + \dot{S}_{gen,t} = \dot{m}_4 s_4$
ExBE: $\dot{m}_3 ex_3 = \dot{W}_t + \dot{m}_4 ex_4 + \dot{E}x_{d,t}$

When we calculate the heat input to the thermofluid, the balance equations of solar collectors in each system can be expressed as:

MBE: $\dot{m}_2 = \dot{m}_3$
EBE: $\dot{m}_2 h_2 + \dot{Q}_{SC} = \dot{m}_3 h_3$
EnBE: $\dot{m}_2 s_2 + \frac{\dot{Q}_{SC}}{T_b} + \dot{S}_{gen,SC} = \dot{m}_3 s_3$
ExBE: $\dot{m}_2 ex_2 + \dot{E}x_{\dot{Q}_{SC}} = \dot{m}_3 ex_3 + \dot{E}x_{d,SC}$

When we calculate the released heat from the condenser, the balance equations of condenser in each system can be written as:

MBE: $\dot{m}_4 = \dot{m}_1$
EBE: $\dot{m}_4 h_4 = \dot{m}_1 h_1 + \dot{Q}_{CON}$

EnBE: $\dot{m}_4 s_4 + \dot{S}_{gen,CON} = \dot{m}_1 s_1 + \frac{\dot{Q}_{CON}}{T_b}$
ExBE: $\dot{m}_4 ex_4 = \dot{m}_1 ex_1 + \dot{E}x_{\dot{Q}_{CON}} + \dot{E}x_{d,CON}$

In order to calculate the total solar energy reaching the solar collectors, first of all, the area of each solar collector needs to be multiplied by the number of solar collectors as follows:

$$A_{SC} = N_{SC} a_{SC}$$

where A_{SC} gives the total area of solar collectors in the plant. The total area of solar collectors in the plant needs to be multiplied by the global horizontal irradiation per unit area ($\dot{q}_{GHI}$), as follows:

$$\dot{Q}_{GHI,SC} = \dot{q}_{GHI} \times A_{SC}$$

On the other hand, for the exergy analysis of the solar energy, the Parrot's model can be used, which can be expressed as follows:

$$\dot{E}x_{\dot{Q}_{GHI,SC}} = \dot{Q}_{GHI,SC}\left(1 - \frac{4}{3}\frac{T_o}{T_{sun}}\left(1 - \cos\delta\right)^{\frac{1}{4}} + \frac{1}{3}\left(\frac{T_o}{T_{sun}}\right)^4\right)$$

where the half angle of the disc of sun is denoted by δ.

To calculate the energy and exergy efficiencies of a non-concentrating solar thermal–based power generation system, electricity can be considered as the useful output and the energy and exergy versions of solar energy reaching the area of solar collectors can be taken as the total energy input. From this aspect, energy and exergy efficiencies of the non-concentrating solar thermal–based power generation system are:

$$\eta_{en,ov} = \frac{\dot{W}_{El,SC}}{\dot{Q}_{GHI,SC}}$$

and

$$\eta_{ex,ov} = \frac{\dot{W}_{El,SC}}{\dot{Q}_{GHI,SC}}$$

Both of these systems are designed to generate 50 MW of electricity. Flat plate collector can heat the steam to 180°C when the solar intensity is 1000 W/m^2. Flat plate collector is able to transfer 55% of the heat that came from the sun. On the other hand, evacuated tube collectors can heat the steam to 220°C and they are able to transfer 60% of the solar radiation into the thermofluid.

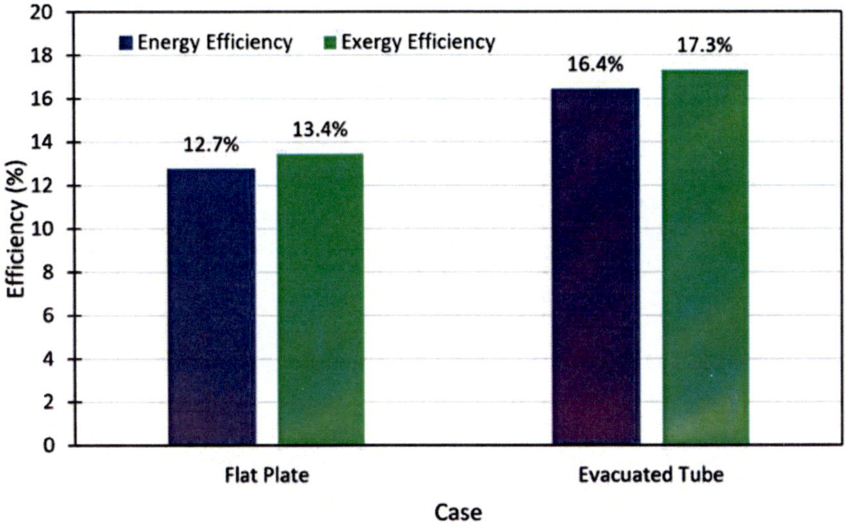

Fig. 5.15 Energy and exergy efficiencies of non-concentrating solar thermal–based power generation systems

In order to generate 50 MW of net electricity, 392,337.6 m^2 of flat plate collector area or 304,414 m^2 of evacuated tube collector area is required.

When we look at the energy and exergy efficiencies of non-concentrating solar thermal–based power generation systems shown in Figs. 5.13 and 5.14; for the flat plate collector, 12.7% of energy and 13.4% of exergy efficiencies are found for the overall system. For the evacuated tube collector, the overall energy and exergy efficiencies are found as 16.4% and 17.3%, as shown in Fig. 5.15.

(b) *Concentrating Solar Thermal Systems*
Concentrating solar thermal systems are more advanced technologies compared to non-concentrating systems. By virtue of the high-temperature heat generation ability of concentrating solar thermal systems, there is a rich option for potential integrations. Suitable processes vary from power generation to hydrogen production. When we look at the currently operating commercial concentrating solar plants, solar towers and parabolic trough systems are popular technologies in order to collect solar radiation and turn it into power. Linear Fresnel, on the other hand, still needs to be improved in terms of cost effectiveness.

Solar tower collectors can concentrate the solar radiation in order to reach higher temperatures compared to other concentrating solar collectors. Solar tower collectors can concentrate the sun to more than 1000 of the sun. Typically, available commercial solar tower–based power plants have process heat temperatures around 250–600 °C.

Parabolic trough systems are also mature technologies and currently operating in many countries. The curved mirror of the parabolic trough collector concentrates the

Table 5.1 The linear Fresnel projects across the world

Field	Location	Power capacity
Augustin Fresnel 1	Occitanie, France	300 kW
Dhursar	Rajasthan, India	125 MW
eLLO	Occitanie, France	9 MW
Huaqiang	Hebei, China	15 MW
Lanzhou Dacheng Dunhuang	Gansu, China	10 MW
Lanzhou Dacheng Dunhuang	Gansu, China	50 MW
Puerto Errado 2	Murcia, Spain	30 MW

Compiled from Ref. [7]

solar radiation and focuses it on the absorber tube line in order to heat the thermofluid, typically molten salt.

Linear Fresnel collectors have flat mirrors which reflect the sun to the receiver line on top of the flat mirrors, typically higher than the receiver line of parabolic trough collectors. A list of commercially operating linear Fresnel projects is given in Table 5.1.

The concentrating solar thermal–based power generation systems are illustrated in Fig. 5.16. The power generation mechanism is the same in different concentrating solar thermal collectors. Heat transferred from solar collectors to thermofluid can be used in the Rankine cycle. In commercial systems, usually, a heat exchanger is added between solar collector and the power generation cycle. When there is more than one loops, thermofluid transfers its heat to steam which runs the Rankine cycle.

The balance equations are the same for three of the concentrating solar systems given in Fig. 5.16. For the pump component in Fig. 5.16, the MBE, EBE, EnBE, and ExBE can be expressed as:

MBE: $\dot{m}_1 = \dot{m}_2$

EBE: $\dot{m}_1 h_1 + \dot{W}_p = \dot{m}_2 h_2$

EnBE: $\dot{m}_1 s_1 + \dot{S}_{gen,p} = \dot{m}_2 s_2$

ExBE: $\dot{m}_1 ex_1 + \dot{W}_p = \dot{m}_2 ex_2 + \dot{E}x_{d,p}$

The balance equations for the turbine in Fig. 5.16 can be written as:

MBE: $\dot{m}_3 = \dot{m}_4$

EBE: $\dot{m}_3 h_3 = \dot{W}_t + \dot{m}_4 h_4$

EnBE: $\dot{m}_3 s_3 + \dot{S}_{gen,t} = \dot{m}_4 s_4$

ExBE: $\dot{m}_3 ex_3 = \dot{W}_t + \dot{m}_4 ex_4 + \dot{E}x_{d,t}$

When we calculate the heat input to the thermofluid, the balance equations of solar collectors in Fig. 5.16 can be expressed as:

MBE: $\dot{m}_2 = \dot{m}_3$

EBE: $\dot{m}_2 h_2 + \dot{Q}_{SC} = \dot{m}_3 h_3$

EnBE: $\dot{m}_2 s_2 + \frac{\dot{Q}_{SC}}{T_b} + \dot{S}_{gen,SC} = \dot{m}_3 s_3$

ExBE: $\dot{m}_2 ex_2 + \dot{E}x_{\dot{Q}_{SC}} = \dot{m}_3 ex_3 + \dot{E}x_{d,SC}$

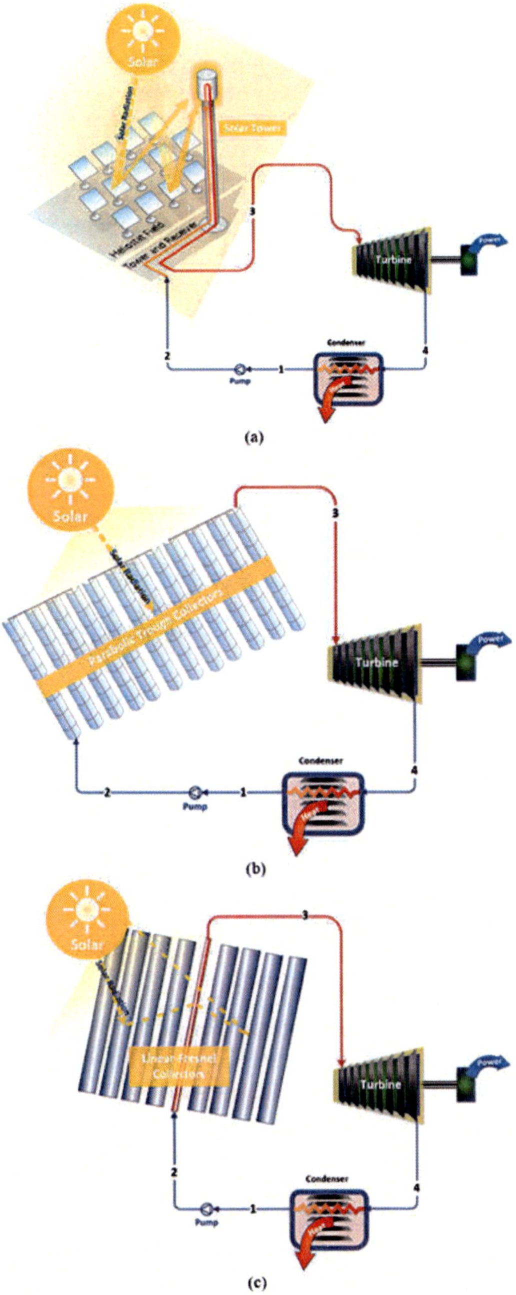

Fig. 5.16 Illustration of concentrating solar thermal–based power generation systems with (**a**) solar tower, (**b**) parabolic trough, and (**c**) linear Fresnel collectors

When we calculate the released heat from the condenser, the balance equations of condenser in Fig. 5.16 can be written as:

MBE: $\dot{m}_4 = \dot{m}_1$

EBE: $\dot{m}_4 h_4 = \dot{m}_1 h_1 + \dot{Q}_{CON}$

EnBE: $\dot{m}_4 s_4 + \dot{S}_{gen,CON} = \dot{m}_1 s_1 + \frac{\dot{Q}_{CON}}{T_b}$

ExBE: $\dot{m}_4 ex_4 = \dot{m}_1 ex_1 + \dot{E}x_{\dot{Q}_{CON}} + \dot{E}x_{d,CON}$

It is important to find how the source is utilized. With better source utilization, sources can be exploited for more useful outputs. Source utilization can be found by using the energy and exergy efficiencies. For the specific system, solar energy is used to generate useful heat and then power. The overall energy and exergy efficiencies of the parabolic trough collector–type concentrated solar power plant can be calculated as follows:

$$\eta_{en} = \frac{\dot{W}_{net}}{\dot{Q}_{DNI,CS}}$$

and

$$\eta_{ex} = \frac{\dot{W}_{net}}{\dot{E}x_{Q_{DNI,CS}}}$$

Illustrative Example 5.3 An illustrative example is comparatively assessed for three different concentrating solar collector technologies. Solar tower, linear Fresnel, and parabolic trough types of solar collectors are considered in order to evaluate their power generation performance by using the data from case study 4.2. Figures 5.16a, b, c depict the concentrating solar systems for power generation. Each system is designed to have the same solar radiation input. According to their system characteristics, power generation will be different. The objective of the current illustrative example is comparatively assessing different types of concentrating solar systems due to their power generation.

The power generation system is added to the simulation carried out in case study 4.2. Similarly, the simulations are carried out by using System Advisor Model for Lat: 34.85 DD and Long: −116.78 DD location, which is in the town of Daggett in California, the United States. Three types of concentrating solar systems: parabolic trough, solar tower, and linear Fresnel are simulated for the same heat generation capacities. According to the obtained data, design points are arranged such as process heat temperatures and freeze protection temperatures. System Advisor Model's weather data shows 2127.95 kWh of global horizontal irradiance, where

Table 5.2 The details of the location for illustrative example 5.3

Parameter	Details
Location	Daggett, CA, the United States
Latitude	34.85 DD
Longitude	−116.78 DD
Elevation	561 m
Meteorological data	National Solar Radiation Database
Weather station ID	91,486
Meteorological data type	Typical meteorological year
Annual global horizontal irradiance	2127.95 kWh/m^2
Annual direct normal irradiance	2799.55 kWh/m^2
Annual diffuse horizontal irradiance	456.25 kWh/m^2

between −3°C and 44°C ambient temperatures are shown in a typical year. Location-related parameters are given in Table 5.2.

Solar tower-type concentrating solar power system is designed with 193.5 m of height, 21.6 m of receiver length, and 17.6 m of receiver width. The coating emittance and absorbances are accounted as 0.88 and 0.94 when AISI316 stainless steel used. For the same total aperture reflective area, parabolic trough and linear Fresnel concentrating solar power systems are designed. The parabolic trough concentrating solar thermal system design parameters are given in Table 5.3.

The simulations showed different heat and power generation results for different technologies. Since the aperture reflective areas are same for three of the concentrating solar power systems that simulated in this case study, the available solar irradiation is found 3384 GWh for all of the systems. The total aperture reflective area is accounted as 1,208,730 m^2. Among the concentrating solar collector technologies, solar tower generated the highest process heat as 1531.5 GWh in a typical year. The maximum process heat temperature is found for 573.98°C. The lowest process heat is generated by the linear-Fresnel concentrating solar power system, where 1021 GWh of process heat is generated in a typical year, with the highest 536.04°C.

The parabolic trough concentrating solar power system generated 1482.8 GWh of process heat in a typical year. The maximum process heat temperature is found as 436.86°C. According to the heat generation, solar tower produced the highest amount, followed by the parabolic trough system. There is a significant difference between the second and third places where the linear Fresnel technology generated 45% less than the parabolic trough and 50% less than the solar tower systems.

When one looks at the exergy versions of collected solar radiance and generated heat, 3210 GWh of exergy of solar radiance is collected by each type of concentrating solar system. Useful thermal exergy outputs of concentrating solar systems are

Table 5.3 The parabolic trough–type concentrated solar thermal system design parameters considered

Parameter	Value
Heat absorption capacity	820.5 MW_t
Collector exit temperature	391°C
Collector inlet temperature	293°C
Number of collector loops	238
Heat transfer fluid	Therminol VP-1
Row spacing	15 m
Stow angle	170°
Deploy angle	10°
Washes per year	63
Total aperture reflective area	1,208,730.0 m^2
Solar field area	772 acres
Total land area	1081 acres
Tracking power	238 kW
Pump efficiency	0.85
Piping thermal loss coefficient	0.45 W/m^2 K
Collector	SkyFuel sky trough
Collector reflective aperture area	656 m^2
Aperture width	6 m
Length of collector assembly	115 m
Number of modules per assembly	10
Receiver	Schott PTR80
Absorber tube inner diameter	0.076 m
Absorber tube outer diameter	0.08 m

found as 992.06 GWh for solar tower, 812.22 GWh for parabolic trough, and 602.9 GWh for linear Fresnel technologies.

Figure 5.17b shows the electricity generation from the concentrating solar thermal–based systems. Among the concentrating collector types, solar tower–based systems produced the highest electricity in a typical year for the selected location, at 520.7 GWh of electricity. On the other hand, the parabolic trough–based system produced slightly lower electricity than the solar tower system, at 504.2 GWh of electricity. Linear Fresnel system produced the lowest electricity among the concentrating solar systems, at 347.1 GWh of electricity in a typical year.

The energy and exergy efficiencies are calculated by dividing the electricity generated by the concentrating solar systems to the total direct normal irradiances collected by the concentrating solar collectors. Among the concentrating solar systems, solar tower type has the highest overall energy and exergy efficiencies at 16.22% and 17.09%. For the parabolic trough concentrating solar system, the overall energy and exergy efficiencies are found as 15.7% and 16.55%, respectively. Overall, energy and exergy efficiencies for the linear Fresnel–type concentrated solar system were found to be 10.81% and 11.39%, as shown in Fig. 5.18.

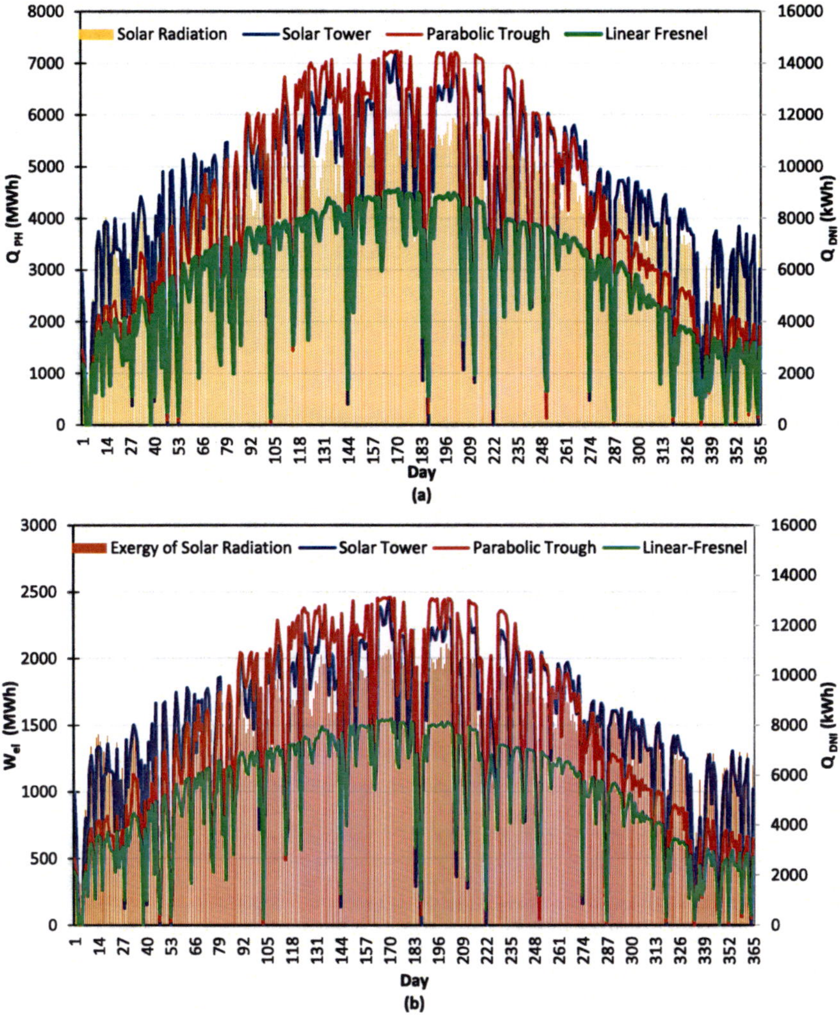

Fig. 5.17 (**a**) Process heat and (**b**) electricity generation from solar tower, parabolic trough, and linear Fresnel concentrating solar systems by energy and exergy of direct normal irradiance

5.4 Wind Power

For the last 20 years, wind is the fastest growing renewable power generation source, due to the cost drops for both onshore and offshore wind farms. Levelized cost of electricity has dropped significantly due to the cost drops in components as well as with the improved supply chain of the components of wind farms. When we look at the levelized cost of electricity for wind farms, offshore wind farms had \$0.18/kWh of levelized cost of electricity in 2010, which dropped to \$0.0752/kWh in 2021. On

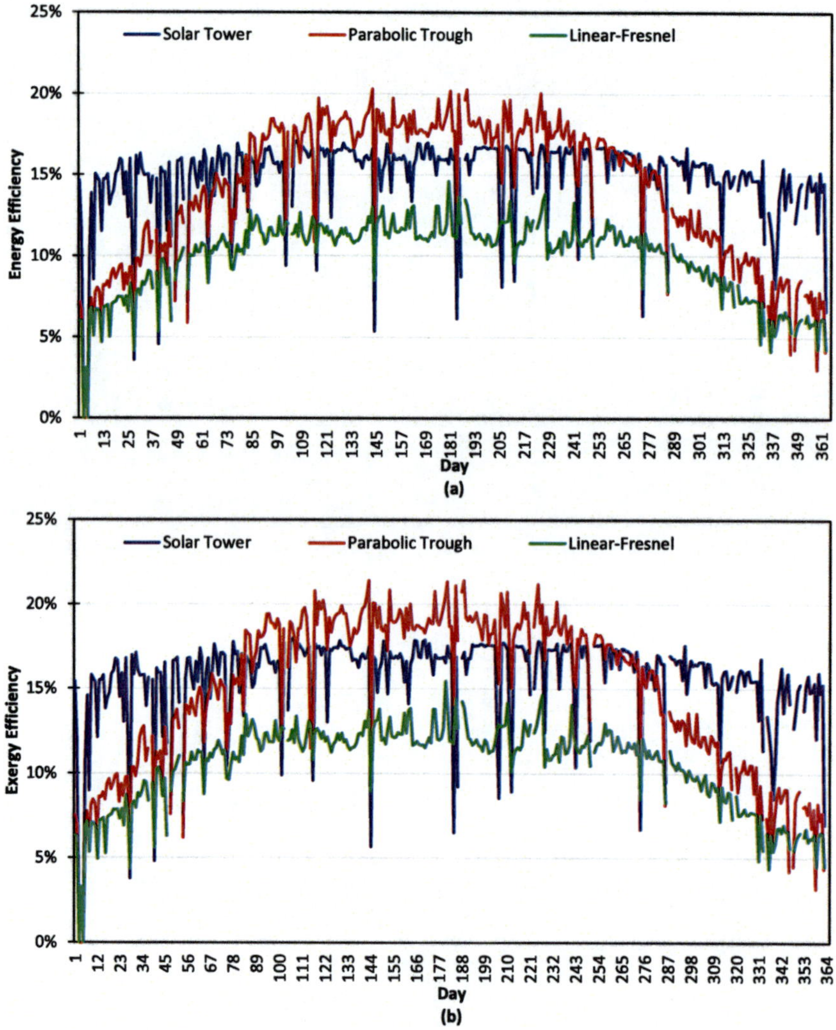

Fig. 5.18 (**a**) Energy and (**b**) exergy efficiencies of solar tower, parabolic trough, and linear Fresnel concentrating solar thermal–based power generation systems

the other hand, onshore wind farms became even less costly. In 2010, onshore wind farms had \$0.1021/kWh of levelized cost of electricity, which dropped to \$0.0331/kWh of electricity.

The main difference between offshore and onshore windfarm besides cost can be mentioned as the intermittency of the source. Offshore wind farms usually have more continuous wind penetration since there are less obstacles in the offshore environment compared to the terrestrial environment.

Besides the offshore and onshore wind farms, wind power is also classified due to the wind turbine. According to the direction of the axis of the rotating shaft, there are vertical and horizontal wind turbines. Horizontal wind turbines currently form the majority of commercial wind farms. On the other hand, commercial and research works are also carried out for vertical wind turbines in order to improve the vertical wind turbines to the level of horizontal wind turbines.

In order to carry out a thermodynamic analysis for a wind turbine, based on the first and second laws of thermodynamics, in other words, energy and exergy aspects, wind energy, and forms need to be clarified. Before defining wind energy, the general energy and exergy balance equations can be expressed as follows:

$$\sum_{i} \dot{m}_i (h + ke + pe)_i = \sum_{o} \dot{Q}_o + \sum_{o} \dot{W}_o + \sum_{o} \dot{m}_o (h + ke + pe)_o$$

$$\sum_{i} \dot{E}x_{\dot{Q}_i} + \sum_{i} \dot{W}_i + \sum_{i} \dot{m}_i ex_i = \sum_{o} \dot{E}x_{\dot{Q}_o} + \sum_{o} \dot{W}_o + \sum_{o} \dot{m}_o ex_o + \dot{E}x_d$$

where i and o subscripts denote input and output, and ke and pe denote kinetic energy and potential energy. Specific exergy for the general balance equations can be expressed as follows:

$$ex = (ke + pe + (h - h_0) - T_0(s - s_0))$$

Wind energy can be expressed as the function of kinetic energy which can be expressed as the air at a certain mass flow rate flows through the wind turbine at a certain speed:

$$E_{k,\text{wind}} = \frac{1}{2} \dot{m}_{\text{air}} V_{\text{air}}^2$$

where mass flow rate can also be expressed as follows:

$$\dot{m}_{\text{air}} = \rho \pi r^2 V_{\text{air}}$$

The wind power can be expressed as follows:

$$P_{\text{wind}} = \frac{1}{2} \rho \pi r^2 V^3$$

Betz limit is the maximum power coefficient that denotes the exploitable mechanical energy within the kinetic energy of the wind. According to the theory exhibited by Albert Betz, wind turbines cannot exploit more than 59.3% of wind kinetic energy for power generation. The Betz Limit expresses that no wind turbine can

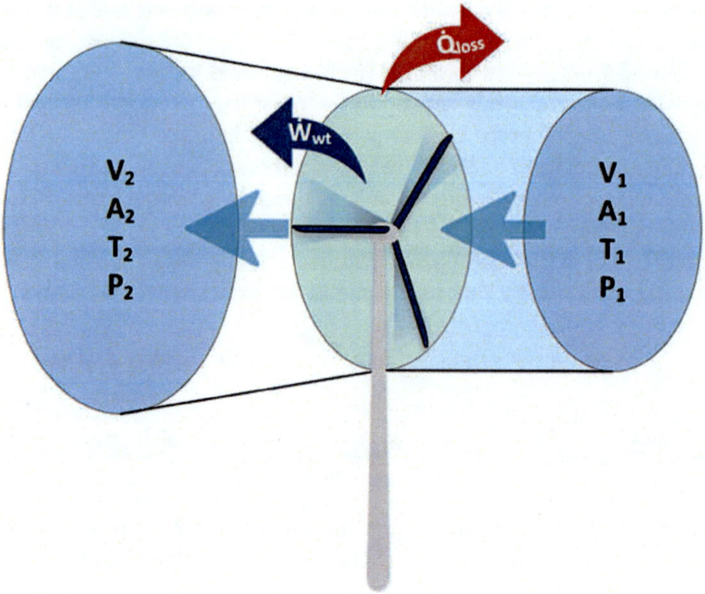

Fig. 5.19 Illustration of wind energy system state points for inlet and exit

generate more power than 59.3% of the kinetic energy of the wind. To account Betz limit, the following expression can be used:

$$\dot{W}_{wt,max} = C_B P_{wind}$$

where $\dot{W}_{wt,\,max}$ denotes the maximum possible work generation of the wind turbine.

In order to carry out calculation according to the second law of thermodynamics, the change in enthalpy needs to be defined as the change of the flowing matter at two different state points which is shown in Fig. 5.19, which can be expressed as follows:

$$\Delta H_{air} = \dot{m}_{air} C_P (T_2 - T_1)$$

where T_1 and T_2 denote the wind chill temperatures at the state points in Fig. 5.19. According to the Joint Action Group for Temperature Indices expression [8], Wind chill temperature can be expressed in °F and mph as follows:

$$T_{wc} = 35.74 + 0.6215\, T_{air} - 35.75\, \left(V^{0.16}\right) + 0.4274\, T_{air}\left(V^{0.16}\right)$$

On the other hand, entropy difference needs to be defined in order to proceed to the exergy definition. The entropy difference can be written as follows:

$$\Delta S_{air} = \dot{m}_{air} T_{ambient}\left(C_P \ln\left(\frac{T_2}{T_1}\right) - R \ln\left(\frac{P_2}{P_1}\right) - \frac{Q_{loss}}{T_{at}}\right)$$

Therefore, exergy of the wind can be calculated as follows:

$$Ex_{\text{wind}} = E_{\text{generated}} + \dot{m}_{\text{air}}\left(C_P(T_2 - T_1)\right.$$
$$\left. + \left(T_{\text{ambient}}\left(C_P \ln\left(\frac{T_2}{T_1}\right) - R\ln\left(\frac{P_2}{P_1}\right) - \frac{Q_{\text{loss}}}{T_{\text{at}}}\right)\right)\right)$$

where $E_{\text{generated}}$ denotes the generated electricity from the total kinetic energy difference.

Illustrative Example 5.4 Offshore wind farm system is designed by using the System Advisor Model. General Electric 1.6 MW wind turbines with 100 m of rotor diameter and 100 m of hub height are used in the wind farm.

To form an offshore wind farm, 63 wind turbines are used for 100.8 MW of total wind-based electricity generation capacity. Wind turbines start to generate electricity from 3.5 m/s and cut out at 25 m/s wind speed, where the rated wind speed is 11 m/s for the current wind turbine model. The mechanical energy from the wind is used via 7854 m^2 of swept area and three blades of each wind turbine.

Coastal water of San Luis Obispo County is considered for the offshore wind farm simulations. Wind direction and average wind speed for the selected location are shown in Fig. 5.20. The cut-in wind speed for the wind turbine is 3.5 m/s, the rated wind speed is 11 m/s, and the cut-out wind speed is 25 m/s. The electricity

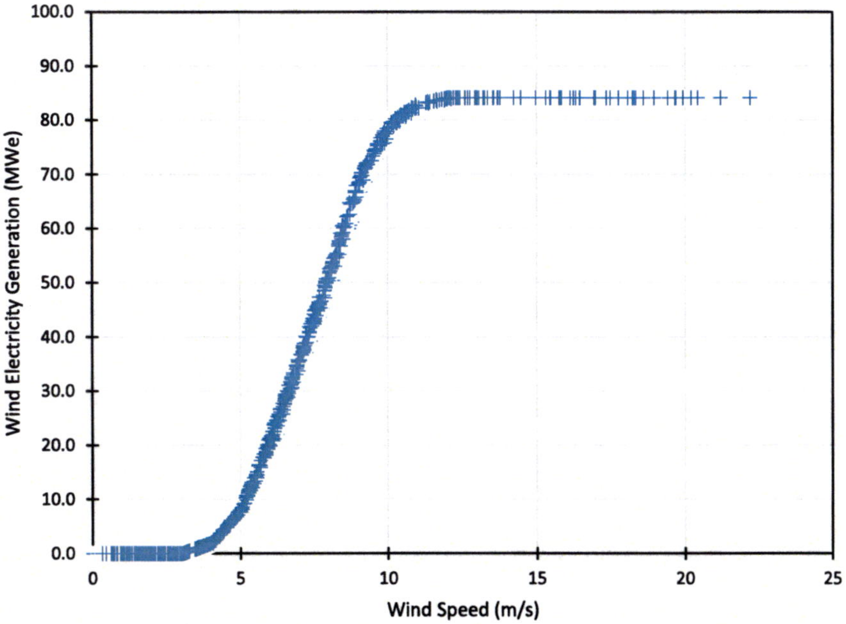

Fig. 5.20 Monthly electricity production from wind farm arrays in different wind speeds

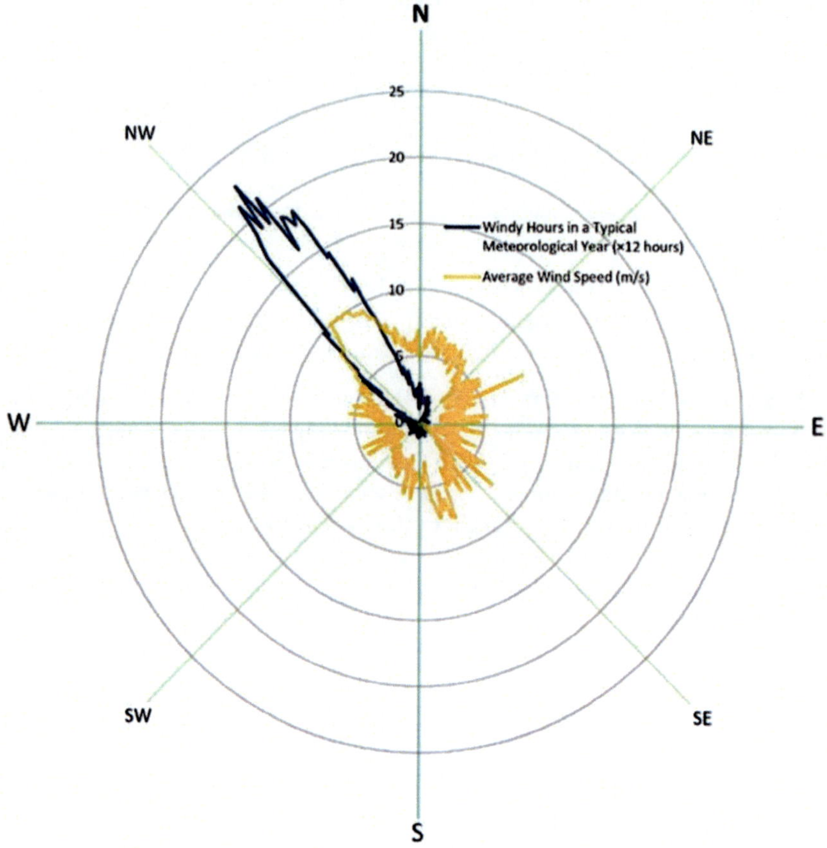

Fig. 5.21 Annual wind speed and direction variations for a typical meteorological year in California

production rates at different wind speeds of the overall wind farm are shown in Fig. 5.21. Annually, 367.5 GWh of electricity is generated by the proposed wind farm.

5.5 Geothermal-Based Power Generation

Geothermal-based power generation system utilizes the thermal energy that comes from the underground. The center of the Earth consists of magma, which is very high temperatures molten or semi-molten rocks. Typically, the plate boundaries of the Earth have more geothermal energy potential compared to the other regions. Countries with high renewable share in their electricity mixture, usually have hydro or geothermal as the dominant source. The main reason behind this is the high- capacity factor. Geothermal energy is a more reliable source compared to other renewables. When we look at the solar and wind, there are high intermittencies and typically capacity factors are lower. From this aspect, geothermal energy is a promising solution in order to meet the baseload.

Power generation is one of the key outputs of geothermal energy. Mainly, there are primary loop and secondary loop systems. If the geothermal fluid is suitable for the power generation process and has contaminations under safe limits, then a primary loop system can be used in order to directly use geothermal fluid to generate power. In some systems, geothermal fluid does not enter the Rankine cycle or another power generation cycle, where a secondary loop is used. Flash steam power systems can be included in the primary loop, where the flashing process occurs by decreasing the pressure of the fluid in order to get steam. According to the temperature and quality of the steam, number of flashing stages varies. Binary steam power systems differ from the first two methods since binary systems use geothermal fluid indirectly with a heat exchanger in order to heat another working fluid within the power generation cycle. Binary cycles can exploit the lower temperatures for power generation since different working fluids can be used, including organic fluids. Organic Rankine cycle and Kalina cycle are two examples of low-temperature geothermal binary cycle power systems. Other than space heating and power generation, application areas and potential services of geothermal-based energy systems have a large variety.

Temperature is always considered the key constraint to select the suitable application. From 10°C, space heating can be done via heat pump systems. When the outside temperature is lower than the ground temperature, the heat pump consumes more energy to upgrade the heat at lower temperatures. Geothermal water of 20–50° C can be used in fish farms, especially the ones that are located in cold climatic conditions. Greenhouse systems also can benefit from the heat from 25°C temperatures.

For the desalination process, heat waste from geothermal-based power generation systems can be exploited under vacuum pressures. Another way to use geothermal energy is using power-driven desalination methods, where geothermal energy can be exploited to generate power and later on power can be used to generate fresh water.

The overall energy and exergy efficiencies are defined for the geothermal-based power generation system with reinjection as:

$$\eta_{en,ov} = \frac{\dot{W}_{El,net}}{\dot{m}_P h_P - \dot{m}_{RI} h_{RI}}$$

and

$$\eta_{ex,ov} = \frac{\dot{W}_{El,net}}{\dot{m}_P ex_P - \dot{m}_{RI} ex_{RI}}$$

The overall energy and exergy efficiencies are now defined for the geothermal-based power generation system without reinjection as:

$$\eta_{en,ov} = \frac{\dot{W}_{El,net}}{\dot{m}_P h_P - \dot{m}_0 h_0}$$

and

$$\eta_{ex,ov} = \frac{\dot{W}_{El,net}}{\dot{m}_P ex_P - \dot{m}_0 ex_0}$$

Illustrative Example 5.5 An illustrative example is carried out for a geothermal source in Kizildere, Denizli, Turkey. A reservoir with a 260°C source temperature is considered in order to carry out the illustrative example. A steam Rankine cycle system is integrated as a secondary loop to the primary geothermal loop. Figure 5.22 is considered the geothermal-driven power generation system with reinjection in Kizildere, Denizli, Turkey.

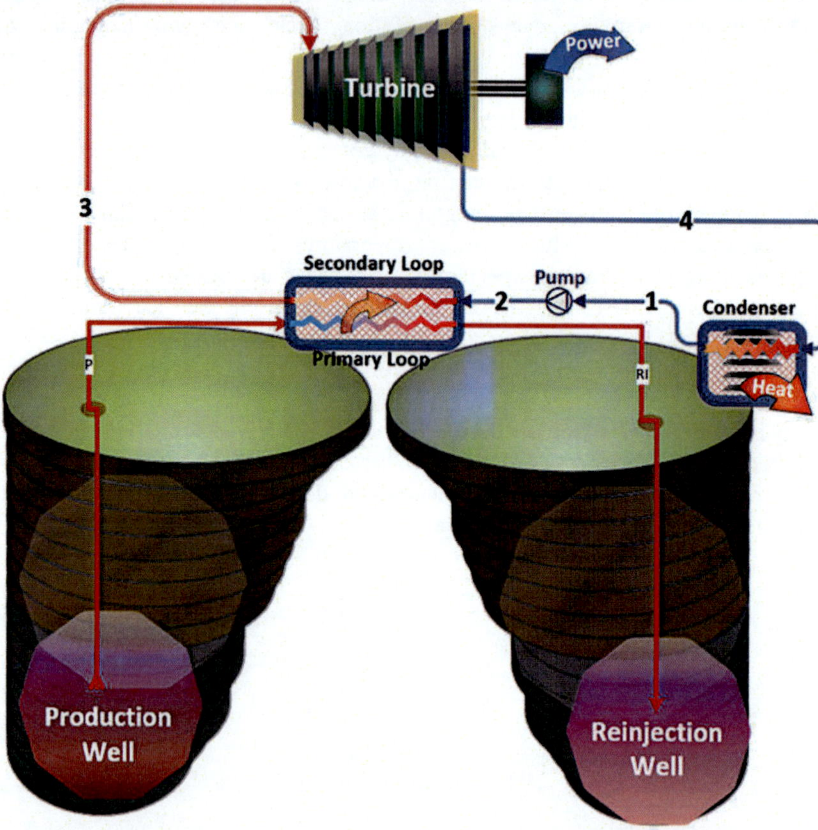

Fig. 5.22 Geothermal-based power generation system via steam Rankine cycle

To calculate the energy and exergy performances of the proposed system, balance equations need to be applied.

For the pump component in Fig. 5.22, the MBE, EBE, EnBE, and ExBE can be expressed as:

MBE: $\dot{m}_1 = \dot{m}_2$
EBE: $\dot{m}_1 h_1 + \dot{W}_p = \dot{m}_2 h_2$
EnBE: $\dot{m}_1 s_1 + \dot{S}_{gen,p} = \dot{m}_2 s_2$
ExBE: $\dot{m}_1 ex_1 + \dot{W}_p = \dot{m}_2 ex_2 + \dot{E}x_{d,p}$

The balance equations for the turbine in Fig. 5.22 can be written as:

MBE: $\dot{m}_3 = \dot{m}_4$
EBE: $\dot{m}_3 h_3 = \dot{W}_t + \dot{m}_4 h_4$
EnBE: $\dot{m}_3 s_3 + \dot{S}_{gen,t} = \dot{m}_4 s_4$
ExBE: $\dot{m}_3 ex_3 = \dot{W}_t + \dot{m}_4 ex_4 + \dot{E}x_{d,t}$

When we calculate the heat input to the thermofluid, the balance equations of the steam generator in Fig. 5.22 can be expressed as:

MBE: $\dot{m}_2 = \dot{m}_3$
EBE: $\dot{m}_2 h_2 + \dot{Q}_{SC} = \dot{m}_3 h_3$
EnBE: $\dot{m}_2 s_2 + \frac{\dot{Q}_{SC}}{T_b} + \dot{S}_{gen,SC} = \dot{m}_3 s_3$
ExBE: $\dot{m}_2 ex_2 + \dot{E}x_{\dot{Q}_{SC}} = \dot{m}_3 ex_3 + \dot{E}x_{d,SC}$

When we calculate the released heat from the condenser, the balance equations of the condenser in Fig. 5.22 can be written as:

MBE: $\dot{m}_4 = \dot{m}_1$
EBE: $\dot{m}_4 h_4 = \dot{m}_1 h_1 + \dot{Q}_{CON}$
EnBE: $\dot{m}_4 s_4 + \dot{S}_{gen,CON} = \dot{m}_1 s_1 + \frac{\dot{Q}_{CON}}{T_b}$
ExBE: $\dot{m}_4 ex_4 = \dot{m}_1 ex_1 + \dot{E}x_{\dot{Q}_{CON}} + \dot{E}x_{d,CON}$

For the energy and exergy efficiencies of the geothermal-based power generation system with reinjection, the respective equations can be used as follows:

$$\eta_{en,ov} = \frac{\dot{W}_{El,net}}{\dot{m}_P h_P - \dot{m}_{RI} h_{RI}}$$

and

$$\eta_{ex,ov} = \frac{\dot{W}_{El,net}}{\dot{m}_P ex_P - \dot{m}_{RI} ex_{RI}}$$

Process heat temperature entered into the Rankine cycle is used as 180°C. According to the calculations, the overall system uses 172.63 MW of heat in order

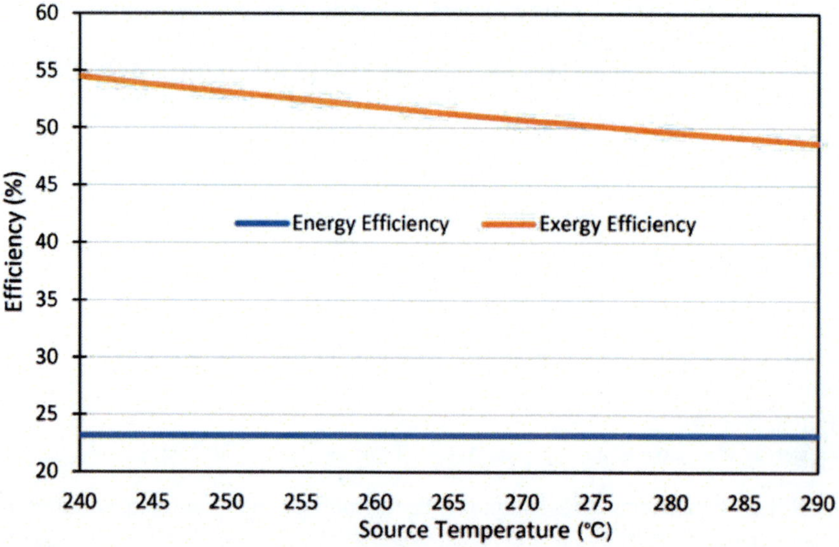

Fig. 5.23 Overall energy and exergy efficiencies of the geothermal-based power generation system in steady-state conditions

to generate 40 MW of net electricity, in steady-state conditions. The overall energy and exergy efficiencies of the geothermal-based power generation system are found to be 23.17% and 52.54%. When the source temperature is changed, the exergy efficiency of the overall system also changes. However, the energy efficiency of the overall system does not change with the change of source temperature. Figure 5.23 depicts the energy and exergy efficiency changes with different source temperatures.

In order to improve efficiencies, additional integrated systems can be considered such as desalination and space heating. The temperature of the excess heat released from the condenser is calculated as 40°C, which is useful for many processes. In the following chapters, integrated energy systems will comprehensively present how to exploit excess or waste energy with additional processes.

5.6 Biomass-Based Power Generation

Biomass refers to the contained energy in chemical form within organic or biologic materials such as wood, plants, and wastes. Biomass-based systems have high-capacity factors compared to the other renewables which is typically around 68%. When we look at the cost and capacity factors, bioenergy is a promising solution. The emissions of bioenergy sources need to be considered by balancing it. The photosynthesis process catches the carbon dioxide in order to generate bioenergy sources. For this reason, net greenhouse gas emissions will be balanced, and there will be no extra carbon dioxide emissions related to biomass.

There are two main generations of biofuels: conventional (first generation) and advanced (second generation). Edible bioenergy sources are utilized, such as corn, sugarcane, sugar beets, and rapeseed, to generate bioethanol or other biofuels, in the conventional biofuel generation process. On the other hand, the advanced biofuel generation process utilizes and treats non-edible bioenergy sources. The reason behind the initiation of the advanced biofuel generation process is to keep edible food products for eating. Pyrolysis, gasification, and anaerobic digestion are among the processes to generate advanced biofuels. There are also third and fourth generations of biofuels which refer to algae-driven biofuel generation (third) and genetically engineered feedstocks and genomically synthesized microorganism-driven biofuel generation (fourth).

Combustion is a popular process for both direct usage of solid biomass and biomass-based liquid fuels (biofuels). For solid wastes, typically a secondary loop is used to utilize heat for power generation. Figure 5.24a depicts a solid biomass combustion system integrated with a steam Rankine cycle and Fig. 5.24b depicts a liquid biofuel system with a gas turbine cycle.

For the biomass combustor, the mass balance equation can be expressed as follows:

$$\sum \dot{n}_R = \sum \dot{n}_P$$

The energy balance equation for the biomass combustor can be expressed as follows:

$$H_R = H_P$$

where

$$H_R = \sum \dot{n}_R \left(\overline{h_f}^{\circ} + \overline{h} - \overline{h^{\circ}} \right)_R$$

and

$$H_P = \sum \dot{n}_P \left(\overline{h_f}^{\circ} + \overline{h} - \overline{h^{\circ}} \right)_P$$

The entropy balance equation for the biomass combustor is then obtained as follows:

$$\dot{S}_R + \dot{S}_{gen} = \dot{S}_P$$

where

$$\dot{S}_R = \sum (\dot{n}\overline{s})_R$$

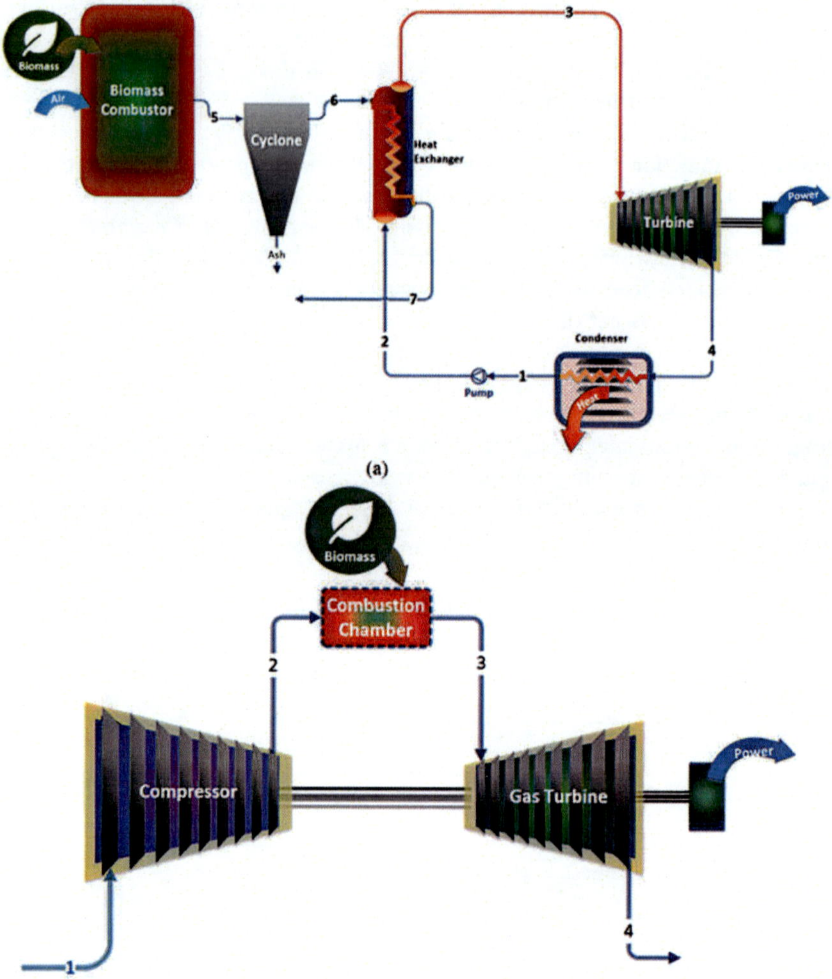

Fig. 5.24 Biomass-based power generation systems (**a**) with steam Rankine cycle for solid wastes and (**b**) with gas turbine cycle for liquid biofuels

and

$$\dot{S}_R = \sum (\dot{n}\bar{s})_P$$

where $\bar{s}$ stands for the absolute entropy and it can be calculated for a specific temperature (T) and pressure (P) as follows:

$$\bar{s}(T,P)=\overline{s^{\circ}}(T_0,P)-R_u\ln\left(\frac{P}{P_0}\right)$$

The exergy balance equation for the biomass combustor is then written as follows:

$$\dot{\mathrm{E}}\mathrm{x}_R=\dot{\mathrm{E}}\mathrm{x}_P+\dot{\mathrm{E}}\mathrm{x}_d$$

where the exergy destruction ($\dot{\mathrm{E}}\mathrm{x}_d$) can be expressed from entropy generation and the reference temperature as:

$$\dot{\mathrm{E}}\mathrm{x}_d=T_0\dot{\mathrm{S}}_{\mathrm{gen}}$$

Moreover, the exergy flows of reactants ($\dot{\mathrm{E}}\mathrm{x}_R$) and products ($\dot{\mathrm{E}}\mathrm{x}_P$) is then expressed as:

$$\dot{\mathrm{E}}\mathrm{x}_R=\sum\dot{n}_R\left(\overline{h_f{}^{\circ}}+\bar{h}-\overline{h^{\circ}}-T_0\bar{s}\right)_R$$

and

$$\dot{\mathrm{E}}\mathrm{x}_P=\sum\dot{n}_P\left(\overline{h_f{}^{\circ}}+\bar{h}-\overline{h^{\circ}}-T_0\bar{s}\right)_P$$

Illustrative Example 5.6 An illustrative example is carried out for the system in Fig. 5.24a. A municipal solid waste processing facility uses municipal solid waste with a combustor in order to produce steam at 220°C. Heating value of the solid waste is reported as 18,000 kJ/kg and exergy content of the solid waste is reported as 16,000 kJ/kg. According to the facility, carbon dioxide emission from municipal solid waste is 1.1 kg of CO_2. In order to generate 500 kW of electricity, the required municipal solid waste mass flow rate and emission rate will be presented.

For the pump component in Fig. 5.24a, the MBE, EBE, EnBE, and ExBE can be expressed as:

MBE: $\dot{\mathrm{m}}_1=\dot{\mathrm{m}}_2$
EBE: $\dot{\mathrm{m}}_1h_1+\dot{\mathrm{W}}_{\mathrm{p}}=\dot{\mathrm{m}}_2h_2$
EnBE: $\dot{\mathrm{m}}_1s_1+\dot{\mathrm{S}}_{\mathrm{gen,p}}=\dot{\mathrm{m}}_2s_2$
ExBE: $\dot{\mathrm{m}}_1\mathrm{ex}_1+\dot{\mathrm{W}}_{\mathrm{p}}=\dot{\mathrm{m}}_2\mathrm{ex}_2+\dot{\mathrm{E}}\mathrm{x}_{\mathrm{d,p}}$

The balance equations for the turbine in Fig. 5.24a can be written as:

MBE: $\dot{\mathrm{m}}_3=\dot{\mathrm{m}}_4$
EBE: $\dot{\mathrm{m}}_3h_3=\dot{\mathrm{W}}_{\mathrm{t}}+\dot{\mathrm{m}}_4h_4$
EnBE: $\dot{\mathrm{m}}_3s_3+\dot{\mathrm{S}}_{\mathrm{gen,t}}=\dot{\mathrm{m}}_4s_4$
ExBE: $\dot{\mathrm{m}}_3\mathrm{ex}_3=\dot{\mathrm{W}}_{\mathrm{t}}+\dot{\mathrm{m}}_4\mathrm{ex}_4+\dot{\mathrm{E}}\mathrm{x}_{\mathrm{d,t}}$

When we calculate the heat input to the thermofluid, the balance equations of steam generator in Fig. 5.24a can be expressed as:

MBE: $\dot{m}_2 = \dot{m}_3$
EBE: $\dot{m}_2 h_2 + \dot{Q}_{SC} = \dot{m}_3 h_3$
EnBE: $\dot{m}_2 s_2 + \frac{\dot{Q}_{SC}}{T_b} + \dot{S}_{gen,SC} = \dot{m}_3 s_3$
ExBE: $\dot{m}_2 ex_2 + \dot{E}x_{\dot{Q}_{SC}} = \dot{m}_3 ex_3 + \dot{E}x_{d,SC}$

When we calculate the released heat from the condenser, the balance equations of the condenser in Fig. 5.24a can be written as:

MBE: $\dot{m}_4 = \dot{m}_1$
EBE: $\dot{m}_4 h_4 = \dot{m}_1 h_1 + \dot{Q}_{CON}$
EnBE: $\dot{m}_4 s_4 + \dot{S}_{gen,CON} = \dot{m}_1 s_1 + \frac{\dot{Q}_{CON}}{T_b}$
ExBE: $\dot{m}_4 ex_4 = \dot{m}_1 ex_1 + \dot{E}x_{\dot{Q}_{CON}} + \dot{E}x_{d,CON}$

For the energy and exergy efficiencies of the municipal solid waste–based power generation system, equations can be used as follows:

$$\eta_{en,ov} = \frac{\dot{W}_{El,net}}{\dot{m}_{SW} HV_{SW}}$$

and

$$\eta_{ex,ov} = \frac{\dot{W}_{El,net}}{\dot{m}_P ex_{SW}}$$

According to the calculations, 220°C of steam is used to drive the Rankine cycle. In order to generate 500 kW of electricity, 1826.5 kW of heat is required from the combustor. According to the heating value, a total of 365.3 kg of solid waste is required hourly to run the system. According to the efficiency assessment, both 27.4% of energy and 30.8% of exergy efficiencies are found.

5.7 Hydropower Systems

Hydropower systems use the potential energy, which is related to the position of the water, specifically the elevation of water. The elevation between the upper water reservoir and lower water reservoir gives the potential energy that can be utilized. According to the generator efficiency and head losses, the final electricity can be calculated as:

$$\dot{W}_{\text{hydro,el}} = \eta_{\text{gen}} \dot{m} g \left(Z_{\text{upper reservoir}} - Z_{\text{lower reservoir}} - \sum \text{Head losses} \right)$$

where the mass flow rate is denoted with $\dot{m}$, the gravitational force is denoted with g, elevation is denoted with z, and generator efficiency is denoted with η_{gen}.

Among the renewables, hydro-based electricity generation represents the highest renewable electricity generation globally. As of 2022, hydro-based electricity generation reached 4326 TWh of electricity, which is even higher than total electricity produced by other renewables.

Illustrative Example 5.7 An illustrative example is carried out by using the data from [9] for Cochrane, Ontario, Canada. A hydropower system is integrated with a floating PV plant in the integrated system in order to store excess energy for later use. According to the geographical characteristics of the location, abandoned mines are considered as the lower reservoir, whereas lakes are considered as the upper reservoir. This is also environmentally friendly for the abandoned mines since oxidation of unmined ore poses a threat to the environment. The integrated system is given in Fig. 5.25.

For the illustrative example, Kam Kotia Mine in Cochrane, Ontario, with 55 m of depth, is considered as the lower reservoir. Kamistokia Lake in Cochrane, Ontario is considered as the upper reservoir for the study. On Kamistokia Lake, there is 450 hectares of water surface is available, which is considered for the floating PV plant installation.

For the hydropower system with a charging pump, the hydropower generation efficiency is given as 80%, whereas the pumping efficiency is given as 84%, respectively.

According to the results, the hydropower system was used more during the summer months for charging purposes, since the available excess energy was higher during the summer months. 4.2 GWh of electricity is generated during the months of June, July, and August, in total. 1.5 GWh of electricity is used to charge Kamistokia Lake with water during the month of May. In December, only 0.627 GWh of electricity is used to charge Kamistokia Lake with water. Figure 5.26a shows the water withdrawal from the lower reservoir and Fig. 5.26b shows the charge and discharge rates as well as available energy in hydropower storage.

The cooling effect of the upper reservoir of the hydropower system is exploited by the PV plant. A floating PV plant designed on the Kamistokia Lake produces 1519 MWh of electricity in a typical year. The cooling effect of the upper reservoir increased the PV electricity generation by 1.58% compared to the terrestrial PV system on average during a typical year. Power generation losses related to the temperature coefficient are given in Fig. 5.27.

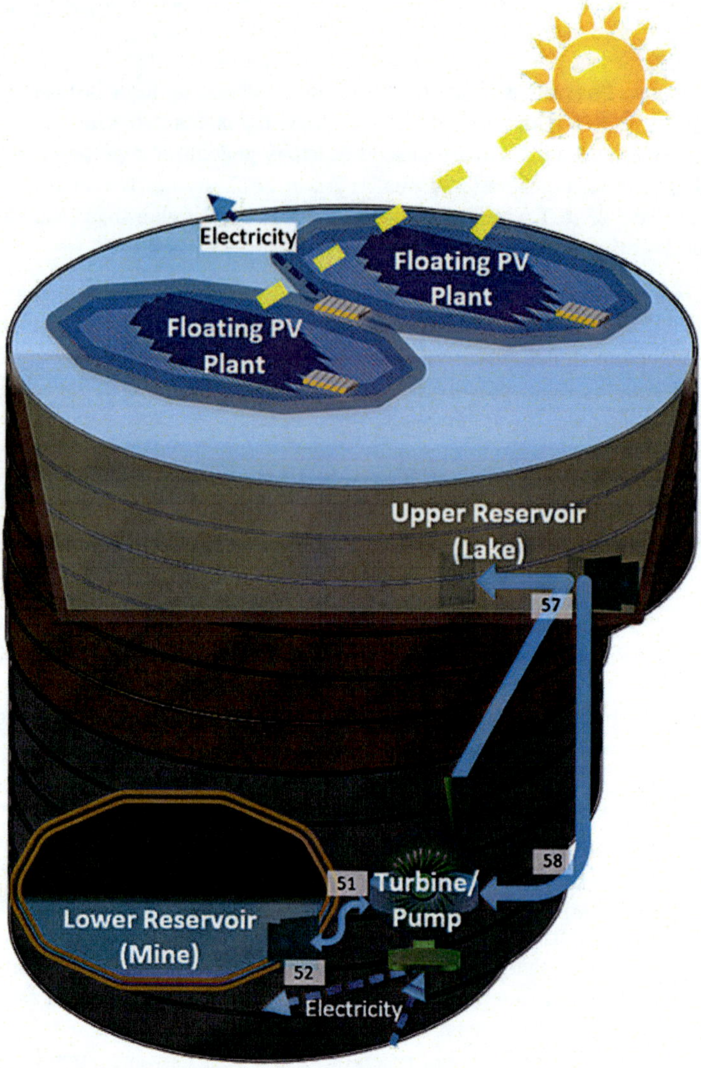

Fig. 5.25 An integrated energy system with solar PV and hydropower

5.8 Marine-Based Power Generation

Marine energy is the potential alternative to currently available fossil fuels. Ocean can be considered as the world's largest thermal energy storage, moreover, due to the heated water, movements occur which can be exploited to generate power. There are five main methods that can be mentioned for marine-based power generation systems. Tides, tidal current, waves, and salinity gradient systems can be directly

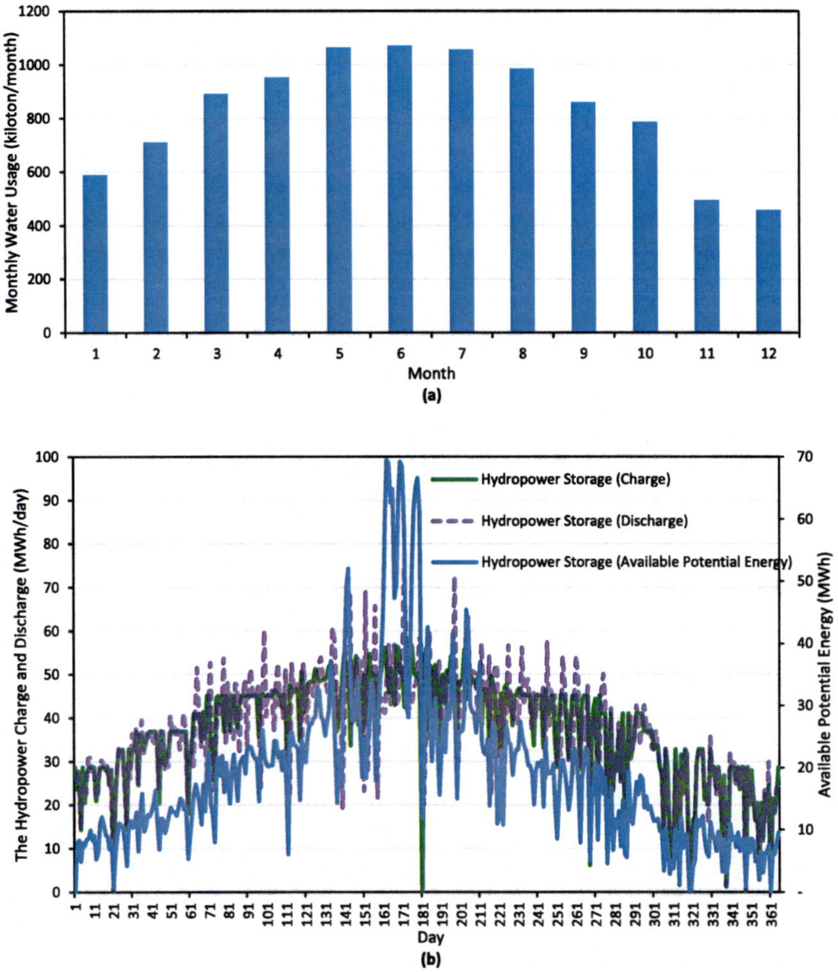

Fig. 5.26 (**a**) Water withdrawal rates during an average year (**b**) the charge and discharge rates of the hydropower storage with available energy

used to generate power without any thermal energy conversion. On the other hand, a thermal gradient needs to be utilized by using a thermal energy conversion system, typically a low-temperature power generation cycle such as the organic Rankine cycle (Fig. 5.28).

When one looks at the conversion of marine energy into power, the majority of marine energy types can be converted directly. Direct conversion typically occurs as the conversion of potential energy or kinetic energy into power. In thermal conversion, the temperature difference between deep water and ambient air is utilized by using low-temperature thermal energy conversion cycles such as organic, Kalina, or trilateral Rankine cycles.

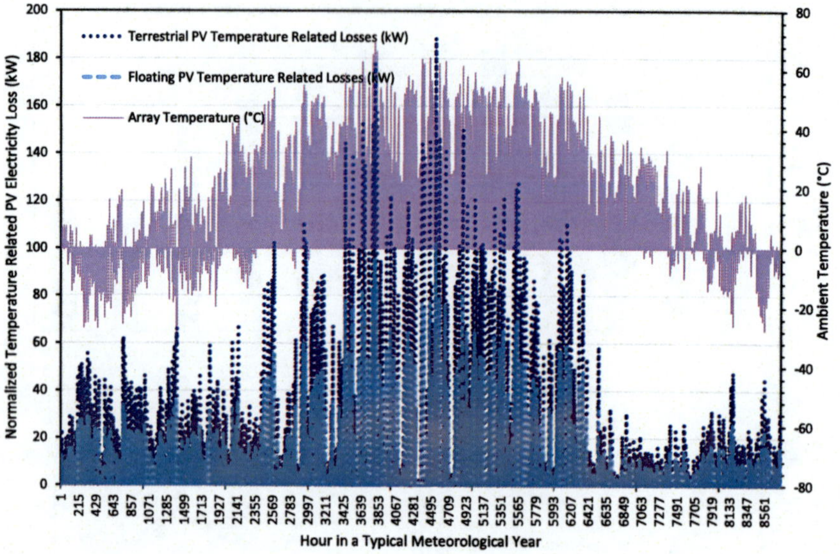

Fig. 5.27 Temperature-related PV electricity losses in floating and terrestrial cases

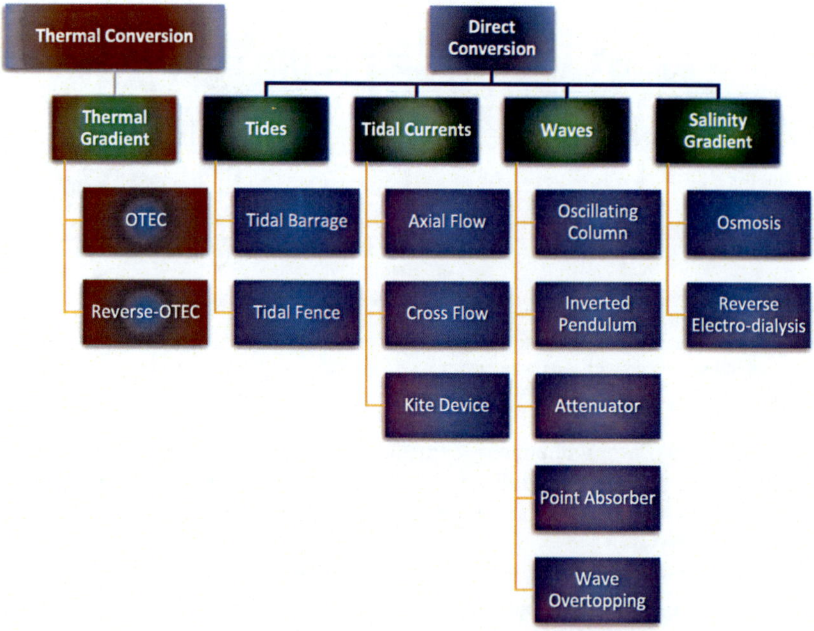

Fig. 5.28 Direct and thermal conversion of marine-based power generation

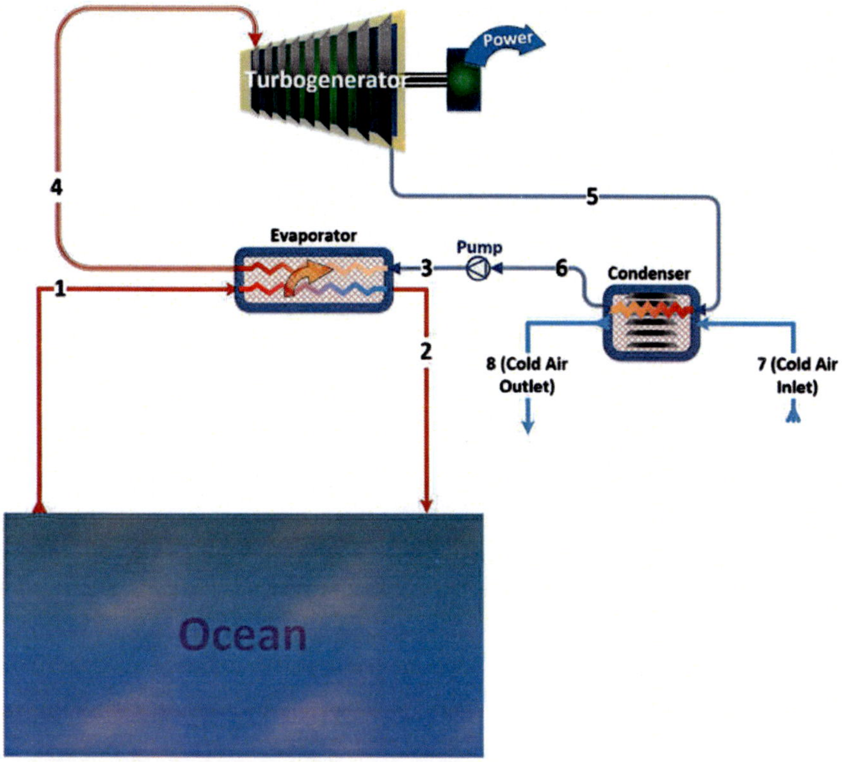

Fig. 5.29 Direct and thermal conversion of marine-based power generation

Illustrative Example 5.8 An illustrative example is presented in order to analyze an ocean thermal energy conversion (OTEC) system presented in Fig. 5.29. High-temperature ocean water is used in order to evaporate ammonia via an evaporator. Then gaseous ammonia is expanded in the turbogenerator to generate power. The balance equations for each component are expressed below.

For the evaporator component, balance equations can be expressed as:

MBE: $\dot{m}_1 = \dot{m}_2$ and $\dot{m}_3 = \dot{m}_4$
EBE: $\dot{m}_1 h_1 + \dot{m}_3 h_3 = \dot{m}_2 h_2 + \dot{m}_4 h_4$
EnBE: $\dot{m}_1 s_1 + \dot{m}_3 s_3 + \dot{S}_{\text{gen,eva}} = \dot{m}_2 s_2 + \dot{m}_4 s_4$
ExBE: $\dot{m}_1 ex_1 + \dot{m}_3 ex_3 = \dot{m}_2 ex_2 + \dot{m}_4 ex_4 + \dot{E}x_{\text{dest,eva}}$

The balance equations for turbogenerator component are:

MBE: $\dot{m}_4 = \dot{m}_5$
EBE: $\dot{m}_4 h_4 = \dot{m}_5 h_5 + \dot{W}_T$

Table 5.4 The thermophysical properties of state points

State	Mass flow rate (kg/s)	Substance	Temperature (°C)	Specific enthalpy (kJ/kg)	Specific exergy (kJ/kg K)
1	2	Water	35	146.7	417.3
2	2	Water	33	138.4	417.1
3	1	Ammonia	−25.2	1409	
4	1	Ammonia	−25.2	1426	
5	1	Ammonia	−35	1418	

EnBE: $\dot{m}_4 s_4 + \dot{S}_{gen,T} = \dot{m}_5 s_5$
ExBE: $\dot{m}_4 ex_4 = \dot{m}_5 ex_5 + \dot{W}_T + \dot{E}x_{dest,T}$

The balance equations for the condenser component are:

MBE: $\dot{m}_5 = \dot{m}_6$ and $\dot{m}_7 = \dot{m}_8$
EBE: $\dot{m}_5 h_5 + \dot{m}_7 h_7 = \dot{m}_6 h_6 + \dot{m}_8 h_8$
EnBE: $\dot{m}_5 s_5 + \dot{m}_7 s_7 + \dot{S}_{gen,con} = \dot{m}_6 s_6 + \dot{m}_8 s_8$
ExBE: $\dot{m}_5 ex_5 + \dot{m}_7 ex_7 = \dot{m}_6 ex_6 + \dot{m}_8 ex_8 + \dot{E}x_{dest,con}$

In order to calculate the electricity consumption of the pump and other parameters related to the pump component, equations can be expressed as:

MBE: $\dot{m}_6 = \dot{m}_3$.
EBE: $\dot{m}_6 h_6 + \dot{W}_P = \dot{m}_3 h_3$
EnBE: $\dot{m}_6 s_6 + \dot{S}_{gen,P} = \dot{m}_3 s_3$
ExBE: $\dot{m}_6 ex_6 + \dot{W}_P = \dot{m}_3 ex_3 + \dot{E}x_{dest,P}$

According to the calculations, Table 5.4 can be created. When we look at the power generation of turbine, 7.43 kW of electricity can be obtained. When we look at the overall energy and exergy efficiencies by considering the energy and exergy inputs and work output, 2.5% of energy and 0.89% of exergy efficiencies are found.

Illustrative Example 5.9 An illustrative example is carried out in order to assess a marine-based power generation system. A tidal barrage is used where the water flow into the bay is allowed. Once the water levels are changed outside of the bay, turbines generate power, as shown in Fig. 5.30. The elevations for points A, B, C, and D are given as 135 m at point A, 45 m at point B, and 30 m at both points C and D. Head losses are given as $\frac{3V_1^2}{2g}$ and $\frac{2V_2^2}{2g}$ for the first and second pipes. Diameters for the first and second pipes are given as 30 cm and 60 cm, respectively. Volumetric flow rate is given as 0.7 m^3/s.

In order to generate electricity from the position difference between the inside (A) and outside (B) of the tidal barrage, potential energy needs to be turned into kinetic energy and then electrical energy. The overall formula for the electricity generation from the tidal barrage is as follows:

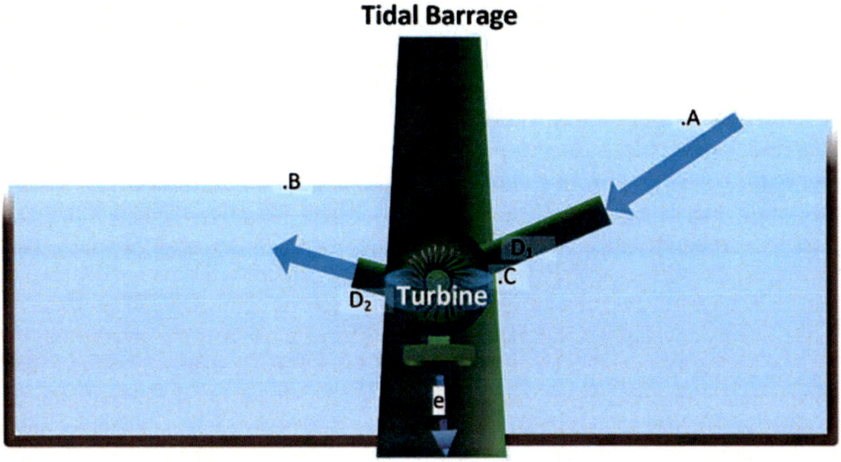

Fig. 5.30 Illustration of a tidal barrage–based power generation system

$$\dot{W}_{\text{tidal barrage}} = \dot{m}g\left(Z_{\text{inside}} - Z_{\text{outside}} - \sum \text{Head losses}\right)$$

Considering the mass flow rate of the system, the following equation can be used to find the mass flow rate from the volumetric flow rate.

$$\dot{m} = \rho\dot{V}$$

According to the given head loss formula, velocities can be calculated as follows:

$$V_n = \frac{\dot{V}}{A_n}$$

An assessment of energy efficiency can be calculated by using the equation as follows:

$$\eta = \frac{\dot{W}_{\text{tidal barrage}}}{\dot{m}g(H_a - H_b)}$$

When the given equations are applied, electricity generation from the given system is found as 511.3 kW. On the other hand, the rate of the potential energy difference between the inside and outside of the reservoirs is calculated as 618.03 kW. From those, the energy efficiency of the tidal barrage system is found as 82.7%.

5.9 Closing Remarks

The current chapter has aimed to cover power generation methods using renewable sources by presenting both technical and non-technical dimensions, such as the historical developments in the power generation area and the current status of renewable power generation globally with local examples and success stories around the world. Especially countries with a high share of renewables in their grid electricity mixture, such as Iceland and Brazil, are more deeply analyzed and presented. In this chapter, renewable power generation is mainly categorized into two: direct conversion of renewables and thermal conversion of renewables. In the direct conversion, solar, marine, hydro, and wind sources are considered. For marine, hydro, and wind sources, position or motion-related energy, in other words, potential and kinetic energies are considered to be utilized. On the other hand, the photoelectric effect is considered to utilize solar energy to generate electric power. Case studies are carried out by using the actual data from different locations. For solar-based case studies, meteorological data is considered with commercially available components. In a geothermal-based power generation example, an actual reservoir is considered. For the example regarding the hydropower system, real geographical characteristics are considered along with the additional cooling effect of the hydropower reservoir. For solar-based systems, simulations are carried out by considering commercial products with actual solar data.

Renewable power generation possesses importance in terms of desalination. Many desalination processes can exploit power to generate fresh water from saline water, which can be mentioned as mechanical vapor compression, reverse osmosis, electrodialysis, and forward osmosis. Moreover, the studied systems have energy wastes such as heat waste in thermal energy conversion cases. It is important to integrate other subsystems to improve the utilization of sources, which will be studied more in the following chapters.

References

1. Ritchie H, Roser M, Rosado P (2022) Energy data. Our world in data. Retrieved from: https://ourworldindata.org/energy (Online Resource)
2. Dincer I (2021) Thermodynamics: a smart approach. Wiley, Chichester
3. IRENA (2023) Global LCOE and auction values. International Renewable Energy Agency, Abu Dhabi
4. The White House (2023) Fact sheet: President Biden to catalyze global climate action through the major economies forum on energy and climate. In: Statements and releases. https://www.whitehouse.gov/briefing-room/statements-releases/2023/04/20/fact-sheet-president-biden-to-catalyze-global-climate-action-through-the-major-economies-forum-on-energy-and-climate/. Accessed 13 Oct 2023
5. Einstein A (1905) Über einem die Erzeugung und Verwandlung des Lichtes betreffenden heuristischen Gesichtspunkt. Ann Phys 4

6. Temiz M, Dincer I (2021) Design and analysis of a new renewable-nuclear hybrid energy system for production of hydrogen, fresh water and power. e-Prime Adv Electr Eng Electron Energy 1: 100021. https://doi.org/10.1016/J.PRIME.2021.100021
7. National Renewable Energy Laboratory (2023) Linear fresnel projects. In: SolarPACES. https://solarpaces.nrel.gov/by-technology/linear-fresnel. Accessed 27 Jun 2023
8. Osczevski RJ (2000) Windward cooling: an overlooked factor in the calculation of wind chill. Bull Am Meteorol Soc 81:2975–2978
9. Temiz M, Dincer I (2022) Design and analysis of a floating photovoltaic based energy system with underground energy storage options for remote communities. J Energy Storage 55:105733. https://doi.org/10.1016/j.est.2022.105733

Chapter 6
Desalination Methods

Nomenclature

3S	Source-system-service
BRICS	Brazil, Russia, India, China and South Africa
en	energy
ex	specific exergy (kJ/kg)
$\dot{E}x$	exergy rate (kW)
h	specific enthalpy (kJ/kg)
$\dot{m}$	mass flow rate (kg/s)
MW	molecular weight
η	efficiency
OECD	Organization for Economic Cooperation and Development
P	pressure (kPa)
PEC	photoelectrochemical
PV	photovoltaic
PR	performance ratio
Q	heat (kJ)
$\dot{Q}$	heat rate (kW)
R	Recovery
S	total entropy (kJ /K)
s	specific entropy (kJ/kg K)
$\dot{S}$	entropy rate (kW/K)
T	temperature (°C)
w	work
$\dot{W}$	power (kW)
y	mass ratio
x	molar ratio

I. Dincer, M. Temiz, *Renewable Energy Options for Power Generation and Desalination*, https://doi.org/10.1007/978-3-031-53437-9_6

Subscripts

act	actual
ASM	anion selective membrane
b	brine
con	condenser
CSM	cation selective membrane
ED	electrodialysis
f	fresh
LF	linear Fresnel
P	production
p	pump
r	ration
RI	reinjection
s	salt
SC	solar collector
t	turbine
tot	total
w	water

6.1 Introduction

Humanity has historically been using primitive salt-removing methods, such as distillation where solar heat was utilized to boil the water and get clean water. In addition, there were basic water treatment and cleaning efforts by simply boiling the water to make them drinkable. The process of removing salt from the saline water from any source, like ocean or sea, in order to turn it into the fresh water, is called desalination. It is recognized as an essential process for all living organisms, which occurs continuously in nature. Natural desalination starts with water evaporation from a saline water source such as sea or ocean. Water vapor does not contain salt, and it accumulates in the atmosphere. When the weather conditions are suitable in terms of pressure and temperature for condensation, the accumulated water vapor in the atmosphere condenses and rains. Suitable geographies on land can accumulate the fresh water as freshwater bodies.

The history of man-made desalination starts with the ancient Greek Philosopher Aristotle's description in their ancient text called Meteorologia in 340 BC. Aristotle described that saline water is actually a mixture of salt and water; however, the water vapor of saline water does not contain salt. Thus, water vapor can be used as a sweet and freshwater source if it gets condensed. Moreover, Aristotle noticed that the wax vessel could be used as a membrane in order to filter salt to get fresh water. Aristotle described the process as if a water-tight vessel was plunged into the sea or ocean, salt would be filtered by the waxen walls, thus fresh water would be poured into the vessel. Especially, sailors from ancient times were troubling to find fresh water during their long journeys. The Greek sailors in the third century BC boiled seawater in order to get fresh water. During the early periods of history, filters to trap the salt were very common, such as the sand filters, wool wicks, and clay filters. In ancient Greece, sand filters were used to treat water in order to remove solid particles during the third and second century BC. The wool wicks were also used by ancient Greeks between the second and first century BC. The Romans developed the clay filters that act as a membrane in order to generate fresh water, in the first century. An ancient Greek philosopher who lived between the second and third centuries in modern-day Aydin, Turkey, described the distillation desalination with a comment and drawing as seawater can be used by sailors by boiling it, where a bronze vessel and large sponges can be used to collect the water vapor as sweet water. Condensing steam in sponges became a common process from the fourth century BC among sailors. In the eighth century, medieval Muslim chemists published the first scientific text about the desalination process with detailed descriptions.

Until 1452, desalination technology was not feasible for large-scale applications. Leonardo da Vinci realized the feasibility of desalination and proposed that in order to make desalination feasible for larger scales, a still attachment can be adapted to a cookstove, which enables the distillation at larger quantities. Before the Industrial Revolution, freshwater supply was one of the biggest concerns of the ships. Sir Richard Hawkins reported that they were able to carry out the distillation process on board and supplied fresh water during their travels in the South Seas between 1562 and 1622. Thomas Jefferson, the third president of the United States from 1801 to 1809, was passionate about applying chemical knowledge to real-life challenges or domestic objects. Thomas Jefferson led some research efforts on desalination processes and publicized practical advice for distillation desalination to all ships registered in the United States. After the development of thermodynamic knowledge about steam-based processes and the development of steam engines, water desalination is accelerated with technological advantages. The first patent was granted in 1869, and the first desalination plant was built in England in the same year. In 1881, the first commercial desalination plant was made operational in Tigne, Malta.

As of 2023, the freshwater need of over 300 million people is met by desalination plants. The historical developments in desalination technology are depicted in Fig. 6.1. Modern society has become more and more reliant on desalinated water due to lack of natural drinkable water resources increasing water scarcity. Resources of natural fresh water are decreasing continuously. There are numerous direct drivers behind water scarcity; contamination, conflicts, poor management of water resources, poor infrastructure, destruction of water bodies to accumulate fresh water, etc. Besides the direct drivers of water scarcity, there are also indirect drivers behind water scarcity, greenhouse gas emissions and global warming are the major ones among the indirect drivers. Water scarcity belongs to the center of climate change since they are inextricably linked. Increasing greenhouse gas concentrations and global temperatures triggers the water crisis due to the changing climate balances, which leads to shrinking ice sheets, rising sea levels, droughts, and floods.

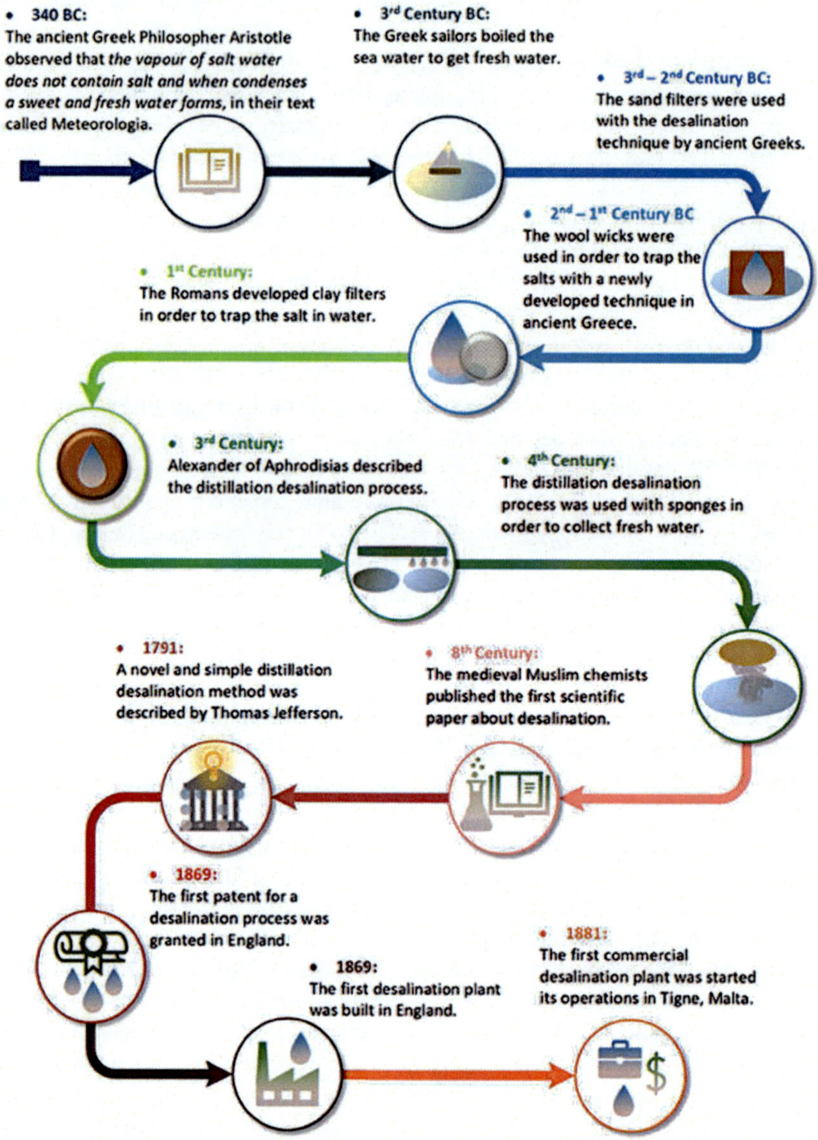

Fig. 6.1 Some historical events of desalination

Water is considered an essential necessity for the existence of life, and it is widely connected with public health as well, besides drinking. Basic hygiene and public health rely on water. In case of water scarcity, sewage systems, cleaning practices, and hygiene practices can stop, which leads to surging diseases. Especially women and children in various underdeveloped regions are under higher stress, since they

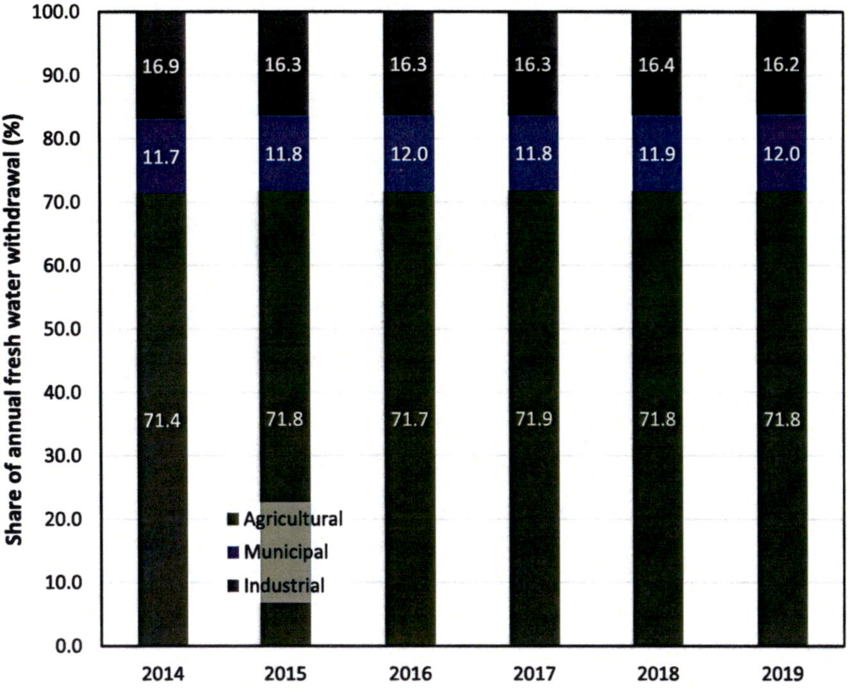

Fig. 6.2 Annual freshwater withdrawal shares by sectors (data from Ref. [1])

are the responsible ones for collecting fresh water. Increasing water scarcity can increase the time spent on to collect and access fresh water, among women and children, which disables them to spend time on other needs such as education, employment, friendship, self-actualization, etc.

Figure 6.2 shows the annual freshwater withdrawal shares by sector Agriculture is the top sector with 71–72% of the water withdrawal share. Industry is the second sector for freshwater withdrawal with 16% of share. Municipal activities, including freshwater use by municipal facilities as well as drinking water or tap water, have 12% of freshwater withdrawal share. This further proves how water is critical to feed the world.

When we look at the sectors for water usage, agricultural activities withdraw the highest share of fresh water, globally. More than 70% of freshwater withdrawals belong to agricultural activities. Therefore, it can be mentioned that agricultural activities highly rely on fresh water.

Figure 6.3 shows a map of annual freshwater withdrawal shares for agricultural activities. Fresh water is essential for global agriculture, which can be supplied from rainfed sources or irrigation via pumps. In order to feed the world and produce livestock, biofuels, or other non-food products, around 2.8 trillion m^3 of fresh water is needed. Water withdrawal shares can be different for different countries. For example, Türkiye uses 87.6% of the freshwater withdrawals for agricultural

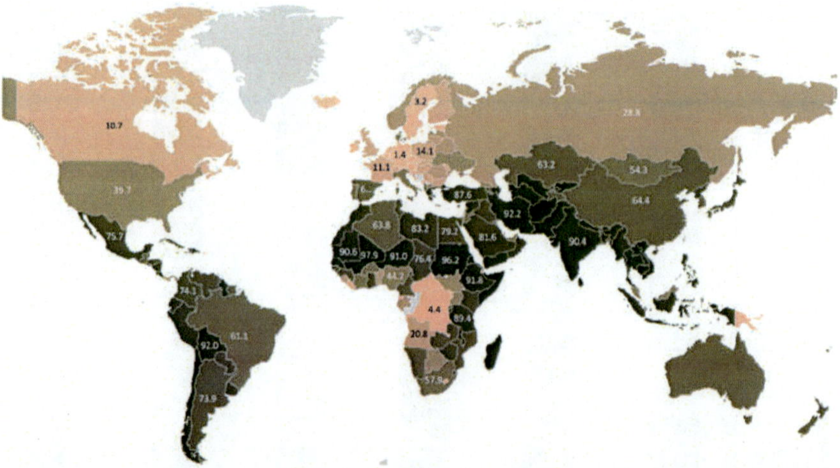

Fig. 6.3 Annual freshwater withdrawal shares for agricultural activities (data from Ref. [1])

purposes, which is higher than the global average. Moreover, Sudan has the highest share among the countries with 96% of the share. Income levels and geographical positions cause large variations. Developed countries typically have lower shares of freshwater withdrawals used in agriculture. Germany has a 1.4% of the share of freshwater withdrawal for agricultural activities.

Industry is considered the second sector for freshwater withdrawal, globally. Many processes, such as dilution, steam generation, cooling, manufacturing, and washing, rely on fresh water. Moreover, the energy industry highly relies on water; power plants require water for condensing and other cooling processes and hydrogen generation is another source for industrial water withdrawals. Especially chemical production, food processing, and paper production industries consume large amounts of fresh water. Figure 6.4 shows the freshwater withdrawal shares for industrial activities. High-income countries typically have a larger share of water consumption for industry, compared to the mid- and low-income countries. The United States withdraws more than 300 billion m^3 of fresh water each year, for industrial processes. Louisiana, Indiana, and Texas are the top states for industrial freshwater consumption. China has the second largest consumption, as 140 billion m^3 of freshwater withdrawal.

Other industrialized countries, such as Canada, Japan and France, follow China for industrial freshwater withdrawal. Further to note that high-income countries consume around 60% of water for industrial activities. Globally, about 16% of annual freshwater consumption occurs due to the industrial activities, which is equal to 650 billion m^3 of freshwater withdrawal. Canada has one of the highest shares, as 75% of its water consumption is related with industrial activities.

Typically, Middle Eastern, North African, and South Asian countries have low shares of water consumption for industrial purposes, which is typically around 1–5%. Estonia has the highest share of water consumption for industry, where

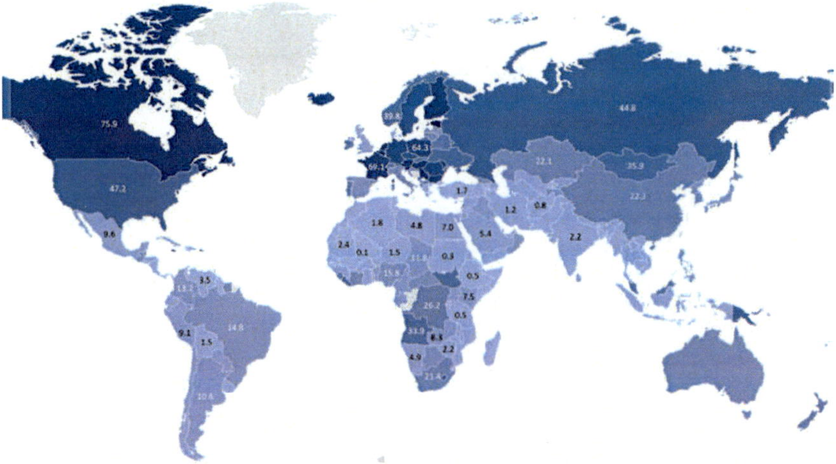

Fig. 6.4 Annual freshwater withdrawal shares for industrial activities (data from Ref. [1])

91.7% of annual water is consumed by the Estonian industries. Slovenia and Belgium are two other countries where 80% of freshwater withdrawals occur due to industrial activities.

When we look at the food-related freshwater withdrawals more deeply, different agricultural products require a large variety of fresh water as shown in Fig. 6.5. Cheese, nuts, and farmed fish have the highest freshwater withdrawals per kg of produce. In order to produce a kg of cheese, 5605 liters of fresh water is needed. Farmed fish requires the highest amount of water requirement among the meat types, as 3691 liters of fresh water per kilogram of farmed fish. Poultry meat requires the lowest fresh water among the meat types, as 660 liters of fresh water per kg of poultry meat. Rice is one of the highest water-intensive agricultural activities, which requires 2248 liters of fresh water in order to produce a kg of rice.

When it comes the household and public service-related (municipal) freshwater withdrawals, there is 12% of the global average, which is the equivalent of 480 billion m^3 of freshwater withdrawal. Although municipal freshwater withdrawal is the most visible freshwater consumption in our daily lives, it represents only 12% of total freshwater withdrawals. Activities can be listed as drinking, cleaning, washing, cooking, etc. Figure 6.6 shows the freshwater withdrawals of municipal activities. Most countries have a low share of municipal freshwater withdrawals with a few exceptions. Especially in countries with high precipitation, agricultural water can be rainfed, thus agricultural-related water withdrawals will be very low. Therefore, countries like the United Kingdom and Ireland have very high municipal freshwater consumptions, where 74% of water withdrawals are for municipal activities in the United Kingdom.

There is a balance between water withdrawals and water replenishment. Rain and river flows replenish renewable freshwater resources. However, there is a global trend of freshwater withdrawal. During the last 60 years, more than half of the

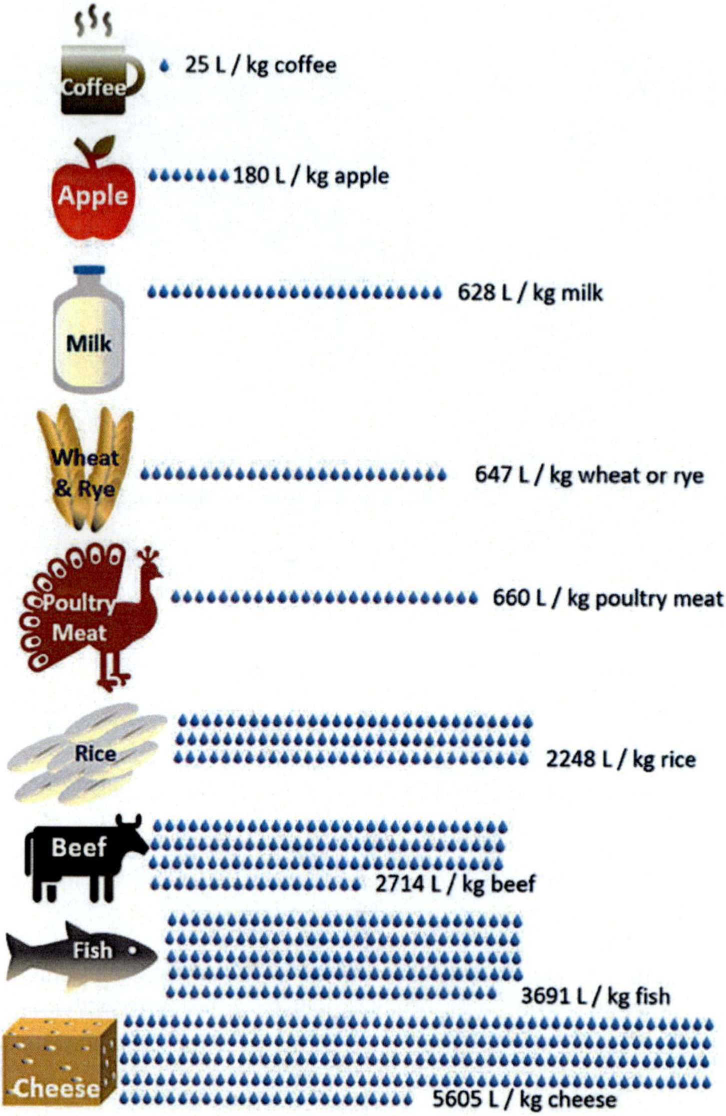

Fig. 6.5 Freshwater requirements in liters per kilogram of food (data from Ref. [2])

freshwater resources per capita are felled. Growing population is also a major factor for declining freshwater resources. Figure 6.7 shows the global renewable water resources per capita. When we look at the regions, South America and Oceania have the highest renewable freshwater resources at around 30000 m^3 per person per year. The total renewable freshwater resource in South America is 12724 km^3, which is followed by North America with 6077 km^3 of total freshwater resources. Northern

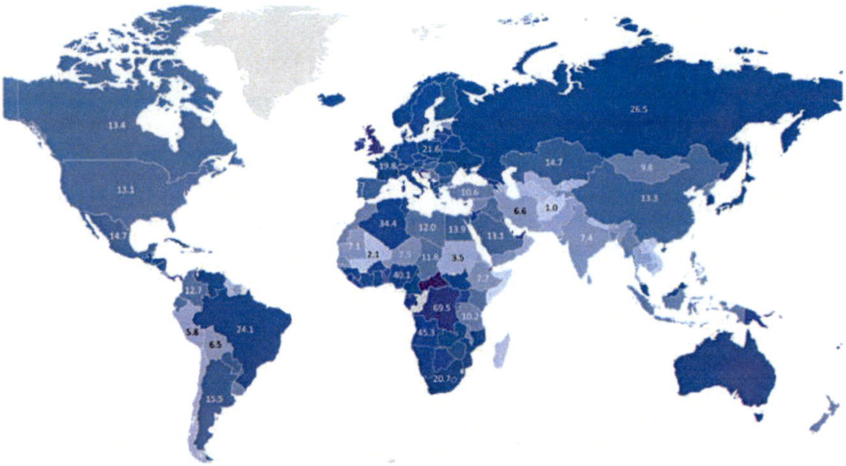

Fig. 6.6 Annual freshwater withdrawal shares for municipal activities (data from Ref. [1])

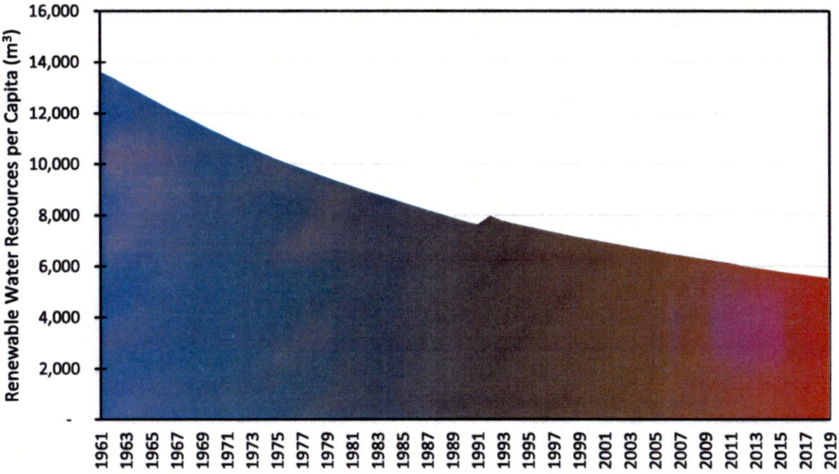

Fig. 6.7 Renewable water resources per capita (data from Ref. [1])

Africa has the lowest freshwater resource with 256 m^3 per person per year. Countries like Libya and Saudi Arabia face high stress since the freshwater withdrawal is eight to ten times higher than the internal freshwater resources.

When we look at the global picture for renewable freshwater resources which is depicted in Fig. 6.8, Canada has one of the largest resources with 75,795 m^3 of fresh water per capita. Especially north African countries have the lowest freshwater resources, which is typically between 9 and 500 m^3 of fresh water per capita. Iceland has the highest rate with 471,485 m^3 of freshwater resources per capita.

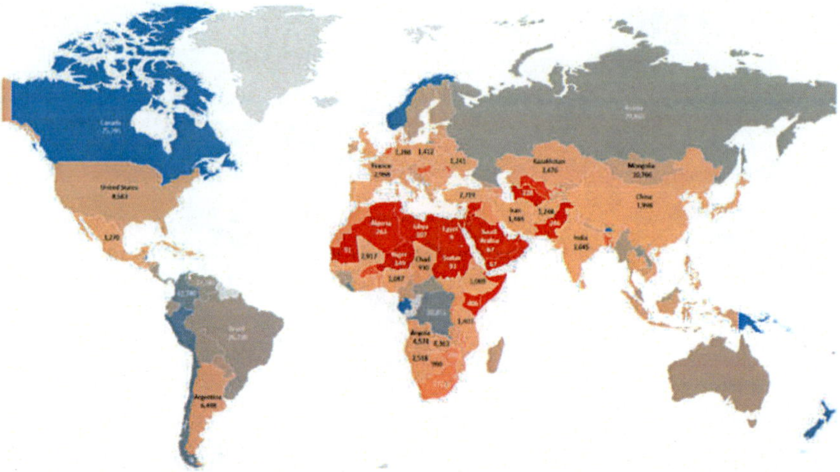

Fig. 6.8 Renewable freshwater resources per capita in m^3 (data from Ref. [1])

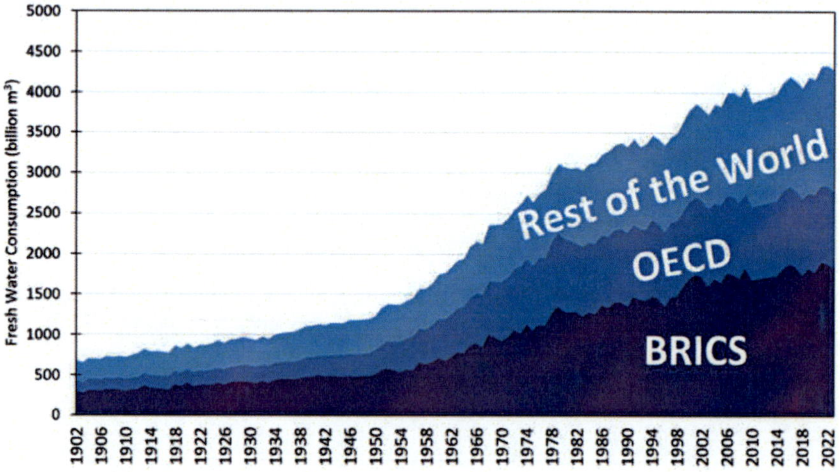

Fig. 6.9 Historical freshwater consumption by different regions (data from Ref. [1])

The population growth and global warming effects are two of the reasons for declining freshwater resources. Water consumption has a growing trend as shown in Fig. 6.9. Global annual freshwater consumption reached 4 trillion m^3, where 20–25% of it is consumed by the Organization for Economic Cooperation and Development (OECD) countries, 45% of it consumed by Brazil, Russia, India, China and South Africa (BRICS), and the remaining part is consumed by the rest of the world.

Fig. 6.10 Illustration of source-system-service (3S) approach for sustainable freshwater generation

When we look at the current situation of water, desalination possesses are of high importance. Especially in countries with high freshwater consumption and low freshwater resources, desalination is essential. North Africa, South Asia, and the Middle East are three regions where the majority of countries consume more fresh water than their freshwater resources. The United Arab Emirates consumes 15 times of their freshwater resources, Saudi Arabia consumes 9.7 times of their freshwater resources, and Qatar consumes 4.3 times of their freshwater resources. For these and other countries, internal freshwater resources are not sufficient to meet their freshwater needs, which puts them at tin critical stress category in terms of freshwater withdrawals. Under the effects of population growth and global warming, global freshwater resources will continue to decrease; thus, water exporting will not be feasible and reliable in future. Therefore, freshwater production is the most promising option for the regions with water stress. Desalination requires an energy source since thermal or mechanical energy drives the desalination processes. It is also critical to drive the desalination process by using a sustainable source, such as renewables [3]. A source system service approach [4] is applied to address sustainable freshwater generation, which is depicted in Fig. 6.10.

6.2 Desalination

Currently, there are more than 12 desalination methods available, which are driven by power and/or heat. Hybrid desalination methods combine power and heat-driven desalination techniques in a synergetic manner to improve the source utilization; however, the complexity is higher in hybrid methods.

The majority of desalination methods use either power or thermal energy. When we look at the pathways, renewable sources can be utilized by generating power or heat. Renewable heat or power can be used to generate a variety of useful outputs, including fresh water. Figure 6.11 depicts the pathways to generate fresh water. Geothermal, biomass, concentrating solar, and ocean thermal energy systems can generate thermal energy, which can be used directly in heat-driven desalination processes (Option A), or heat can be converted into power for power-driven desalination methods (Option B). Solar photovoltaic (PV), marine, wind, and hydroenergy systems can directly generate power, which can be supplied to power-driven desalination systems (Option C).

The classification of power-driven, heat-driven and hybrid desalination methods is presented in Fig. 6.12. Reverse osmosis is one of the most common desalination methods when we look at the currently available desalination plants. No heat source

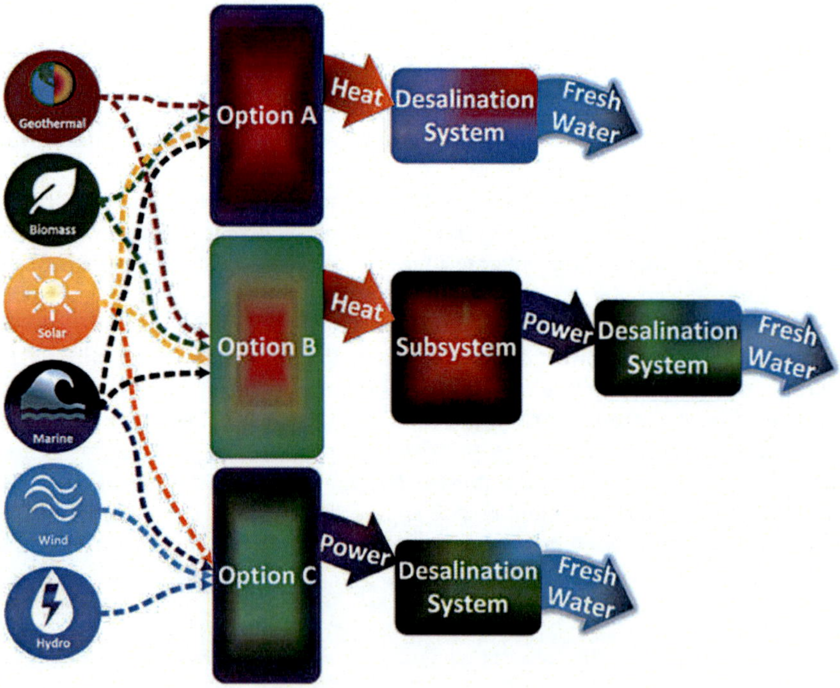

Fig. 6.11 Sustainable pathways of renewable desalination

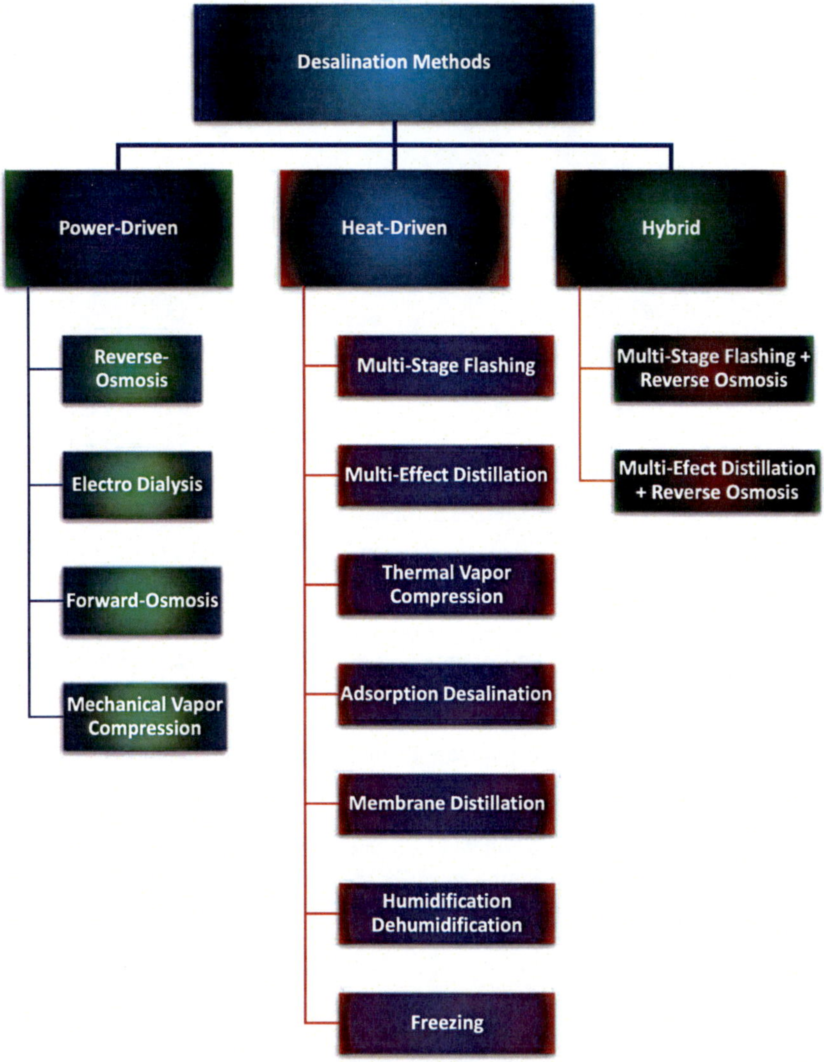

Fig. 6.12 Classification of desalination methods

is required in reverse osmosis processes where pressurized saline water is separated from the salt content via the membrane. In the electrodialysis method, ion exchange membranes are used in order to transfer salt ions under the effect of electric potential difference from one side to another. Typically, multiple electrodialysis cells form an electrodialysis stack in applications. Forward osmosis is another osmosis process; however, pressurization is not required, unlike the reverse osmosis process; therefore, less energy is required compared to the reverse osmosis process. Moreover,

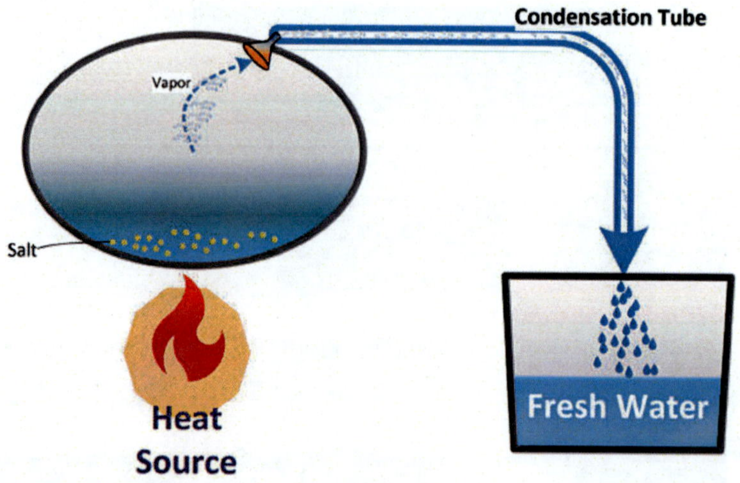

Fig. 6.13 Illustration of the basic distillation desalination

heat can help to facilitate the forward osmosis process, however, power is the main driver. In mechanical vapor compression desalination, compression of vapor increases both pressure and temperature of vapor, which increases the enthalpy; therefore, higher heat content makes the vapor suitable for a distillation process. The base method of heat-driven desalination process is the distillation process, which is not included in the classification since it is not considered as one of the particular desalination systems, but a simple process. It is the fundamental process which used as the main method in the early history of desalination. Figure 6.13 depicts the distillation desalination process, where a heat source evaporates the saline water, and vapor enters the condensation tube in order to become liquid. The vapor that is condensed can be collected in a separate container as fresh water.

Heat-driven desalination processes mainly rely on the distillation of water by evaporating and condensing it to remove the salt content. In a multi-stage flashing process, desalination is carried out with the evaporation of the saline water by dropping the pressure with a flashing process in a series of stages. Multi-effect distillation process also uses multiple stages, unlike the multi-stage flashing, where each stage is heated by steam. In the first stage, the steam used for heating is not collected; however, steam in the following stages is actually the steam generated from the previous process. While the steam is condensing to become liquid freshwater at one stage, rejected heat from the condensing process is used to carry out the flashing at the next stage. Each stage is carried out in lower temperatures and pressures than the previous stage; however, a higher heat transfer surface is required compared to the previous stages. Thermal vapor compression is a similar process to mechanical vapor compression, where the mechanical compressor is replaced by the thermal compression process where lower velocity higher pressure steam can be obtained. In the adsorption desalination method, an adsorbent is used such as silica gel or zeolite, which possess a high affinity to adsorb water. During the evaporation process, continuous collection of steam reduces the pressure, which leads to more evaporation with less thermal energy requirement. The saturated adsorbent is heated after in order to desorb the fresh water from the adsorbent. Membrane distillation is another heat-driven process which is currently at low maturity. A hydrophobic membrane is used which only allows the vapor to pass the membrane in the membrane distillation method. Humidification dehumidification method is one of the simplest desalination methods where seawater is humidified and then dehumidified and collected as fresh water. Freezing is another thermal process, where heat in the saline water should be removed to freeze, usually spraying or low-pressure components are used in order to facilitate the freezing process. In energy systems, integrations can bring additional benefits [5]. Also, there are alternative approaches from researchers where a power- and a heat-driven desalination methods are hybridized for synergetic benefits.

6.3 Power-Driven Desalination Methods

In power-driven desalination systems, pressure plays a critical role in the process. In reverse osmosis and forward osmosis methods, a membrane is used to allow only fresh water to pass. Mechanical vapor compression and reverse osmosis methods pressurize the water, whereas mechanical vapor compression heats the vapor by compressing it; on the other hand, pressurized water improves the movement through membrane in the reverse osmosis method. In an electrodialysis system, seawater circulates between anion and cation-selective membranes. Positive and negative ions pass through the membranes, which desalinate the water.

In order to power the desalination systems by using renewable sources, Fig. 6.14 shows the options. Geothermal, biomass, solar thermal, and ocean thermal energy systems can be used to utilize renewable sources to generate heat. The generated heat from renewables can be used to generate power in Option A. Solar PV, marine, wind, and hydro energy systems can be used to utilize renewable sources to generate power directly in Option B. Power generated from renewable sources can be used in the desalination system for renewable freshwater generation.

For the thermodynamic analysis of the power-driven desalination systems, energy, and exergy inputs are considered as the actual work input to drive the system. The minimum work required to separate the fresh water can be considered the useful output. It can be formulated as the energy or exergy difference between the seawater input and the freshwater output. In the subsections below, each power-driven process will be presented with details.

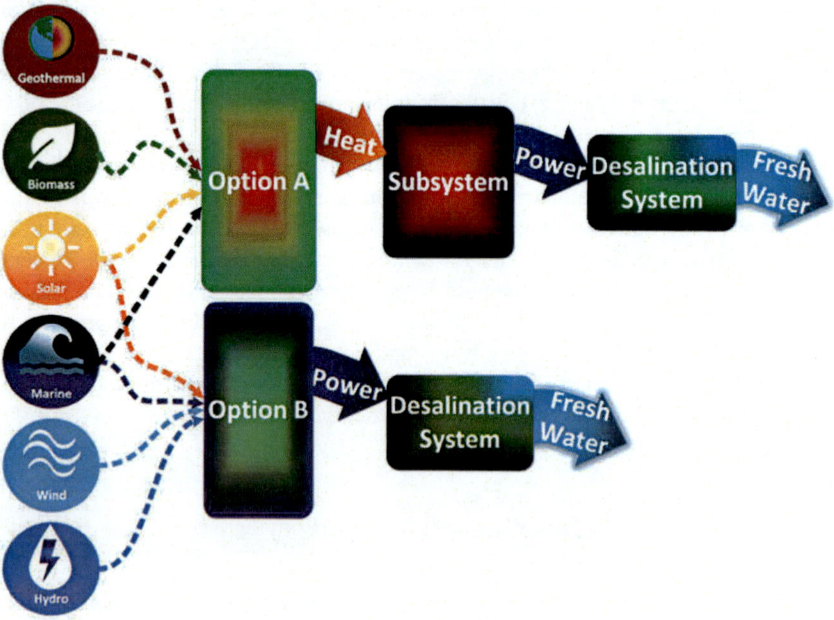

Fig. 6.14 Illustration of the power-driven desalination options

6.3.1 Reverse Osmosis

In reverse osmosis systems, seawater is typically pumped for the pretreatment, where chemical and physical filtrations are used. Later on, a high-pressure pump can be used to pressurize the water. Water enters the reverse osmosis component and circulates. The membrane allows only fresh water to pass to the next section, where fresh water is collected. Brine product, which has a higher salinity than the seawater, leaves from the first section, and it can be reinjected into the sea or can be used for other applications. Figure 6.15 depicts a reverse osmosis system. The reverse osmosis process uses electricity as the energy input, mainly at the low-pressure and high-pressure pumps. Typically, low-pressure pump uses a pressure ratio of around 6.5, and high-pressure pump raises the water pressure to 60 bars.

Reverse osmosis process is one of the most common desalination methods which is commonly used in residential, industrial, commercial, and military applications. Moreover, some of the chemical processes employ reverse osmosis process in order to purify the process water to prevent harmful effects of minerals within the water. Note that hydrogen production is one of the processes, where reverse osmosis is preferred.

In reverse osmosis, desalination process simply separates the salt content from the water. The seawater is fed to the system; once the fresh water is separated as a useful product, brine remains as the by-product from the process. The product water

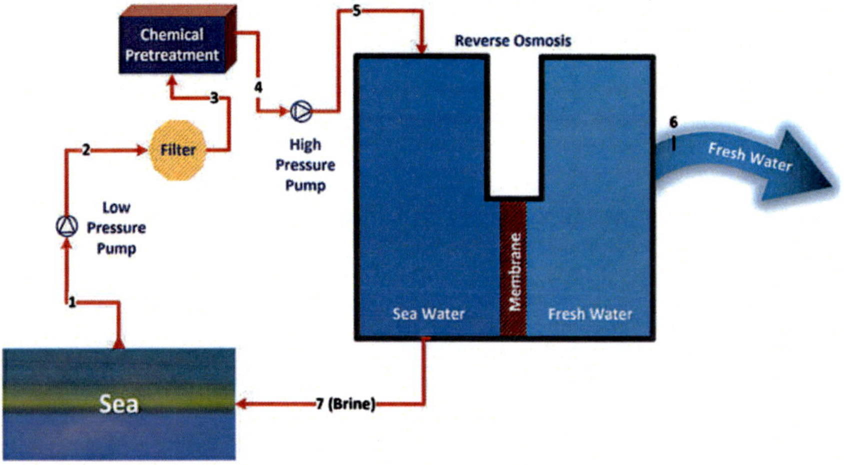

Fig. 6.15 Illustration of the reverse osmosis desalination system

contains the lowest salt concentration, whereas the seawater has the middle salt concentration and the brine has the highest salt concentration.

The recovery ratio can be considered as the ratio of the mass flow rate of the seawater to the fresh water as follows:

$$r_r = \frac{\dot{m}_{fw}}{\dot{m}_{sw}}$$

where fw denotes the fresh water, sw denotes the salt water, and r_r represents the recovery ratio.

When we look at the molar fraction of the salt and water within the seawater content, calculation can be made as follows:

$$x_{s,sw} = \frac{MW_w}{MW_w + MW_s \times \left(\left(\frac{1}{y_{s,sw}}\right) - 1\right)}$$

and

$$x_{w,sw} = 1 - x_{s,sw}$$

where $y_{s,sw}$ denotes the mass fraction of the salt in the seawater.

In the freshwater product, a low concentration of salt remains, which can be calculated as follows:

$$\dot{m}_{s,fw} = y_{s,fw} \times \dot{m}_{fw}$$

For the molar fraction of the salt within the fresh water, calculation can be made as follows:

$$x_{s,fw} = \frac{MW_w}{MW_w + MW_s \times \left(\left(\frac{1}{y_{s,fw}}\right) - 1\right)}$$

and the mass flow rate of the water content within the fresh water can be calculated as follows:

$$\dot{m}_{w,fw} = \dot{m}_{fw} - \dot{m}_{s,fw}$$

Moreover, the mass fraction and molar fraction of the water within the fresh water can be written as follows:

$$y_{w,fw} = 1 - y_{s,fw}$$

and

$$x_{w,fw} = 1 - x_{s,fw}$$

For the by-product, brine, the mass flow rate can be calculated as follows:

$$\dot{m}_b = \dot{m}_{sw} - \dot{m}_{fw}$$

The salinity content of brine is:

$$x_{s,b} = \frac{MW_w}{MW_w + MW_s \times \left(\left(\frac{1}{y_{s,b}}\right) - 1\right)}$$

The total energy and exergy inputs to the reverse osmosis desalination process are in mechanical power, which can also be considered as the total actual work of high-pressure and low-pressure pumps:

$$\dot{W}_{act,total} = \dot{W}_{act,p,LP} + \dot{W}_{act,p,HP}$$

In order to calculate the energy and exergy efficiencies of the reverse osmosis desalination system, the ratio of the minimum work requirement to obtain fresh water via reverse osmosis process to the actual work consumed by reverse osmosis process can be considered. Moreover, it can be reformulated as the ratio of the total enthalpy of the freshwater output to the actual work input for the energy efficiency, and the total exergy of the freshwater output to the actual work input for the exergy efficiency as follows:

$$\eta_{en} = \frac{\dot{m}_{fw} h_{fw}}{\dot{W}_{act,total}}$$

and

$$\eta_{ex} = \frac{\dot{m}_{fw} ex_{fw}}{\dot{W}_{act,total}}$$

The exergy and energy efficiencies of the reverse osmosis process are similar to the other power-driven processes, since the input is the work and the output is the fresh water.

Furthermore, a performance ratio can be considered in order to compare the desalination processes with a more sound parameter. Since the output of the desalination process will be used usually at room temperatures, which brings the final product to a zero energy and zero exergy. Therefore, the energy and exergy content do not possess high importance when we look at the final useful product. Therefore, a performance ratio can be expressed as follows:

$$PR = \frac{\dot{m}_{fw} h_{fg}}{\dot{W}_{act,total}}$$

where h_{fg} is the latent heat of evaporation of water.

Illustrative Example 6.1 An illustrative example is carried out by considering a wind farm with 63 wind turbines. The offshore wind farm system is designed by using the System Advisor Model [6]. Within the software package, General Electric 1.6 MW wind turbines with 100 m of rotor diameter and 100 m of hub height are used in the wind farm.

The proposed system is given in Fig. 6.16 where the wind farm is connected to a reverse osmosis system. The kinetic energy of the wind is converted into electric power via wind farms. The electric power is supplied to the reverse osmosis system. The proposed reverse osmosis system has a capacity of 500,000 kilo-tonnes of daily freshwater generation.

The total power generation capacity of the wind farm is designed as 100.8 MW at peak conditions. The proposed wind turbine can generate peak power at wind speeds higher than 11m/s. Lower wind speeds than 3.5 m/s are not suitable for power generation, as well as freshwater generation in this case. In order to protect the wind turbines, 25 m/s and higher wind speeds are not exploited by the system, which is also called as the cutoff speed.

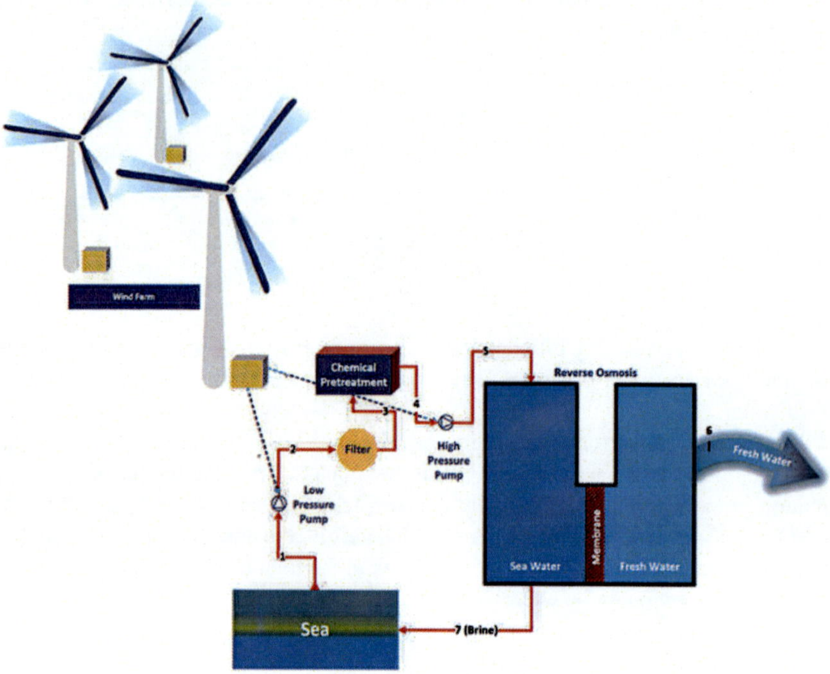

Fig. 6.16 Layout of the wind-based reverse osmosis freshwater generation system

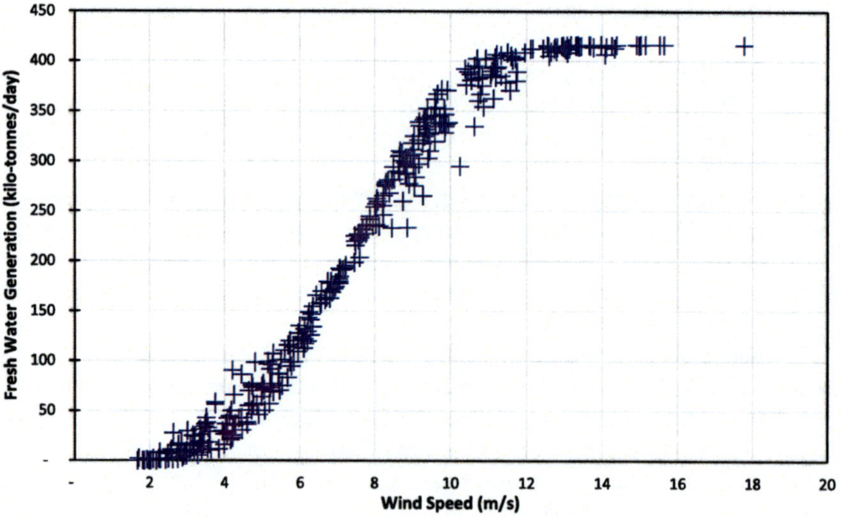

Fig. 6.17 Freshwater generation variations at different wind speeds

Simulations are carried out for California in the United States. As shown in Fig. 6.17, different wind speeds can lead to different daily freshwater generations. In a typical year, 75774 kilo-tonnes of fresh water is generated from the system by using 367.5 GWh of electricity from the wind farm.

6.3.2 *Electrodialysis*

Electrodialysis was invented at the end of the nineteenth century and became implemented on large scale after the 1960s. It is similar to the reverse osmosis in terms of membrane usage. However, the motion of the molecules is promoted by the electric field created by using the external electric power source. Anion-selective and cation-selective membranes pass negative and positive ions, respectively, from the seawater or another feed saline water. Therefore, sodium ions and chloride ions move through the membranes and fresh water remains as shown in Fig. 6.18. Electrodialysis process can generate fresh water at high purities. Also, it is suitable for the intermittent characteristics of renewables. However, compared to the other desalination processes, the cost is higher.

In the electrodialysis process, the electric field facilitates the movement of negative and positive ions through the cation- and anion-selective membranes. The migrative flux of the ions can be modeled mathematically as follows:

$$J_{\mathrm{mig}} = \left(\mathrm{t}_{\mathrm{CSM}}^{\mathrm{counter}} - \left(1 - \mathrm{t}_{\mathrm{ASM}}^{\mathrm{counter}}\right)\right) \frac{\mathrm{i}}{\mathrm{F}}$$

where i denotes the current density, F represents the Faraday constant, and t denotes the transport number.

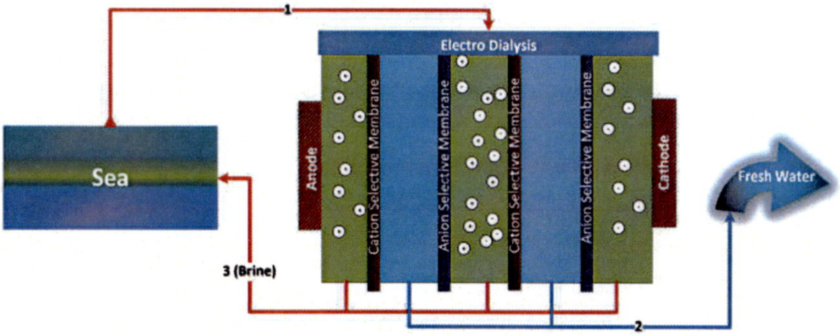

Fig. 6.18 Illustration of the electrodialysis desalination system

The perm-selectivity of cation or anion-selective membranes can be calculated as follows:

$$\alpha_{SM} = \frac{t_{SM}^{counter} - t_{solution}^{counter}}{t_{solution}^{co-ion}}$$

which is the ratio of the difference between counter-ion transport numbers in the selective membrane and solution to the counter-ion transport number in the solution.

The global molar flux is the algebraic sum of the diffusion flux and migrative flux of salt ions.

When we look at the molar fraction of the salt and water within the seawater content, calculation can be made as follows:

$$x_{s,sw} = \frac{MW_w}{MW_w + MW_s \times \left(\left(\frac{1}{y_{s,sw}}\right) - 1\right)}$$

and

$$x_{w,sw} = 1 - x_{s,sw}$$

where $y_{s,sw}$ denotes the mass fraction of the salt in the seawater.

In the freshwater product, a low concentration of salt remains, which can be calculated as follows:

$$\dot{m}_{s,fw} = y_{s,fw} \times \dot{m}_{fw}$$

For the molar fraction of the salt within the fresh water, calculation can be made as follows:

$$x_{s,fw} = \frac{MW_w}{MW_w + MW_s \times \left(\left(\frac{1}{y_{s,fw}}\right) - 1\right)}$$

and the mass flow rate of the water content within the fresh water can be calculated as follows:

$$\dot{m}_{w,fw} = \dot{m}_{fw} - \dot{m}_{s,fw}$$

In addition, the mass fraction and molar fraction of the water within the fresh water can be written as follows:

$$y_{w,fw} = 1 - y_{s,fw}$$

and

$$x_{w,fw} = 1 - x_{s,fw}$$

For the by-product, which is brine, the mass flow rate can be calculated as follows:

$$\dot{m}_b = \dot{m}_{sw} - \dot{m}_{fw}$$

The salinity content of brine is defined as follows:

$$x_{s,b} = \frac{MW_w}{MW_w + MW_s \times \left(\left(\frac{1}{y_{s,b}}\right) - 1\right)}$$

The energy and exergy efficiencies of the electric dialysis can be calculated by considering the electric power provided to the system as the input. The energy efficiency of the electrodialysis process can be expressed as the total enthalpy of the freshwater output to the electric power input of the electrodialysis, and the exergy efficiency of the electrodialysis process is the ratio of the total exergy of the freshwater output to the electric power input of the electrodialysis as follows:

$$\eta_{en} = \frac{\dot{m}_{fw} h_{fw}}{P_{ED}}$$

and

$$\eta_{ex} = \frac{\dot{m}_{fw} ex_{fw}}{P_{ED}}$$

where P_{ED} denotes the electric power input to the electrodialysis process.

The performance ratio can also be used for electrodialysis process due to the low energy content of the useful output from the system. The energy and exergy content of the useful output is replaced with the latent heat of evaporation as follows:

$$PR = \frac{\dot{m}_{fw} h_{fg}}{P_{ED}}$$

where h_{fg} is the latent heat of evaporation of water.

Case Study 6.1 A case study is investigated based on [7]. A wastewater treatment plan was designed with a capacity of 1000 kg of hydrogen and 222 m^3 of fresh water. A photoelectrochemical reactor was employed where heliostat-concentrator combination was used with a 10:1 solar intensity ratio. Photovoltaic (PV) modules are also included in the PEC reactor. The electrolyte reservoir was pressurized to 20.7 bar

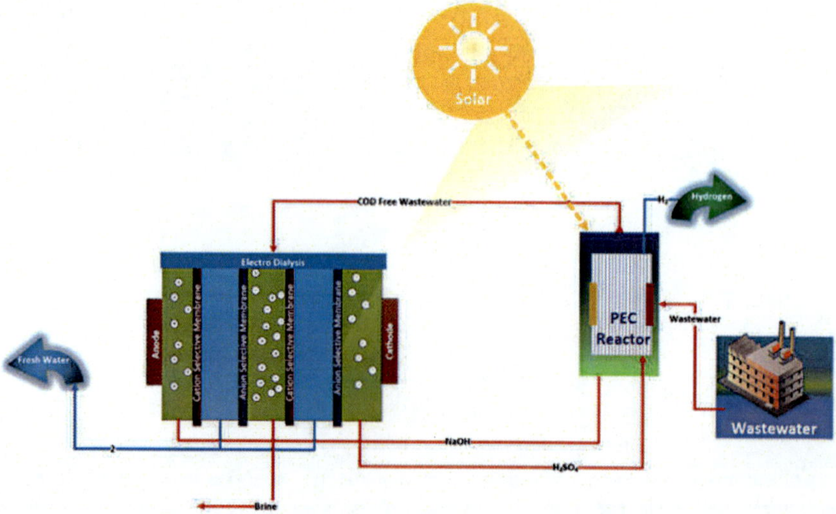

Fig. 6.19 Layout of the solar-based freshwater and hydrogen generation system

Table 6.1 Balance sheet of 81030 tonnes of Na_2SO_4 solution via bipolar membrane electrodialysis

Process	Costs ($)
Treatment cost of Na_2SO_4 solution	444,832.80
Revenue from H_2SO_4 and NaOH recovery	625,105.00
Net revenue	180,272.00

inside the reactor. The idea behind the PEC reactor was to process wastewater for the next process and to generate hydrogen. Textile industry was considered as the wastewater source. The process in PEC reactor was the decolorization and chemical oxygen demand reduction along with hydrogen production. An offset parabolic cylinder array was used to focus the sun's rays on a linear PEC receiver.

In their proposed system which is shown in Fig. 6.19, each concentrator array was 6 to 3 meters which has a 15% hydrogen generation efficiency, where 1982 of these arrays were connected to form a 35654 m^2 of total collector area. Their design parameters are based on the experimental study [8]. At the desalination part, they proposed a bipolar membrane electrodialysis component, which was found synergetic to use with their proposed PEC, since the outputs of bipolar membrane electrodialysis are used in PEC.

They carried out a techno-economic assessment for their system. For their specific wastewater, Na_2SO_4 should be treated specifically. Apart from the freshwater production, their bipolar membrane electrodialysis design recovers H_2SO_4 and NaOH. For their techno-economic, they considered the treatment of Na_2SO_4 solution in order to recover H_2SO_4 and NaOH. The balance sheet of the process is given in Table 6.1. In the case study, 81030 tonnes of Na_2SO_4 solution were used for the textile dye process, annually, which needs to be treated. According to their calculations, the operating costs of the bipolar membrane electrodialysis system are \$444,832.8, annually. The expected benefit from the recovery of H_2SO_4 and NaOH is \$625,105.0, annually. In balance, there will be a net revenue of \$180,272.0, annually. Moreover, the desalination process can bring additional benefits as \$81,030.0, annually, which is based on a rate of \$1.0/m^3 for fresh water.

6.3.3 *Forward Osmosis*

Forward osmosis is another power-driven desalination technology which uses a membrane to separate salt from the feed water. Forward osmosis is a process similar to the reverse osmosis. The movement of the fresh water occurs due to the pressure provided by the pumps, on the other hand, osmotic pressure is used to move fresh water in the reverse osmosis process. The membrane design is very critical in forward osmosis systems. There are cellulosic, thin-film composite, and chemically reformed membrane types used for the forward osmosis process. The circulation of the liquid both from sea to the forward osmosis system and within the forward osmosis system consumes energy. Especially in large-scale applications, forward osmosis faces challenges such as membrane fouling, and reverse solute diffusion. In order to overcome these challenges, membrane design and concentration ratios are important. Figure 6.20 shows the forward osmosis system, where seawater is

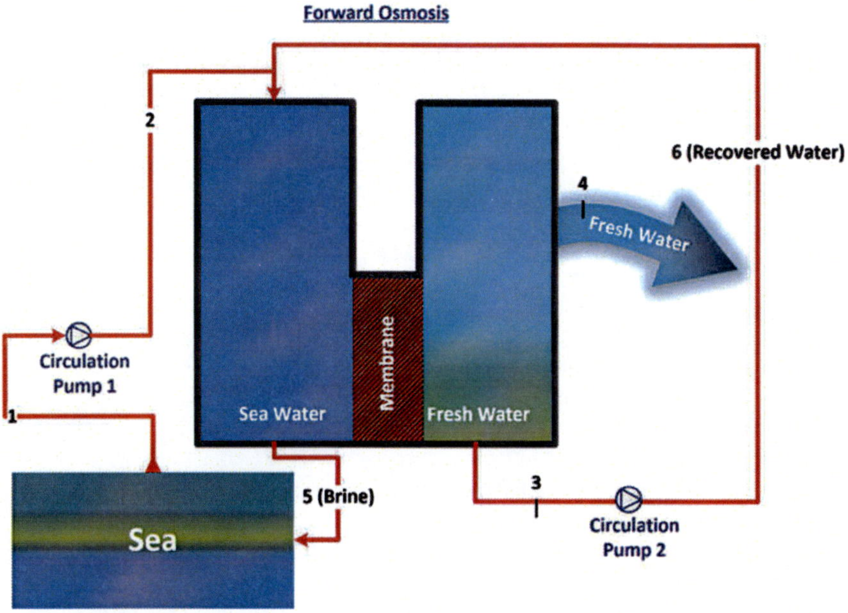

Fig. 6.20 Illustration of the forward osmosis desalination system

circulated into the feed water (seawater) section, fresh water passed to the second section (also called the draw section), where fresh water is separated, and the remaining water is recovered to the first section. Brine solution can be sent back to the sea or it can be used for other purposes.

In the forward osmosis unit, the desalination process starts by pumping the seawater into the system for circulation. The membrane allows the water to pass to the second section where brine with high saline content is left as the remaining by-product. In the second section, fresh water circulates within the draw solution where a major portion of fresh water is taken, and the leftover is recirculated to the first section.

When we look at the recovery ratio, it can be calculated as:

$$r_r = \frac{\dot{m}_{fw}}{\dot{m}_{sw}}$$

The molar ratio of the salt and water content within the seawater is:

$$x_{s,sw} = \frac{MW_w}{MW_w + MW_s \times \left(\left(\frac{1}{y_{s,sw}}\right) - 1\right)}$$

and

$$x_{w,sw} = 1 - x_{s,sw}$$

where $y_{s,sw}$ denotes the mass fraction of the salt in the seawater.

The mass flow rate of the salt content within the fresh water can be expressed as:

$$\dot{m}_{s,fw} = y_{s,fw} \times \dot{m}_{fw}$$

The molar fraction of the salt within the fresh water is:

$$x_{s,fw} = \frac{MW_w}{MW_w + MW_s \times \left(\left(\frac{1}{y_{s,fw}}\right) - 1\right)}$$

and the mass flow rate of the water content within the fresh water is:

$$\dot{m}_{w,fw} = \dot{m}_{fw} - \dot{m}_{s,fw}$$

Moreover, the mass and molar fractions of the water within the generated fresh water can be written as follows:

$$y_{w,fw} = 1 - y_{s,fw}$$

and

$$x_{w,fw} = 1 - x_{s,fw}$$

For the by-product, brine, the mass flow rate can be calculated as follows:

$$\dot{m}_b = \dot{m}_{sw} - \dot{m}_{fw}$$

The molar fraction of the salt in brine can be expressed as:

$$x_{s,b} = \frac{MW_w}{MW_w + MW_s \times \left(\left(\frac{1}{y_{s,b}} \right) - 1 \right)}$$

The energy and exergy efficiencies can be calculated by considering the energy and exergy outputs of the desalination process as follows:

$$\eta_{en} = \frac{\dot{m}_{fw} h_{fw}}{\dot{W}_{act,total}}$$

and

$$\eta_{ex} = \frac{\dot{m}_{fw} ex_{fw}}{\dot{W}_{act,total}}$$

As explained earlier, the energy and exergy contents of the final useful products are significantly low. Therefore, a performance ratio can be calculated by replacing the energy and exergy contents of final products with the latent heat of evaporation as follows:

$$PR = \frac{\dot{m}_{fw} h_{fg}}{\dot{W}_{act,total}}$$

where the total energy inputs the total energy input is the total work from the first and second circulation pumps as follows:

$$\dot{W}_{act,total} = \dot{W}_{act,CP1} + \dot{W}_{act,CP2}$$

and h_{fg} is the latent heat of evaporation.

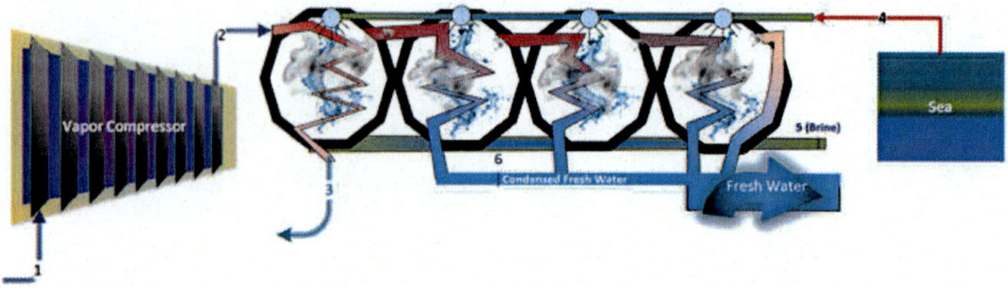

Fig. 6.21 Illustration of the mechanical vapor compression desalination system

6.3.4 Mechanical Vapor Compression

The mechanical vapor compression process use mechanical power to heat up the vapor by compressing it. The heat from the compressed vapor is utilized to drive the desalination stages; from this aspect it aparts from other power-driven desalination methods. Since the power is actually converted into heat by compressing the vapor, which makes the vapor molecules move more rapidly that leads to increasing temperatures. The heat from the compressed vapor is extracted from the first stage in order to evaporate the sea water which is sprayed on heat exchanger tubes. According to the design, the mechanical vapor compression desalination can be carried out either in a single stage or multi-stage. If the process is designed as a multi-stage system, the remaining sea water from the first stage can be sprayed again in the following stages. The next stages, evaporation process is carried out typically at lower pressures to facilitate the evaporation. A multi-stage mechanical vapor compression desalination system is depicted in Fig. 6.21.

The water is separated from the salt content by using the heat energy from the compressed water. Although the evaporation is carried out by using the heat of vapor, the overall system uses mechanical energy or work as the input. The recovery ratio can be calculated as follows:

$$r_r = \frac{\dot{m}_{fw}}{\dot{m}_{sw}}$$

The molar fractions of the salt and water within the feed water or seawater can be expressed as follows:

$$x_{s,sw} = \frac{MW_w}{MW_w + MW_s \times \left(\left(\frac{1}{y_{s,sw}}\right) - 1\right)}$$

and

$$x_{w,sw} = 1 - x_{s,sw}$$

The salt mass within the fresh water can be expressed as:

$$\dot{m}_{s,fw} = y_{s,fw} \times \dot{m}_{fw}$$

and the molar fraction of the salt within the fresh water is:

$$x_{s,fw} = \frac{MW_w}{MW_w + MW_s \times \left(\left(\frac{1}{y_{s,fw}}\right) - 1\right)}$$

Moreover, the water content as well as mass and molar fractions of the water can be expressed as:

$$\dot{m}_{w,fw} = \dot{m}_{fw} - \dot{m}_{s,fw},$$

$$y_{w,fw} = 1 - y_{s,fw}$$

and

$$x_{w,fw} = 1 - x_{s,fw}$$

When we look at the performances of the vapor compression system, the actual compressor work can be considered as the total energy and exergy inputs. The ratio of the energy and exergy of the useful output to the actual work of the compressor gives the energy and exergy efficiencies as:

$$\eta_{en} = \frac{\dot{m}_{fw} h_{fw}}{\dot{W}_{act,c}}$$

and

$$\eta_{ex} = \frac{\dot{m}_{fw} ex_{fw}}{\dot{W}_{act,c}}$$

The performance ratio of the mechanical vapor compression desalination process can be calculated as:

$$PR = \frac{\dot{m}_{fw} h_{fg}}{\dot{W}_{act,c}}$$

which is the total latent heat of evaporation for the mass of fresh water generated from the system to the total work input.

6.4 Heat-Driven Desalination Methods

In the heat-driven desalination, heat source is mainly used in order to evaporate the seawater by using different techniques and conditions, later on to condense the vapor to obtain the fresh water. For renewable heat-driven desalination, renewable sources are utilized, mainly with geothermal, biomass, solar thermal, and ocean thermal

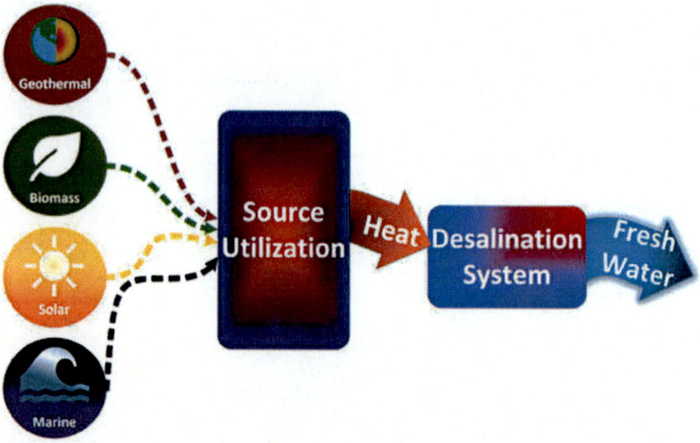

Fig. 6.22 Illustration of the heat-driven desalination options

energy systems, in order to generate heat. The heat generated from renewables is the driver of the desalination process. Figure 6.22 shows what are the renewable energy source options and the pathway to generate fresh water. Geothermal and solar energy sources are already in the heat form in the nature. Geothermal systems can be utilized by extracting the geothermal water, if suitable underground water can be used directly as a primary loop geothermal system to run the processes, or heat can be extracted from the geothermal water indirectly by using a secondary loop geothermal system. For solar heat-driven desalination systems, heat energy from solar radiation can be collected via concentrating or non-concentrating solar collectors and transferred into the desalination process. For biomass-driven systems, liquid or solid biomass can be burned and thermal energy can be transferred to the desalination process. In marine energy systems, low-temperature heat can be used to remove the heat from the desalination system for freezing desalination. Or in some cases, when the sea or ocean temperature is higher than the ambient temperature, it can be used as the heat source for the desalination process, such as in adsorption desalination process.

There are mainly two types of heat-driven desalination methods due to the membrane use. The majority of the heat-driven desalination methods do not require a membrane, since the distillation is the core process to separate water from the salt; therefore, an additional membrane requirement is not necessary.

Multi-stage flashing, multi-effect distillation, thermal vapor compression, adsorption desalination, humidification dehumidification, and freezing desalination methods do not require a membrane to separate salt and water. On the other hand, membrane distillation method requires a selective membrane.

6.4.1 Multi-Stage Flashing

Multi-stage flashing system is the distillation desalination system where seawater is evaporated and then condensed through multiple stages. The number of stages can be

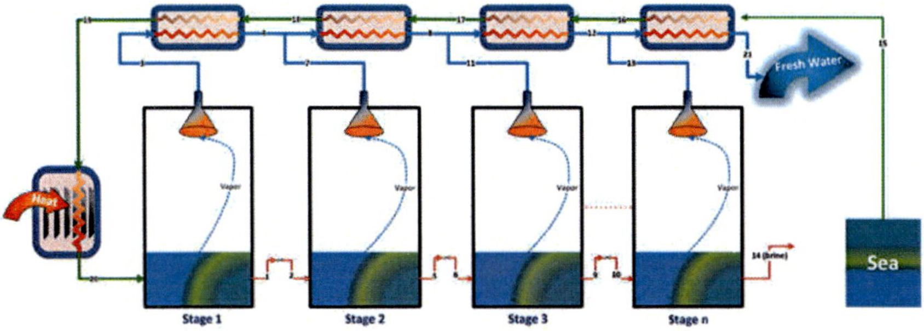

Fig. 6.23 Illustration of the multi-stage flashing desalination system

as high as 30 stages. Figure 6.23 shows a multi-stage flashing system which can be considered effective solution. The working principle of the multi-stage desalination system starts with circulating seawater and heating it before introducing to the stages. There are pre-heaters placed before the main heater, where after each stage, heat energy from the vapor is transferred into the seawater. At the main heater, seawater is getting heated to its maximum temperature between 60 and 120°C depending on the design. Then the heated seawater enters the first stage, and a portion of it evaporates. The remaining brine moves toward the second stage after the flashing process. Flashing is the key in order to proceed to the process, where pressure is decreased in order to improve the evaporation process. The following stages have lower pressures than the previous stages. The vapor enters the pre-heaters where the vapor is condensed by transferring its heat to the seawater.

In the multi-stage flashing system, brine is used as the seawater input to the stages from 2 to n. For the first stage, mass flow rate of the brine can be calculated as follows:

$$\dot{m}_{sw,1} = \dot{m}_{b,1} + \dot{m}_{fw,1}$$

or

$$\dot{m}_{20} = \dot{m}_1 + \dot{m}_3$$

The molar ratio of salinity of brine can be expressed as:

$$x_{s,b,1} = \frac{MW_w}{MW_w + MW_s \times \left(\left(\frac{1}{y_{s,b,1}}\right) - 1\right)}$$

When we look at the stages from 2 to n, the mass balance equations can be expressed as follows:

$$\dot{m}_{b,i-1} = \dot{m}_{b,i} + \dot{m}_{fw,i}$$

where i specifies the stage number. It can be reformulated for the second stage according to Fig. 6.21 as:

$$\dot{m}_2 = \dot{m}_5 + \dot{m}_7$$

The salt from the seawater entering the stages from 2 to n can be calculated as follows:

$$\dot{m}_{s,sw,i} = y_{s,sw} \times \dot{m}_{sw,i}$$

where $y_{s,sw}$ denotes the mass fraction of the salt in the seawater.

The molar fraction of the salt within the seawater can be expressed as:

$$x_{s,sw} = \frac{MW_w}{MW_w + MW_s \times \left(\left(\frac{1}{y_{s,sw}} \right) - 1 \right)}$$

and

$$x_{w,sw} = 1 - x_{s,sw}$$

The recovery ratio of the multi-stage flashing process can be written as:

$$r_r = \frac{\dot{m}_{fw}}{\dot{m}_{sw}}$$

It is really important to study the performances of the multi-stage flashing desalination system, and the total heat energy and its exergy version can be considered as the inputs. The ratio of the energy and exergy of the water to the inputs gives the energy and exergy efficiencies as:

$$\eta_{en} = \frac{\dot{m}_{21} h_{21}}{\dot{Q}_{MSF}}$$

and

$$\eta_{ex} = \frac{\dot{m}_{21} ex_{21}}{\dot{E}x_{Q_{MSF}}}$$

where state points are written according to Fig. 6.19.

The performance ratio of the multi-stage flashing desalination process can be calculated as:

$$PR = \frac{\dot{m}_{21} h_{fg}}{\dot{Q}_{MSF}}$$

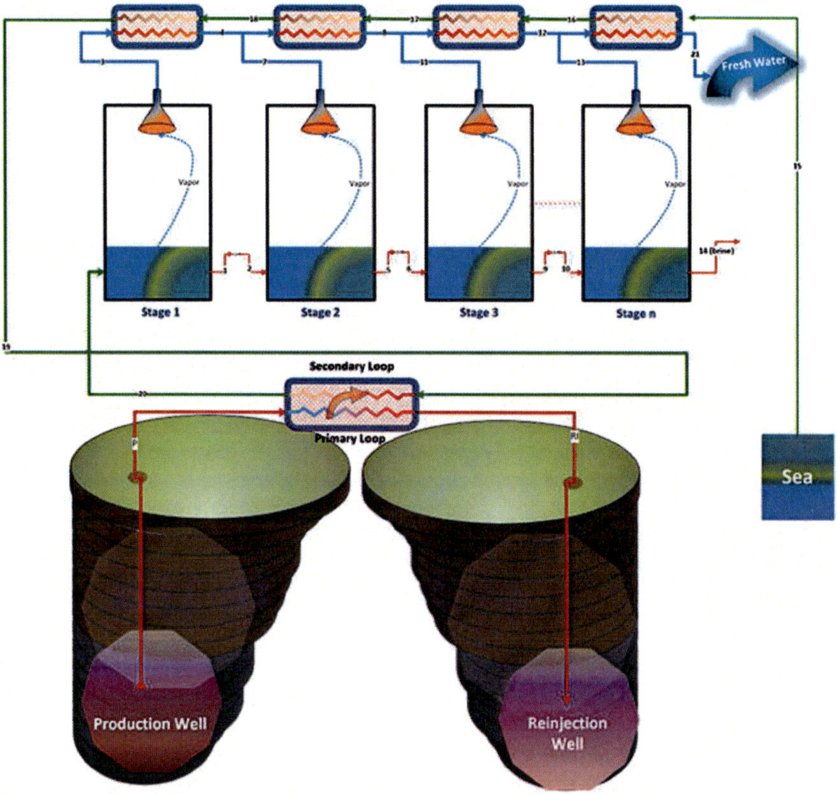

Fig. 6.24 A geothermal-based freshwater generation system via multi-stage flashing

which is the total latent heat of evaporation for the mass of fresh water generated from the system to the total enthalpy input according to the state points in Fig. 6.19.

Illustrative Example 6.2 An illustrative example is carried out for a geothermal source in Kizildere, Denizli, Turkey. A reservoir with 260°C source temperature is considered in order to carry out the illustrative example. The proposed system is illustrated in Fig. 6.24. A multi-stage flashing desalination system is integrated to a geothermal system. A secondary loop is used to exploit to heat. The seawater heated by the primary loop of the geothermal water is sent to the first stage of the multi-stage flashing desalination system.

According to the calculations, 172.63 MW of heat is transferred from the primary loop to the secondary loop. In a steady state condition, 2.48 tonnes of fresh water is generated hourly from the multi-stage flashing desalination system.

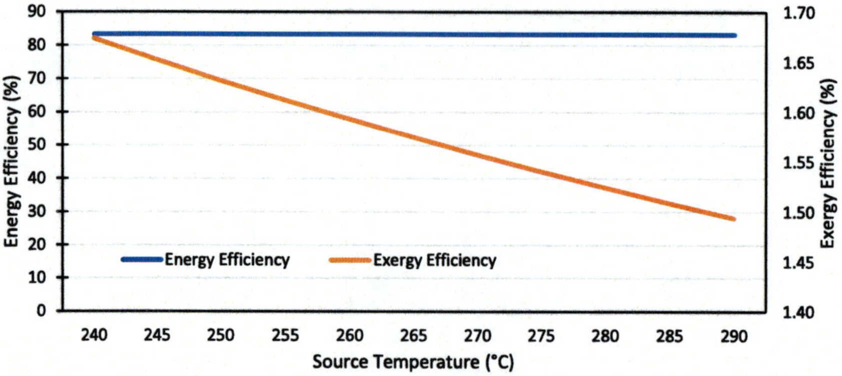

Fig. 6.25 Overall energy and exergy efficiencies of the geothermal-based power generation system in steady state conditions

For the energy and exergy efficiencies of the geothermal-based power generation system with reinjection, equations can be used as follows:

$$\eta_{en,ov} = \frac{\dot{m}_{21}h_{21}}{\dot{m}_P h_P - \dot{m}_{RI}h_{RI}}$$

$$\eta_{ex,ov} = \frac{\dot{m}_{21}ex_{21}}{\dot{m}_P ex_P - \dot{m}_{RI}ex_{RI}}$$

The overall energy and exergy efficiencies of the geothermal-based power generation system are found to be 83.5% and 1.7%. When the source temperature is changed, exergy efficiency of the overall system also changes. However, energy efficiency of the overall system does not change with the change of source temperature. Figure 6.25 depicts the energy and exergy efficiency changes with different source temperatures.

6.4.2 *Multi-Effect Distillation*

Multi-effect distillation desalination method is one of the most popular methods, where stages or effects are used in order to improve the utilization of the heat source. Figure 6.26 depicts the multi-effect distillation desalination system. At the first stage, a low-pressure and low-temperature vapor, which is around 70°C and 40 kPa, enters and transfers its heat to the feed water or seawater. Simultaneously, seawater is usually sprayed to the first stage on the heat exchanger coils. The external vapor enters the first stage and leaves the system, however, the generated vapor from the seawater is moves toward the second stage, remaining seawater or brine also leaves from the system. The following stages have lower temperatures and pressures than the previous stages. Stages between 2 to n use the vapor generated from the previous

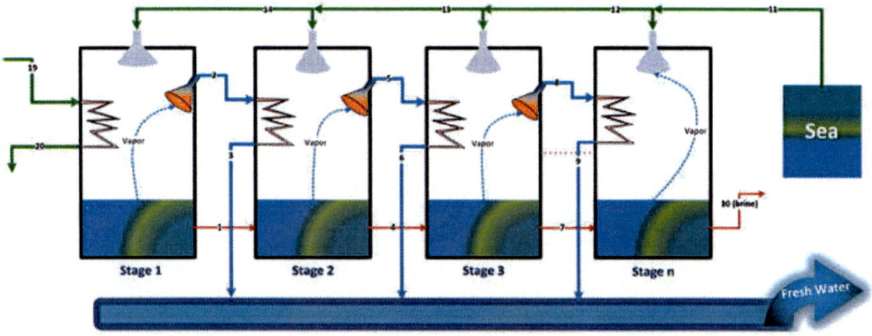

Fig. 6.26 Illustration of the multi-effect distillation desalination system

stage as the heater. Seawater is sprayed again on those stages also in order to generate vapor. The condensed vapor is collected as the fresh water.

In order to calculate the mass flow rates to the system, stage-by-stage approach would be more practical. For the first stage:

$$\dot{m}_{b,1} = \dot{m}_{sw,1} - \dot{m}_{fw,1}$$

and

$$\dot{m}_{vapor,in} = \dot{m}_{vapor,out}; \dot{m}_{19} = \dot{m}_{20}$$

The molar ratio of salinity of brine can be expressed as:

$$x_{s,b,1} = \frac{MW_w}{MW_w + MW_s \times \left(\left(\frac{1}{y_{s,b,1}}\right) - 1\right)}$$

When we look at the stages from 2 to n, the mass balance equations can be expressed as follows:

$$\dot{m}_{b,i-1} + \dot{m}_{b,i} = \dot{m}_{sw,i} - \dot{m}_{fw,i}$$

where i specifies the stage number.

The salt from the seawater entering the stages from 2 to n can be calculated as follows:

$$\dot{m}_{s,sw,i} = y_{s,sw} \times \dot{m}_{sw,i}$$

where $y_{s,sw}$ denotes the mass fraction of the salt in the seawater.

In order to find the total salt within the brine content for stages between 2 and n, the calculation can be made as:

$$\dot{m}_{s,b,i} = \dot{m}_{s,sw,i} + \dot{m}_{s,b,i-1}$$

which can be reformulated as:

$$\dot{m}_{s,b,i} = y_{s,sw} \times \dot{m}_{sw,i} + \dot{m}_{b,i-1}$$

The molar fraction of the salt within the seawater can be expressed as:

$$x_{s,sw} = \frac{MW_w}{MW_w + MW_s \times \left(\left(\frac{1}{y_{s,sw}} \right) - 1 \right)}$$

and

$$x_{w,sw} = 1 - x_{s,sw}$$

The recovery ratio of the multi-effect distillation process can be written as:

$$r_r = \frac{\dot{m}_{fw}}{\dot{m}_{sw}}$$

When we look at the performances of the multi-effect desalination system, the total enthalpy and total exergy can be considered as the inputs. The ratio of the energy and exergy of the water to the total inputs gives the energy and exergy efficiencies as:

$$\eta_{en} = \frac{\dot{m}_{fw} h_{fw}}{\dot{m}_{19} h_{19} - \dot{m}_{20} h_{20}}$$

and

$$\eta_{ex} = \frac{\dot{m}_{fw} ex_{fw}}{\dot{m}_{19} ex_{19} - \dot{m}_{20} ex_{20}}$$

where state points are written according to Fig. 6.26.

The performance ratio of the multi-effect distillation desalination process can be calculated as:

$$PR = \frac{\dot{m}_{fw} h_{fg}}{\dot{m}_{19} h_{19} - \dot{m}_{20} h_{20}}$$

which is the total latent heat of evaporation for the mass of fresh water generated from the system to the total enthalpy input according to the state points in Fig. 6.26.

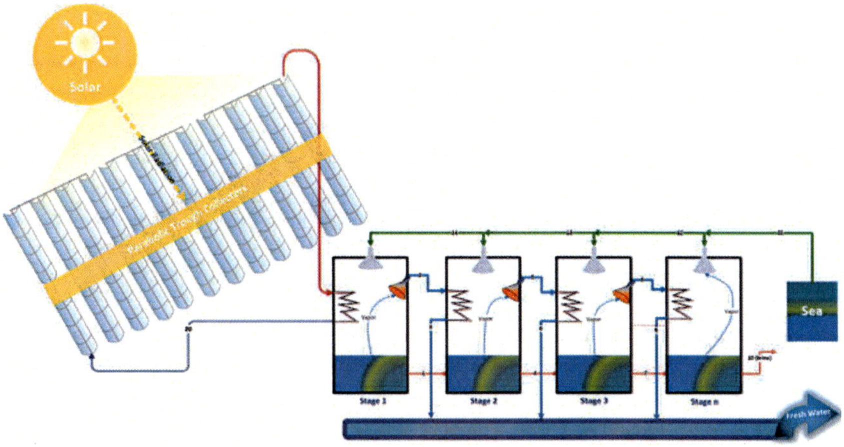

Fig. 6.27 Illustration of the multi-effect distillation integrated parabolic trough concentrating solar plant

Illustrative Example 6.3 An illustrative example is carried out for a parabolic trough type of solar collector with a multi-effect distillation desalination system as shown in Fig. 6.27. Meteorological data is used for Daggett, California, United States, where 2.8 MWh of annual irradiation is reached annual for each m^2 of area. A parabolic trough collector system is simulated with a 1208730 m^2 of total aperture reflective area on 1081 acres of land. SkyFuel Sky Trough collectors and Schott PTR 80 receivers are selected for further simulations. A multi-effect distillation desalination system is designed along with the parabolic trough concentrating solar plant which has a 134.8 kilo-tonnes of daily water production capacity.

According to the simulations, 3384 GWh of solar radiance via 1208730 m^2 of total aperture reflective and 3210 GWh of exergy version of solar radiance are collected, in a typical year. Parabolic trough–type concentrating solar thermal system generated 1482.8 GWh of process heat in a typical year. Maximum process heat temperature is found as 436.86°C.

In a typical meteorological year, between 0 and 134 kilo-tonnes of fresh water in each day, is generated by concentrating solar heat-driven multi-effect distillation desalination system, which makes 28734.4 kilo-tonnes of fresh water per year, as shown in Fig. 6.28.

6.4.3 *Thermal Vapor Compression*

The thermal vapor compression process can employ multi-stage flashing or multi-effect desalination in the distillation section. On the other hand, the vapor that is used as the heat content needs to be generated by using a thermal vapor compressor.

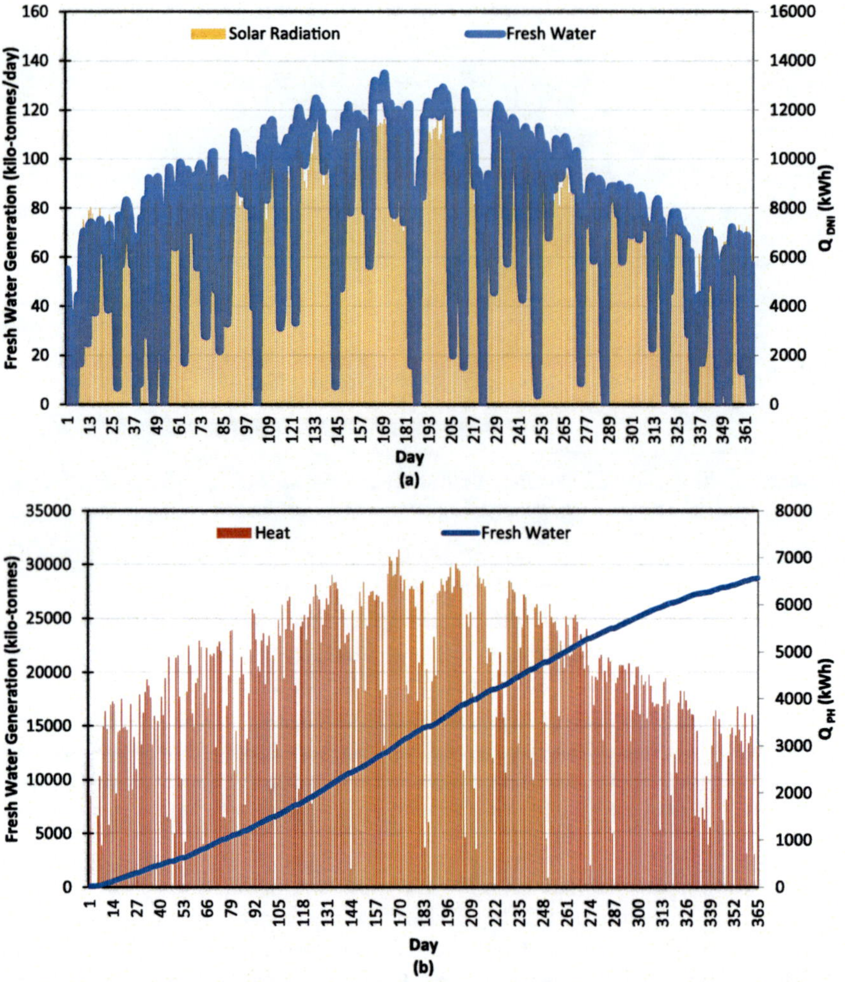

Fig. 6.28 (a) Daily freshwater generation and the available solar radiation variations, (b) cumulative freshwater generation and process heat variations for each day in a typical year

Thermal vapor compressor is the similar to the mechanical vapor compressor; however, the thermal vapor compressor is powered by the thermal energy. Figure 6.29 shows a thermal vapor compression desalination system, where a multi-effect desalination system used in the distillation section.

The calculations are similar to the multi-effect desalination system. In the thermal vapor compression desalination system, brine and seawater inputs are considered to the stages from 2 to n. However, only seawater enters the first stage:

$$\dot{m}_{b,1} = \dot{m}_{sw,1} - \dot{m}_{fw,1}$$

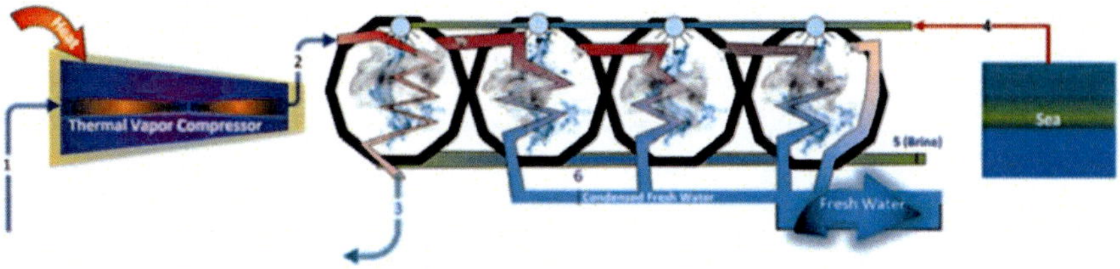

Fig. 6.29 Illustration of the thermal vapor compression desalination system

and

$$\dot{m}_{vapor,in} = \dot{m}_{vapor,out}; \dot{m}_2 = \dot{m}_3$$

The molar ratio of salinity of brine can be expressed for the first stage as:

$$x_{s,b,1} = \frac{MW_w}{MW_w + MW_s \times \left(\left(\frac{1}{y_{s,b,1}}\right) - 1\right)}$$

Considering the stages from 2 to n, the mass balance equation can be expressed as follows:

$$\dot{m}_{b,i-1} + \dot{m}_{b,i} = \dot{m}_{sw,i} - \dot{m}_{fw,i}$$

where i specifies the stage number.

The salt from the seawater entering the stages from 2 to n can be calculated as follows:

$$\dot{m}_{s,sw,i} = y_{s,sw} \times \dot{m}_{sw,i}$$

where $y_{s,sw}$ denotes the mass fraction of the salt in the seawater.

In order to find the total salt within the brine content for stages between 2 and n, the calculation can be made as:

$$\dot{m}_{s,b,i} = \dot{m}_{s,sw,i} + \dot{m}_{s,b,i-1}$$

which can be reformulated as:

$$\dot{m}_{s,b,i} = y_{s,sw} \times \dot{m}_{sw,i} + \dot{m}_{b,i-1}$$

The molar fraction of the salt within the seawater can be expressed as:

$$x_{s,sw} = \frac{MW_w}{MW_w + MW_s \times \left(\left(\frac{1}{y_{s,sw}}\right) - 1\right)}$$

and

$$x_{w,sw} = 1 - x_{s,sw}$$

The recovery ratio of the thermal vapor compression desalination process can be written as:

$$r_r = \frac{\dot{m}_{fw}}{\dot{m}_{sw}}$$

Considering the performances (through efficiencies) of the thermal vapor compression desalination system, the total heat energy and its exergy version can be considered as the inputs. The ratio of the energy and exergy of the water to the inputs gives the energy and exergy efficiencies as:

$$\eta_{en} = \frac{\dot{m}_6 h_6}{\dot{Q}_{TVC}}$$

and

$$\eta_{ex} = \frac{\dot{m}_6 ex_6}{\dot{E}x_{Q_{TVC}}}$$

where state points are written according to Fig. 6.29.

The performance ratio of the thermal vapor compression desalination process can be calculated as:

$$PR = \frac{\dot{m}_6 h_{fg}}{\dot{Q}_{TVC}}$$

which is the total latent heat of evaporation for the mass of fresh water generated from the system to the total enthalpy input according to the state points in Fig. 6.29.

6.4.4 Adsorption Desalination

The adsorption desalination system desalinates water through evaporation by using assistive evaporation with adsorption. The cooling effect of adsorbers was introduced in Japan when the oil crisis peaked. Later on, this technology is adapted for the desalination process for the first time in Singapore. Silica gels were the first adsorbers used with the adsorption desalination method. When the adsorption of water vapor is carried out in a closed system, the partial pressure of the water vapor

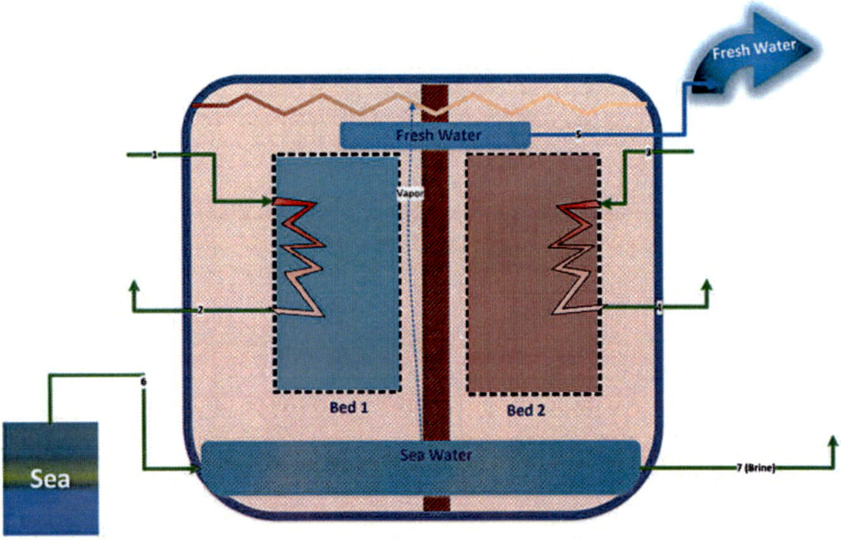

Fig. 6.30 Illustration of the adsorption desalination system

lowers, which facilitates the evaporation process. Therefore, lower-temperature heat sources can be used to carry out evaporation of seawater with an adsorber.

Figure 6.30 shows an adsorption desalination system. The seawater is fed to the closed system, and simultaneously the adsorber adsorbs water vapor which continuously drops the pressure. After the adsorber is saturated, a heat source can be used to heat the adsorber, which starts desorption and facilitates the condensation of the freshwater process.

The recovery ratio can be considered as the ratio of the mass flow rate of the seawater to the fresh water as follows:

$$r_r = \frac{\dot{m}_{fw}}{\dot{m}_{sw}}$$

Considering the molar fraction of the salt and water within the seawater content, the specific molar fraction calculation is made as follows:

$$x_{s,sw} = \frac{MW_w}{MW_w + MW_s \times \left(\left(\frac{1}{y_{s,sw}}\right) - 1\right)}$$

and

$$x_{w,sw} = 1 - x_{s,sw}$$

The fresh water does not contain any salt at the end of the process. However, small quantities can be added to improve the quality of the drinkable water.

For the by-product, brine, the mass flow rate can be calculated as follows:

$$\dot{m}_b = \dot{m}_{sw} - \dot{m}_{fw}$$

The salinity content of brine is:

$$x_{s,b} = \frac{MW_w}{MW_w + MW_s \times \left(\left(\frac{1}{y_{s,b}}\right) - 1\right)}$$

In studying the performance values of the adsorption desalination system, the total enthalpy and total exergy can be considered as the inputs. The ratio of the energy and exergy of the water to the inputs gives the energy and exergy efficiencies as:

$$\eta_{en} = \frac{\dot{m}_5 h_5}{(\dot{m}_1 h_1 - \dot{m}_2 h_2) + (\dot{m}_3 h_3 - \dot{m}_4 h_4)}$$

and

$$\eta_{ex} = \frac{\dot{m}_5 ex_5}{(\dot{m}_1 ex_1 - \dot{m}_2 ex_2) + (\dot{m}_3 ex_3 - \dot{m}_4 ex_4)}$$

where state points are written according to Fig. 6.30.

The performance ratio of the adsorption desalination process can be calculated as:

$$PR = \frac{\dot{m}_5 h_{fg}}{(\dot{m}_1 h_1 - \dot{m}_2 h_2) + (\dot{m}_3 h_3 - \dot{m}_4 h_4)}$$

which is the total latent heat of evaporation for the mass of fresh water generated from the system to the total enthalpy input according to the state points in Fig. 6.30.

6.4.5 Membrane Distillation

In the membrane distillation desalination process, water evaporation is at the core of the process. Unlike other heat-driven desalination processes, a membrane is used in this method. Figure 6.31 depicts the membrane distillation desalination system. Hydrophobic membrane only allows water vapor to pass to the permeate cold side. The seawater is heated by an external heat source and then enters the feed side of the system. The seawater evaporates, and the evaporated water passes through the hydrophobic membrane. In the permeate cold side, the water vapor is condensed and fresh water is obtained.

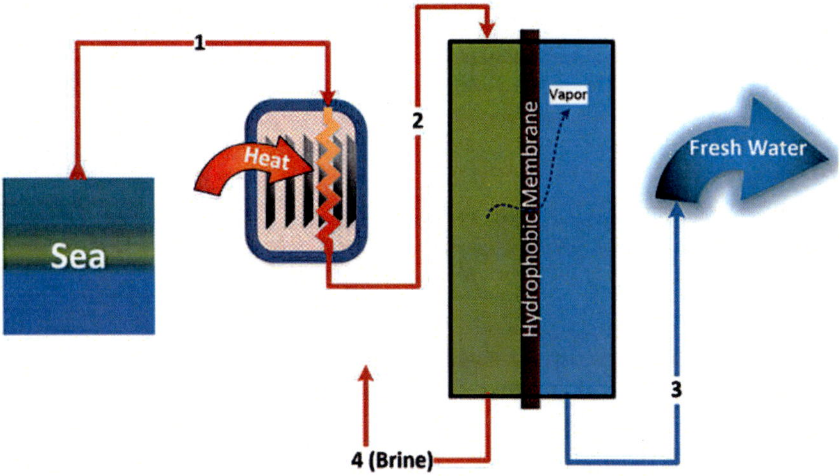

Fig. 6.31 Illustration of the membrane distillation desalination system

In the membrane distillation system, the recovery ratio can be calculated as:

$$r_r = \frac{\dot{m}_{fw}}{\dot{m}_{sw}}$$

The molar ratio of the salt and water content within the seawater is:

$$x_{s,sw} = \frac{MW_w}{MW_w + MW_s \times \left(\left(\frac{1}{y_{s,sw}}\right) - 1\right)}$$

and

$$x_{w,sw} = 1 - x_{s,sw}$$

where $y_{s,sw}$ denotes the mass fraction of the salt in the seawater.

For the by-product, brine, the mass flow rate can be calculated as follows:

$$\dot{m}_b = \dot{m}_{sw} - \dot{m}_{fw}$$

The molar fraction of the salt in brine can be expressed as:

$$x_{s,b} = \frac{MW_w}{MW_w + MW_s \times \left(\left(\frac{1}{y_{s,b}}\right) - 1\right)}$$

In studying the performance values of the membrane distillation desalination system, the total heat energy and its exergy content can be considered as the inputs.

The ratios of the energy and exergy of the water to the inputs technically give the energy and exergy efficiencies as:

$$\eta_{en} = \frac{\dot{m}_3 h_3}{\dot{Q}_{MD}}$$

and

$$\eta_{ex} = \frac{\dot{m}_3 ex_3}{\dot{E}x_{Q_{MD}}}$$

where state points are written according to Fig. 6.31.

The performance ratio of the membrane distillation desalination process can be calculated as:

$$PR = \frac{\dot{m}_3 h_{fg}}{\dot{Q}_{MD}}$$

which is the total latent heat of evaporation for the mass of fresh water generated from the system to the total enthalpy input according to the state points in Fig. 6.31.

6.4.6 Humidification Dehumidification

Humidification dehumidification method is one of the simplest desalination methods where seawater is humidified and then dehumidified via a carrier gas, then collected as fresh water. A simple humidification dehumidification system consists of a humidifier, dehumidifier and a heater section.

Figure 6.32 shows a simple humidification dehumidification desalination system. Air or carrier gas enters the humidifier section along with the heated seawater. The cold air and hot seawater mixes. Temperature and the water content of the air increases. Air with a higher temperature and higher moisture than the entry point moves to the dehumidifier section. Seawater is used as the coolant inside the coils of the dehumidifier. The air loses its heat and the water content becomes condensed. The condensed water can be collected as the fresh water. During the dehumidification process, seawater is getting preheated. Later on, seawater enters the heater and continues to the humidification and dehumidification process.

According to the given figure, the recovery ratio can be calculated as:

$$r_r = \frac{\dot{m}_5}{\dot{m}_4}$$

The molar ratio of the salt and water content within the seawater is:

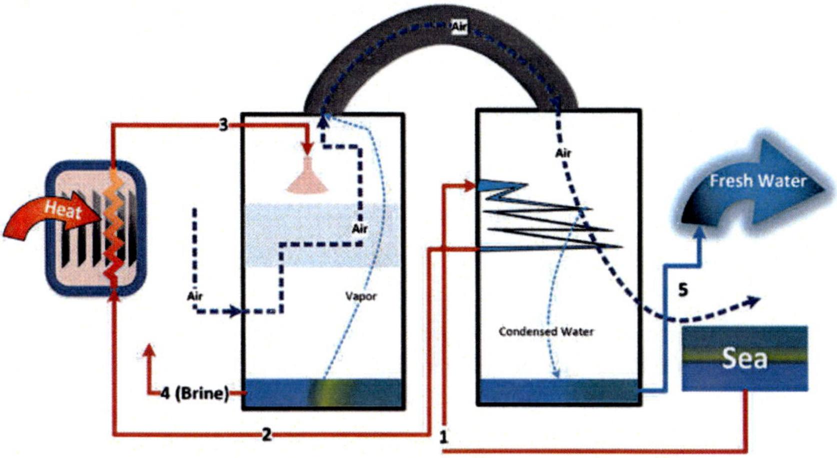

Fig. 6.32 Illustration of the humidification dehumidification desalination system

$$x_{s,sw} = \frac{MW_w}{MW_w + MW_s \times \left(\left(\frac{1}{y_{s,sw}}\right) - 1\right)}$$

and

$$x_{w,sw} = 1 - x_{s,sw}$$

For the by-product, brine, the mass flow rate can be calculated as follows:

$$\dot{m}_b = \dot{m}_{sw} - \dot{m}_{fw}$$

The molar fraction of the salt in brine can be expressed as:

$$x_{s,b} = \frac{MW_w}{MW_w + MW_s \times \left(\left(\frac{1}{y_{s,b}}\right) - 1\right)}$$

The performances of the humidification dehumidification desalination system are required for system assessment, and in this regard the total heat energy and its exergy version can be considered as the inputs. The ratio of the energy and exergy of the water to the inputs gives the energy and exergy efficiencies as:

$$\eta_{en} = \frac{\dot{m}_5 h_5}{\dot{Q}_{HD}}$$

and

$$\eta_{ex} = \frac{\dot{m}_5 ex_5}{\dot{E}x_{Q_{HD}}}$$

where state points are written according to Fig. 6.32.

The performance ratio of the humidification dehumidification desalination process can be calculated as:

$$PR = \frac{\dot{m}_5 h_{fg}}{\dot{Q}_{HD}}$$

which is the total latent heat of evaporation for the mass of fresh water generated from the system to the total enthalpy input according to the state points in Fig. 6.32.

6.4.7 Freezing

Freezing is a unique desalination method, where freezing is exploited to generate fresh water. There are simple and more complicated freezing desalination systems tested or used.

One of the available freezing desalination systems, as shown in Fig. 6.33 uses liquid refrigerant released into the section along with the seawater. Liquid refrigerant

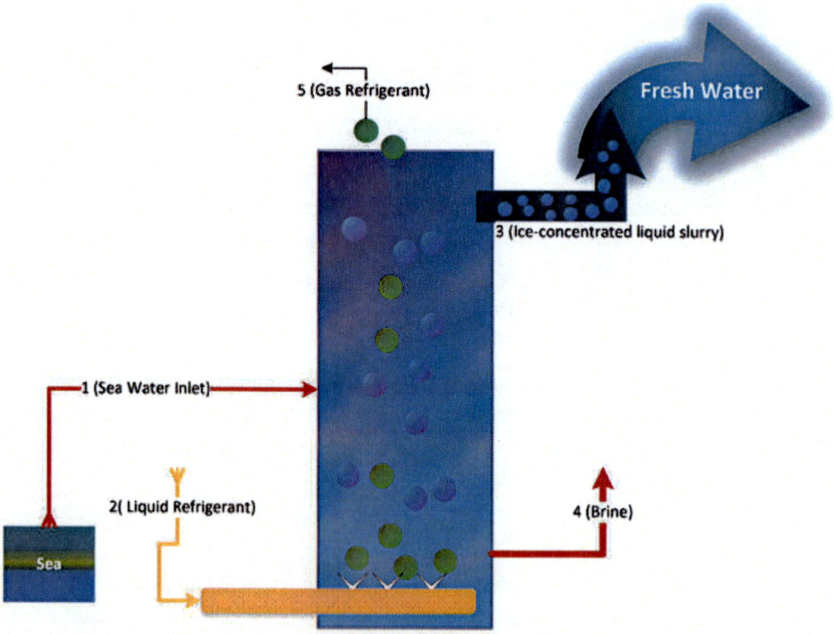

Fig. 6.33 Illustration of the freezing desalination system

evaporates and collects heat from the seawater, which leads to dropping temperatures of the seawater. Therefore, frozen parts become on top of the section. In order to move the frozen part, the form of ice-concentrated liquid slurry is used. The refrigerant leaves in a gas form and the remaining by-product, brine, leaves as a cold liquid.

6.5 Hybrid Desalination Methods

Hybrid desalination methods refer to the desalination where more than one system is employed together, in order to improve the utilization of the source with the synergy of hybridization. The combination of a power-driven desalination method along with the heat-driven desalination method can be considered as the hybrid desalination method.

Figure 6.34 shows the options for hybrid desalination methods. In Fig. 6.34a, the pathway option consists of a heat generation step, where a subsystem is used to utilize the heat from renewable energy, and another subsystem is used to utilize the heat for power generation. Later on, generated heat and power are used in one of the hybrid desalination systems. In the second pathway in Fig. 6.34b, some of the resources are directly used to generate power and directly generate heat. The generated heat and power are used in the hybrid desalination system to generate fresh water.

Sorgulu and Dincer [9] used a reverse osmosis and multi-effect distillation-based hybrid desalination system where they employed biomass energy as the main source. In addition to the fresh water, they also considered the power aimed to be produced. A synthesis gas and biogas were obtained by using an environmentally friendly method.

According to their calculation, 83.29 kg/s of freshwater generation rate is obtained from the reverse osmosis unit when 897.6 kW of electricity was consumed by the unit. On the other hand, 9.1 kW of electrical power and 5.6 MW of heat were utilized by the multi-effect distillation desalination unit in order to generate 9 kg/s of fresh water.

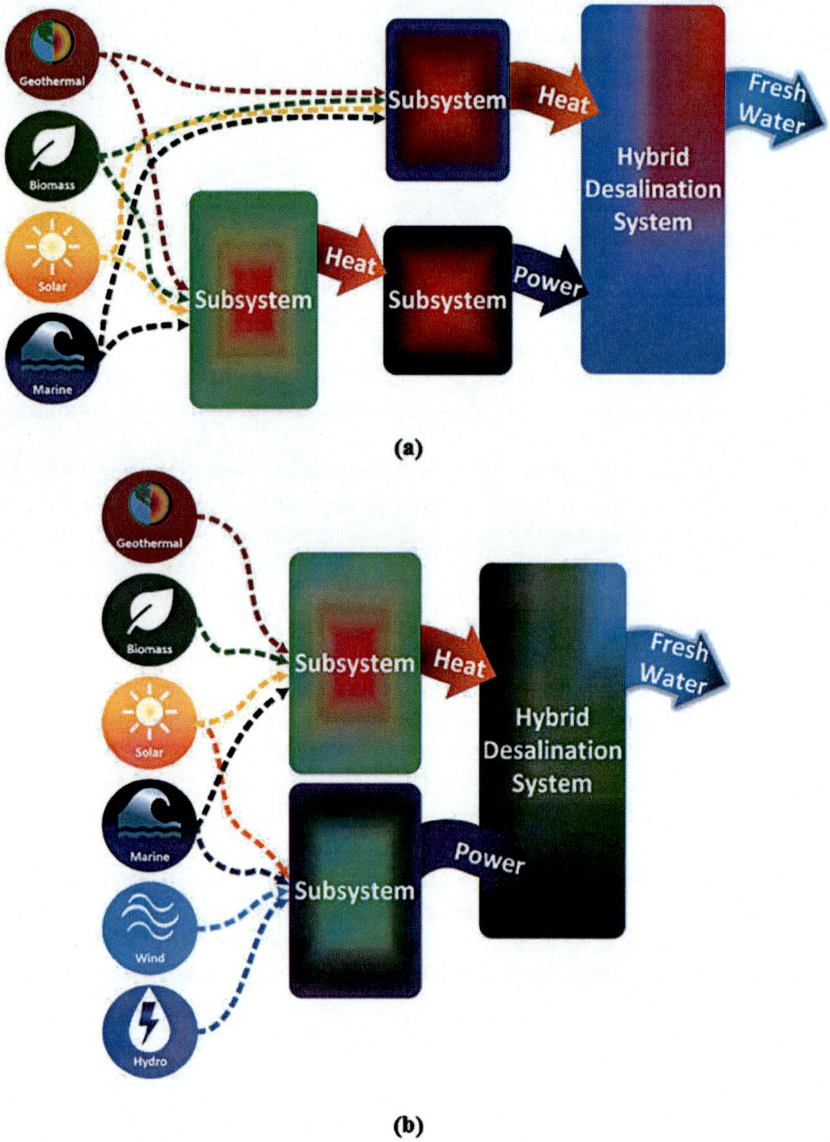

Fig. 6.34 Hybrid renewable desalination pathways (**a**) with indirect (**b**) direct power generation steps

6.6 Closing Remarks

The current chapter has presented some key information to cover renewable options to carry out the desalination process. At the beginning of this chapter, it is clearly shown that the current water resources are at highly critical levels. Especially, North Africa, Middle East, and South Asia regions, countries already consume higher than what they have. Desalination is already essential for many countries. According to the trend, it will be also essential for many more countries. Global warming and global population growth are two major contributors to the worsening conditions of the water situation. Fossil fuel–driven desalination methods are more harmful than the solution. Therefore, the importance of the renewable desalination is clear. These are presented with details in the first two sections.

As shown historically, developments in thermodynamics have accelerated the developments in desalination. The knowledge of steam processing is one of the core parts of thermodynamics, which is also critical for desalination. Therefore, the thermodynamic aspects of desalination are investigated with details for both heat and power-driven desalination systems. Moreover, critical methods are studied with illustrative examples. However, it is critical to improve the source utilization; therefore, integrated desalination systems can improve the feasibility of such systems, which are discussed in the following chapters.

References

1. Hannah Ritchie, Max Roser (2023) Water use and stress. Our World in Data Retrieved from: https://ourworldindata.org/water-use-stress (Online Resource)
2. Poore J, Nemecek T (2018) Reducing food's environmental impacts through producers and consumers. Science 360:987–992
3. Dincer I, Abu-Rayash A (2020) Energy sustainability. Elsevier, New York
4. Dincer I (2021) Thermodynamics: a smart approach. Wiley, Chichester
5. Dincer I, Bicer Y (2020) Integrated energy systems for multigeneration. Elsevier
6. System Advisor Model Version 2020.2.29 (2020) National Renewable Energy Laboratory. Golden, CO
7. Demir ME, Dincer I (2022) An integrated solar energy, wastewater treatment and desalination plant for hydrogen and freshwater production. Energy Convers Manag 267:115894. https://doi.org/10.1016/J.ENCONMAN.2022.115894
8. Demir ME, Chehade G, Dincer I et al (2019) Design and analysis of a new system for photoelectrochemical hydrogen production from wastewater. Energy Convers Manag 199: 111903. https://doi.org/10.1016/j.enconman.2019.111903
9. Sorgulu F, Dincer I (2021) Development and assessment of a biomass-based cogeneration system with desalination. Appl. Therm. Eng. 185:116432. https://doi.org/10.1016/j.applthermaleng.2020.116432

Chapter 7
Traditional Energy Systems

Nomenclature

A	area (m^2)
EBE	energy balance equation
EnBE	entropy balance equation
ExBE	exergy balance equation
ex	specific exergy (kJ/kg)
$\dot{E}x$	exergy rate (kW)
g	gravitational acceleration (m/s^2)
h	specific enthalpy (kJ/kg)
$\bar{h}$	absolute enthalpy (kJ/kmol)
HP	heat pump
LHV	lower heating value (kJ/kg)
$\dot{m}$	mass flow rate (kg/s)
MBE	mass balance equation
$\dot{N}$	molar flow rate (mol/s)
P	pressure (kPa)
Q	heat (kJ)
$\dot{Q}$	heat rate (kW)
S	total entropy (kJ /K)
s	specific entropy (kJ/kg K)
$\dot{S}$	entropy rate (kW/K)
T	temperature (°C)
V	overpotential (V)
w	water, work
$\dot{W}$	power (kW)
X	mass ratio of component

I. Dincer, M. Temiz, *Renewable Energy Options for Power Generation and Desalination*, https://doi.org/10.1007/978-3-031-53437-9_7

Subscripts

c	complete, condenser
comb	combustion
con	condenser
f	formation
gt	gas turbine
HX	heat exchanger
in	injection
P	product
PH	process heat
R	reactant
RC	residential cooling
RH	residential heating
re-in	re-injection
s	source
SG	steam generator
st	steam turbine

7.1 Introduction

In this section, traditional energy systems are presented as conventional practices to primarily cover energy production and desalination options using conventional methods. Traditional energy systems can be described as systems where thermal energy or fuel as a single input is used to generate useful outputs, mainly power. Historically, wood was the primary energy input for traditional energy systems. Then, it was replaced by other inputs, particularly coal, oil, and gas. Whale oil was also used as one of the most popular fuels, especially between 1830 and 1860. After the discovery of the mineral oils in 1860s, whale oil was essentially replaced by kerosene. The primary usages of wood, whale oil, kerosene, or camphine were particularly for illumination and heating purposes. The major milestone in traditional energy systems was the invention of converting the thermal energy into the mechanical power or movement during the Industrial Revolution in the 1760s. Initial attempts to turn thermal energy into mechanical power were started with heating water and the displacement of water or the expansion of water was turned into rotational or tangential movement. This major milestone of the traditional energy systems can be considered as the major turning point of the human history. Especially the developments in the steam engine were the main driver of the industrial revolution, which leads an invention boom that can be treated as a global-level transition of human society and economy toward more efficient, stable, and a larger scale manufacturing and production capability. The interconnection between power generation and desalination can be highlighted here since one of the major sectors that directly impacted by the industrial revolution was the global trade; however, freshwater need was the major challenge for ships which travel long distances. The development of knowledge in thermodynamics of water and steam with the outputs from research on power generation from thermal energy enabled the studies and developments in desalination technologies.

As mentioned earlier, it can be stated that the history of traditional energy systems starts with the developments and works of steam and its capability to create movement. Figure 7.1 depicts the historical developments of traditional energy systems. The earliest device in history based on steam expansion to create movement is a device called an aeolipile. It is a bladeless turbine with two nozzles which are curved to create rotation on the axis. Vitruvius mentioned the aeolipile device for the first time as a device which has two openings and can be filled with water; once the device is heated, water starts to boil and released from openings; therefore, movement can be created. About 50 years later, Heron brought a more practical approach and described the aeolipile with more details; the tangential ejection of steam from the nozzles to rotate the pivoted ball was described in his manuscript called Spiritalia seu Pneumatica. The energy and exergy efficiencies of the first device to turn thermal energy to movement, the aeolipile, were very low. In the tenth century, Professor Gerbert of Aurillac (also known as Pope Sylvester II) built the device to run the organ with heated water in Reims, France. In the last quarter of the fifteenth century, Leonardo da Vinci introduced the steam-powered cannon, which was attributed to the works of Archimedes and named as The Architonnerre. The first steam turbine was described for the first time by Taqi al-Din, an Ottoman polymath who lived in the sixteenth century in Ottoman Egypt. Taqi-al-Din described a rudimentary impact steam turbine in order to rotate a spit. In 1648, A similar rudimentary impact steam turbine was also described by John Wilkins. In 1605, another military application was discovered by Florence Rivault, a mathematician from Laval, France. Florence Rivault introduced that if a bombshell filled with water is heated, it can explode the shells, which can be used in battles.

Further note that the very first patent was granted to Spaniard Jeronimo de Ayanz y Beaumont for the steam-powered pump, which was developed in order to drain flooded mines. Between 1657 and 1659, Robert Boyle carried out a series of experiments on the air pump with the assistance of Robert Hooke. The initial design of the air pump was made by Otto von Guericke, which required the labor of two strong men. Robert Boyle worked on the air pump and built a practical one which could be operated on a desktop. Robert Boyle published these experiments in the New Experiments Physico-Mechanical, Touching the Spring of the Air, and its Effects. In 1712, Thomas Newcomen, a merchant in Staffordshire, England, built a steam engine, where water is heated and connected with a piston mechanism. The Newcomen's steam engine used steam to expand the piston, later on, cold water sprayed inside the piston which created vacuum, which is used to pump water from the mines, which was a popular application, especially in Europe.

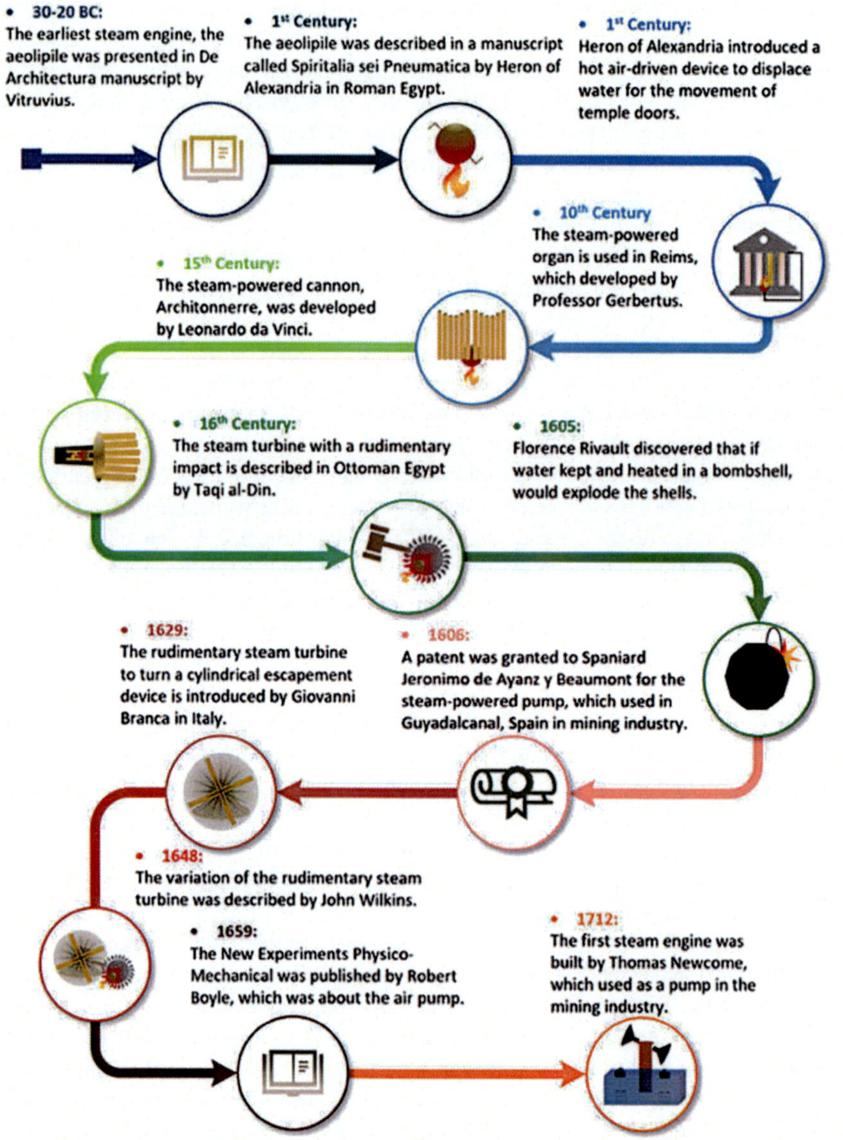

Fig. 7.1 A flowchart about historical events of desalination

As mentioned previously in power generation methods, human power can be considered as the first prime mover. Animals and renewable sources such as wind and hydro sources replaced the human muscle to create motion. Wood, whale oil, and kerosene became popular energy sources one by one. During the pre-industrial period, the only secondary fuel was charcoal, which was generated in locations where an abundance of wood was available, such as communities around the Harz Mountains of Germany. Pyrolysis was the main method to produce charcoal.

Knowledge of thermodynamics unlocked many inventions, including the steam engine, and in a broader aspect, the industrial revolution. As a result of the industrial revolution, the booming demand for thermal energy accelerated the efforts to meet these demands. Fossil fuels were the potential solution to meet these demands; within a short period, coal, oil, and gas became the dominant energy source globally. The modern society highly relies on fossil fuels. When we look at the global primary energy consumption, fossil fuels represent more than 80%. Oil is the most consumed primary energy source globally, which represents the one-third of global primary energy consumption. After oil, coal comes as the second primary energy source, which meets higher than a quarter of global primary energy consumption. Among the fossil fuels, gas comes third, which meets slightly lower than a quarter of global primary energy consumption. In 2022, total fossil fuel consumption was 137.2 PWh, globally, as shown in Fig. 7.2, which also presents the historical trend in fossil fuels. Coal was started as the dominant fuel from the beginning of the nineteenth century, after the 1920s, oil took the dominance and has kept it until today. Especially the transportation sector intends to keep oil as the highest consumed energy source globally.

Although there is a growing trend in fossil fuel consumption, as depicted in Fig. 7.2, the share of fossil fuels in primary energy sources has a declining trend, as shown in Fig. 7.3, with the contributions from nuclear and renewables. The ultimate global target is to remove all fossil fuels from the global energy mix. However, countries have very different energy mix under the effect of numerous factors.

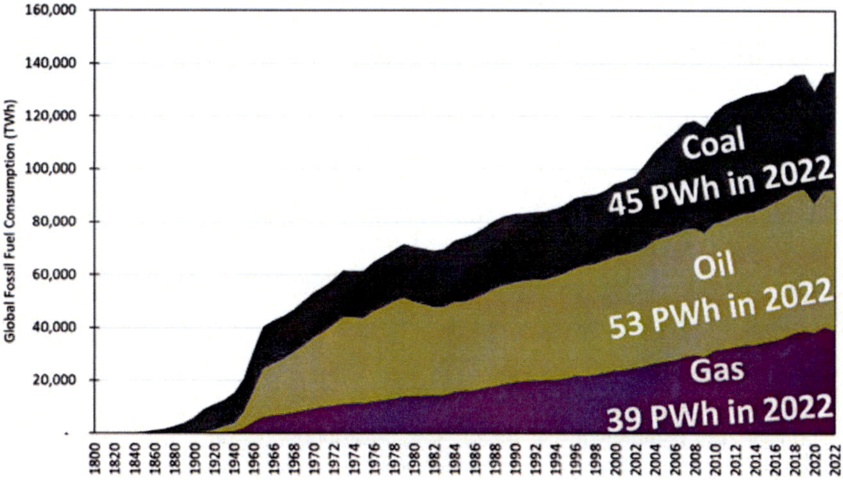

Fig. 7.2 Global fossil fuel consumption as primary energy source (data from Ref. [1])

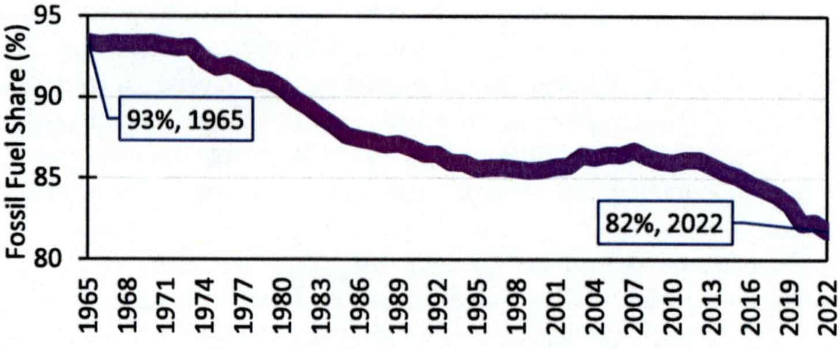

Fig. 7.3 Declining trend of fossil fuel share in primary energy sources (data from Ref. [1])

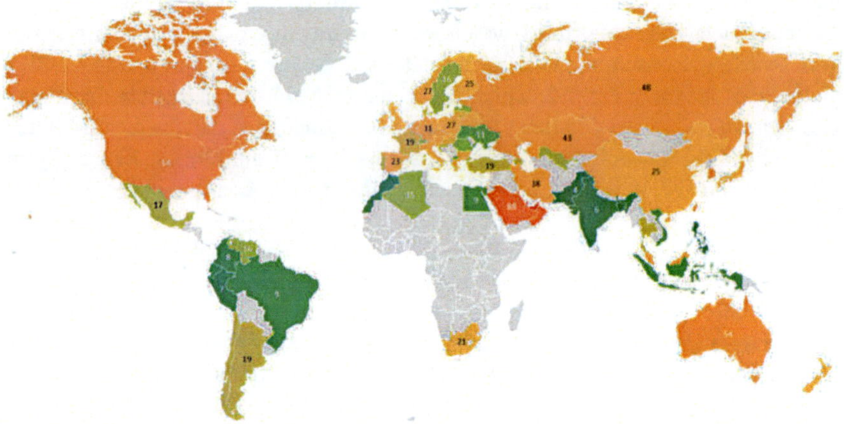

Fig. 7.4 Fossil fuel consumptions of countries in 2022 (unit: PWh) (modified from Ref. [1])

Several countries have very different fossil fuel consumption per capita, as shown in Fig. 7.4. Level of industrial activities, energy efficiency, overall life quality, and source availability are some of the factors that affect the fossil fuel consumption of countries. In 2022, Qatar has the highest fossil fuel consumption per capita with 193.6 MWh equivalent of fossil fuel according to the available data [1]. Singapore comes second with 149.3 MWh of fossil fuel consumption per capita. Among the countries with available data, Bangladesh has the lowest fossil fuel consumption per capita with 2.9 MWh. With a quick math, a person in Qatar consumes equal fossil fuel to 67 people in Bangladesh.

When we look at the trends between fossil fuels, a growing trend in natural gas is very significant. Oil shows a more significant declining trend; however, coal shows mixed trends for the last 50 years as shown in Fig. 7.5.

One of the parameters that affects the consumption is the source availability. Different regions have natural fossil fuel reserves, which are utilized according to the production plans of jurisdictions and the capability to utilize it. Figure 7.6 shows

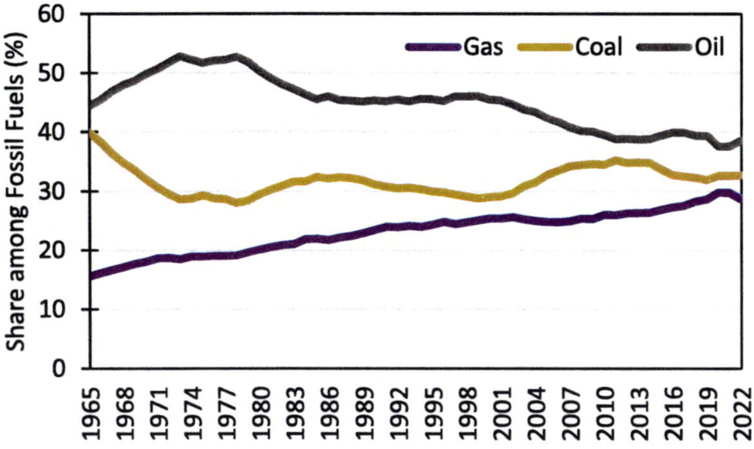

Fig. 7.5 Shares of gas, coal, and oil consumptions among fossil fuels (data from Ref. [1])

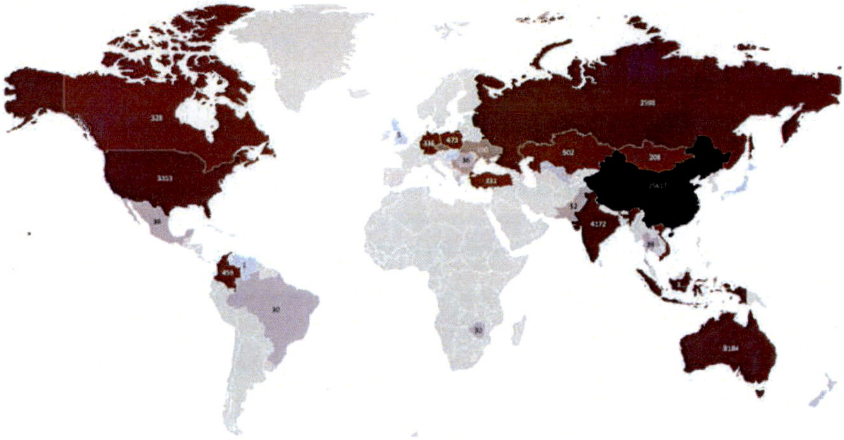

Fig. 7.6 Global map of the coal production in 2022 (unit: TWh) (modified from Ref. [1])

countries and their coal productions in TWh. In 2022, 45 PWh equivalent of coal is consumed and 46.3 PWh equivalent of coal is produced. China is the leader country in terms of coal production, where more than half of the coal is produced only by China. India, Indonesia, the United States, Australia, and Russia follow China, with more than 2.5 PWh equivalent of coal production.

Figure 7.7 depicts a global map for oil production in TWh. In 2022, the global oil consumption was reached to 53 PWh equivalent; on the other hand, 51.3 PWh equivalent of oil produced in the same year. A deeper look to the countries shows that the leader in oil production was the United States in 2022 with 8833 TWh

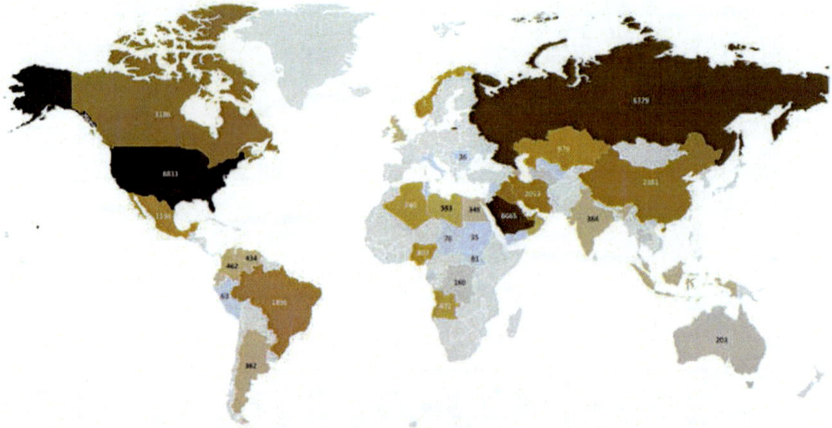

Fig. 7.7 Global map of the oil production in 2022 (unit: TWh) (modified from Ref. [1])

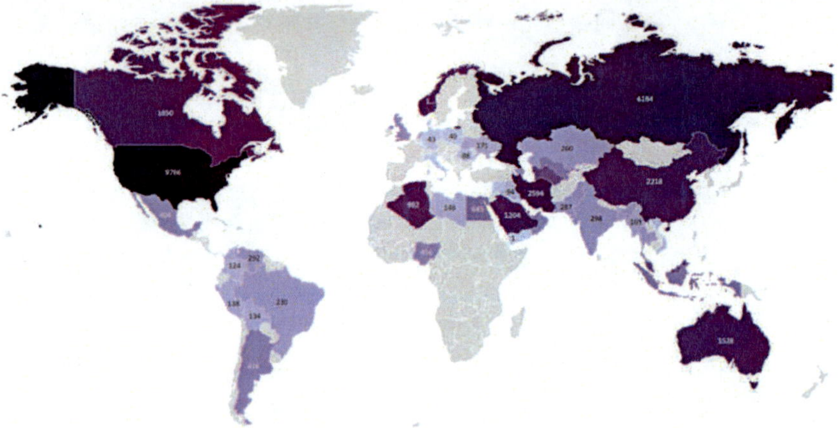

Fig. 7.8 Global map of the gas production in 2022 (unit: TWh) (modified from Ref. [1])

equivalent of oil production. Saudi Arabia and Russia are followed by the United States with more than 6000 TWh equivalent of oil production in each country.

The United States was the leading country also in global gas production in 2022 with 9.8 PWh equivalent of gas. Russia was the second country with 6.2 PWh equivalent of gas production in the same year. Iran, China, Canada, and Qatar follow these countries with more than 1.5 TWh equivalent of gas production. Globally, 40.4 PWh equivalent of gas was produced in 2022; however, 39 PWh equivalent of gas was consumed. The global map for gas production of countries is given in Fig. 7.8.

As mentioned earlier, jurisdictions have the ultimate target as to phase-out from fossil fuels. The United Nations Framework Convention on Climate Change

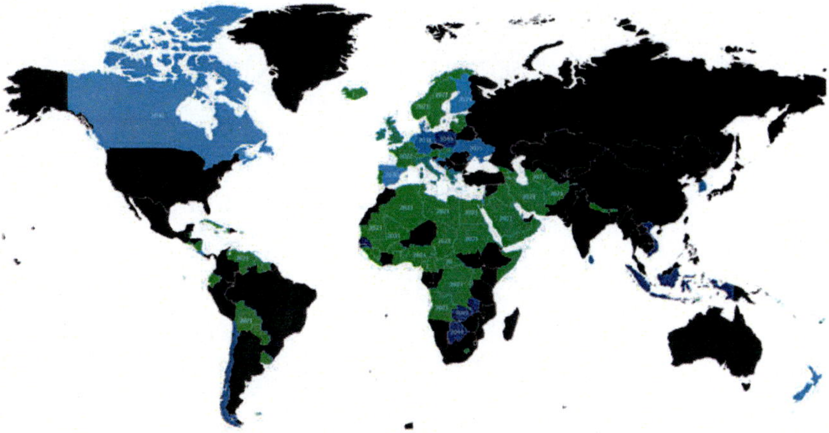

Fig. 7.9 Targeted timelines for coal phase-out (modified from Ref. [1])

includes the majority of the countries globally, which established many treaties such as Kyoto Protocol and Paris Agreement [2]. Series of acts and measures against the climate change are taken by jurisdictions. Phase-out from coal can be highlighted as an example, since the coal power plants emit the highest intensity of carbon dioxide per kWh of electricity generated. There are many countries that are already phased out from coal, such as Azerbaijan, France, Norway, Portugal, Egypt, and Sweden. Among the countries which gave the timeline for phase-out, the latest deadline is 2049. Canada, Spain, and New Zealand are targeting 2030 for their coal phase-out timeline. Figure 7.9 also shows the timelines of different countries for coal-phase-out.

7.2 Traditional Energy Systems

There are many ways to categorize traditional energy systems. In the current book, traditional energy systems are considered conventional methods to turn thermal energy into power and other useful outputs. From this aspect, a fuel-based categorization and a cycle-based categorization can be made. Coal, oil, and gas are the main fuels for traditional energy systems, which are discussed in the previous section. Figure 7.10 illustrates the traditional energy systems with source-system-service (3S) approach, as detailed elsewhere in [3].

When we look at the cycles for traditional energy systems, two main types are available, as vapor cycles and gas cycles. In the vapor cycles, typically a thermofluid is used in a closed system, where basically the thermofluid is evaporated in the boiler, expanded in the turbine, condensed in the condenser, and pressurized in the pump. In typical vapor cycles, the thermal energy is injected to the thermofluid in the boiler component, and the power is generated in the turbine component. Steam

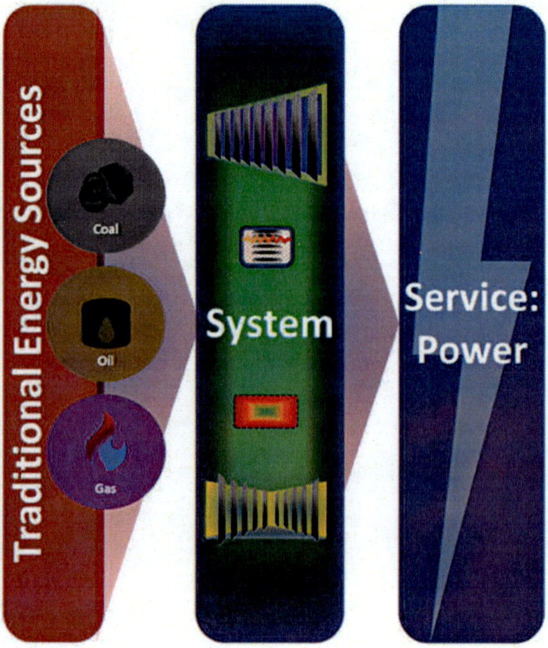

Fig. 7.10 Illustration of source-system-service (3S) approach for traditional energy systems

Rankine cycle, organic Rankine cycle, and trilateral cycles are the most popular ones among the vapor cycles. Especially the steam Rankine cycle is the most popular method to generate power from thermal energy, among conventional power generation methods. Regardless of the fuel type, vapor cycles can be used by utilizing the thermal energy generated by the fuel.

In contrast to the vapor cycles, the gas cycles do not involve a process where the thermofluid/fuel is evaporated or condensed. The fuel remains in the gaseous form throughout the entire cycle, where heat input, heat output, work input, and work output steps occur in order to obtain net power generation, in the gas cycles. There are reversible and irreversible cycles that can be considered in the gas cycles. Reversible cycles are the theoretical cycles, which determine the limits of the processes, in other words, reversible cycles cannot be achieved in practical applications. Carnot cycle is one of the reversible processes that theoretically exist, however, the proposed Carnot efficiency is the limit for heat engines which cannot be exceeded in practice. Brayton, Otto, and Diesel cycles are other gas cycles, which exist in practice and are considered as irreversible processes.

Vapor and gas cycles can be combined in an integrated fashion. In a combined cycle, at least two or more cycles are integrated. Primary cycle utilizes the heat, where a higher temperature process occurs. After the expansion of the thermofluid, excess heat is directed to the heat exchanger, which is connected to the secondary cycle. Thermofluid of the secondary cycle gets heat from the heat exchanger and expands to generate more power. Moreover, combined cycles can be formed as the

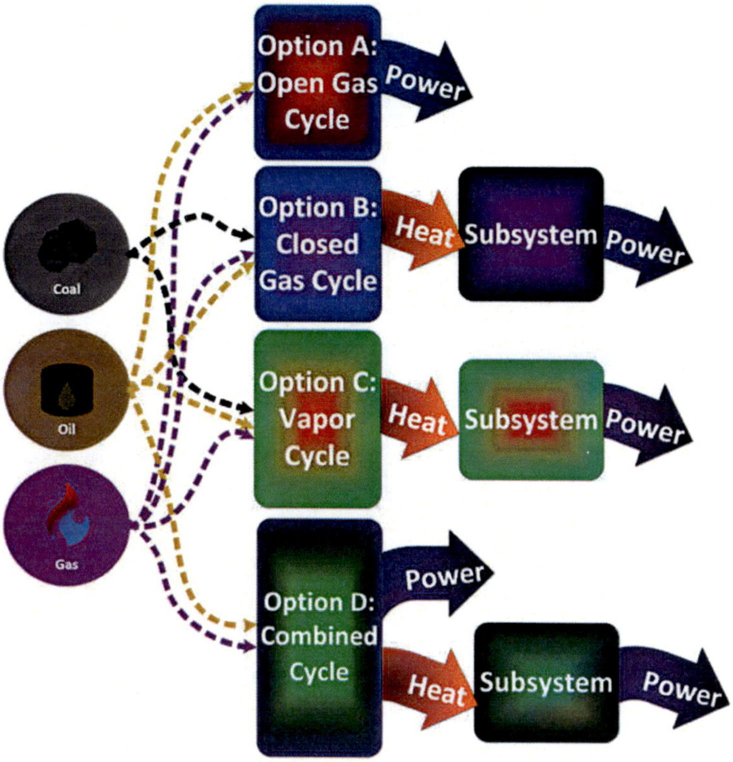

Fig. 7.11 Pathways to utilize fossil fuels to generate power with traditional energy systems

combination of the vapor and gas cycles. Typically, gas cycles occur at higher temperatures than the vapor cycles. For this reason, the excess heat or waste heat from the gas cycles can be exploited via one of the vapor cycles. Temperature of the excess/waste heat is very critical in order to design a combined cycle. Higher temperatures of the excess/waste heat can be exploited via steam Rankine cycles. On the other hand, lower temperatures can be utilized by using low-temperature cycles such as the trilateral cycle or the organic Rankine cycle.

Figure 7.11 shows the pathways to utilize fossil fuels in order to generate power in a traditional fashion. Open gas cycles use fuels directly within the cycle along with air, it can be considered as the fuel is directly utilized by the open gas cycles. Open gas cycles can only use liquid fuels or gaseous fuels. Coal cannot be used directly within the open gas cycles; it needs to be turned into coal gas with certain processes. On the other hand, vapor and closed gas cycles keep the thermofluid within the cycle; therefore, no mass transfer occurs in or out of the system. The earlier steam engines, steam locomotives, and steam vessels were exhausting the steam to the atmosphere from chimneys. This approach is abandoned later after the introduction of the condensing process. Current standard vapor cycles keep the thermofluid

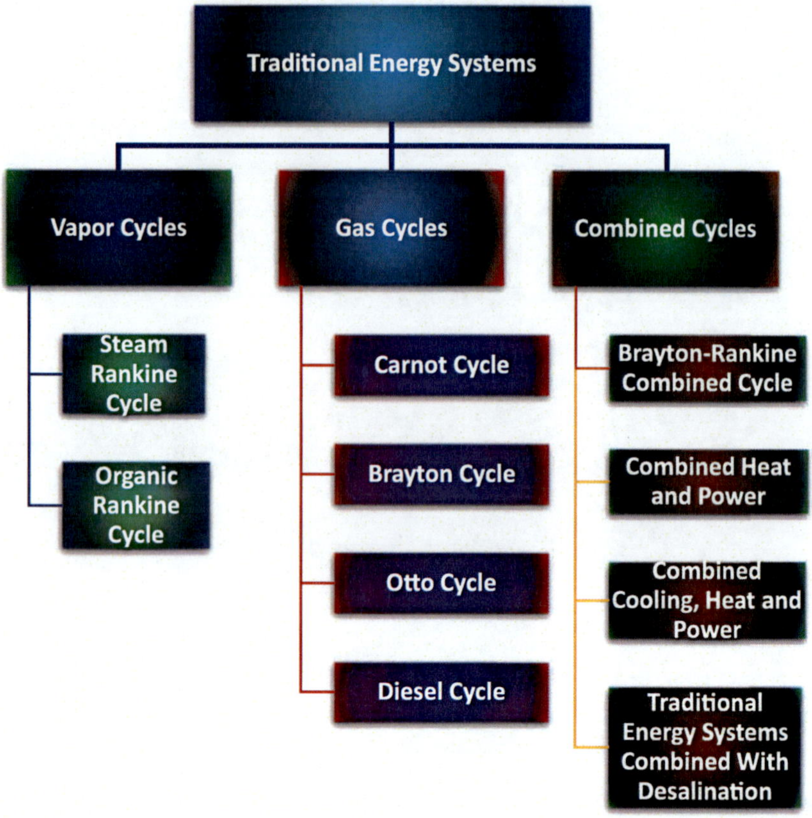

Fig. 7.12 Classification of traditional power generation methods

within. Thermal energy utilized from the fuel is then used in both closed gas cycles and vapor cycles, in order to generate power.

Traditional power generation methods are classified in Fig. 7.12. Steam-based cycles are the processes where evaporation and condensation of the thermofluid occur. Two phases of the thermofluid, liquid and gaseous forms, are used in the vapor cycles. Steam Rankine cycles are the most popular methods, currently used in the industry. Steam Rankine cycles typically require higher source temperatures compared to the other vapor cycles. For lower temperatures, organic Rankine cycles or other cycles such as Kalina or trilateral cycles can be considered. Gas cycles consist of open and closed cycles, which are mainly the Carnot cycle, Brayton cycle, Otto cycle, and Diesel cycle. Combined cycles consist of the combination of two cycles, typically the integration of a gas cycle as a primary cycle and a vapor cycle as a secondary cycle. Some detailed system coverage and analysis of traditional power generation methods are given in the following sections.

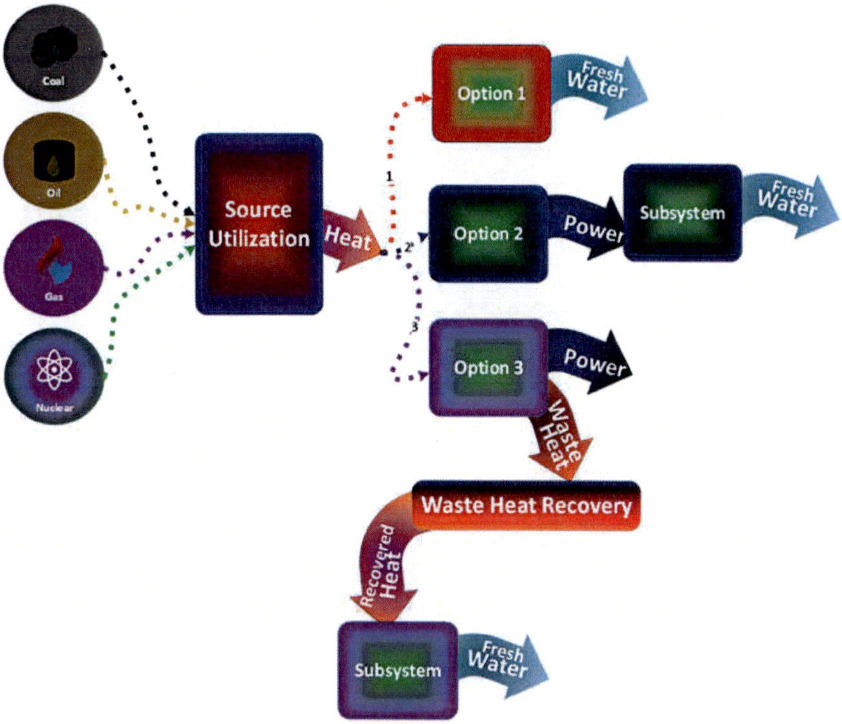

Fig. 7.13 Pathways to generate fresh water with traditional energy systems

There are three main options for generating fresh water by using traditional energy systems. Heat or power can be directly used with conventional systems. Fuels can be used directly to generate the heat required for heat-driven desalination methods. Another option is to utilize heat for power generation with traditional energy systems, and a power-driven desalination method can be used by utilizing the power. Another option is to use waste heat from the power generation system and use a heat-driven desalination method. Three options with conventional energy systems are truly depicted with details in Fig. 7.13.

7.3 Steam Cycles

Currently, steam-based cycles are the dominant technology in the industry to generate power. Historically, vapor cycles were developed as an open cycle where water is heated, expanded to generate power, and exhausted from chimneys in the earlier periods from the first century until the nineteenth century.

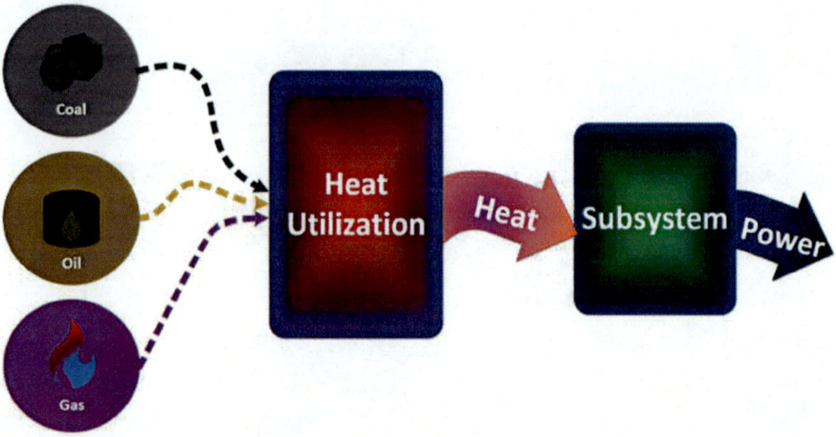

Fig. 7.14 Pathway to generate power by utilizing traditional fuels via vapor cycles

Figure 7.14 shows the pathway to utilize traditional fuels to generate power by using vapor cycles. Conventional power plants typically include a steam Rankine cycle process where fossil fuels are utilized to generate heat in order to drive the cycle. Later on, organic Rankine cycle technology is developed, especially for specific applications, where the source temperatures are lower to run steam Rankine cycles in a feasible manner. Waste heat and renewable heat are typical examples of low-temperature heat sources which are suitable sources for organic Rankine cycle applications. Carbon-containing thermofluids are the organic ones such as octane or fluoromethane, which are used in organic Rankine cycles. There are also other vapor cycles which are not as mature as commercial technologies, such as trilateral cycles or ocean-thermal energy conversion systems, where lower temperatures are exploited by using a variety of thermofluids. Ammonia is one of the popular thermofluids which is used in ammonia trilateral cycles and ocean-thermal energy conversion cycles.

7.3.1 Steam Rankine Cycle

Currently, the steam Rankine cycle is responsible for the majority of power generation globally. Steam Rankine cycle is a thermodynamic cycle where no mass transfer in or out occurs in the process, only heat and work inputs and outputs occur over the system boundary.

The simple form of the steam Rankine cycle is illustrated in Fig. 7.15. In its simple configuration, there are four steps available, which occur in four different components. Since it is a cycle, any of the steps can be considered as the first step. Since the state points are numerated by starting the inlet of the pump in the simple Rankine cycle drawing, the first step is considered as the pressurization.

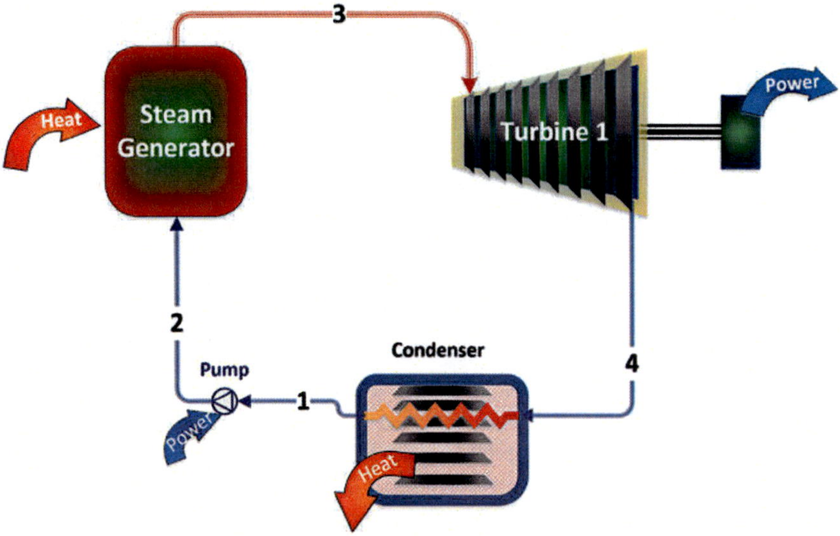

Fig. 7.15 Illustration of a simple steam Rankine cycle

- The pump between state points 1 and 2, pressurizes the steam to the steam generation pressure. In the ideal Rankine cycle, this process is considered as isentropic, where the inlet and outlet entropies are equal. However, in practical applications, actual Rankine cycles have entropy generation in this step, which can be formalized as the isentropic efficiency. The pressurized water flows to the steam generator after the pump.
- In the steam generation step, between state points 2 and 3, the pressurized water is superheated. At state point 2, the pressurized water is typically in a subcritical phase. Later on, water quickly starts to the phase change process and reaches to the saturated vapor state. After that, it is further heated and becomes a superheated steam.
- The superheated steam at state point 3 enters the steam turbine, where it expands to the saturated vapor state (ideally). In the ideal Rankine cycle, the expansion process of the superheated steam is considered as an isentropic process, where no entropy change occurs. However, in practical applications, actual Rankine cycle consists of entropy generation in the during the expansion process, which can be defined with isentropic efficiency.

In order to calculate the heat and work inputs and outputs along with the entropy generations and exergy destructions for the components and the overall steam Rankine cycle, the mass, energy, entropy, and exergy balance equations can be used.

The mass balance equation (MBE), energy balance equation (EBE), entropy balance equation (EnBE), and exergy balance equation (ExBE) can be expressed for the pump component between state points 1 and 2 in Fig. 7.15, as follows:

MBE: $\dot{m}_1 = \dot{m}_2$
EBE: $\dot{m}_1 h_1 + \dot{W}_p = \dot{m}_2 h_2$
EnBE: $\dot{m}_1 s_1 + \dot{S}_{gen,p} = \dot{m}_2 s_2$
ExBE: $\dot{m}_1 ex_1 + \dot{W}_p = \dot{m}_2 ex_2 + \dot{E}x_{d,p}$

In the second step, heat enters to superheat the steam. The steam generation process can be balanced as follows:

MBE: $\dot{m}_2 = \dot{m}_3$
EBE: $\dot{m}_2 h_2 + \dot{Q}_{SG} = \dot{m}_3 h_3$
EnBE: $\dot{m}_2 s_2 + \frac{\dot{Q}_{SG}}{T_b} + \dot{S}_{gen,SC} = \dot{m}_3 s_3$
ExBE: $\dot{m}_2 ex_2 + \dot{E}x^{\dot{Q}_{SG}} = \dot{m}_3 ex_3 + \dot{E}x_{d,SG}$

In the turbine component, the work output along with other parameters according to Fig. 7.15 can be written as:

MBE: $\dot{m}_3 = \dot{m}_4$
EBE: $\dot{m}_3 h_3 = \dot{W}_t + \dot{m}_4 h_4$
EnBE: $\dot{m}_3 s_3 + \dot{S}_{gen,t} = \dot{m}_4 s_4$
ExBE: $\dot{m}_3 ex_3 = \dot{W}_t + \dot{m}_4 ex_4 + \dot{E}x_{d,t}$

The heat output from the condenser along with other parameters can be balanced as:

MBE: $\dot{m}_4 = \dot{m}_1$
EBE: $\dot{m}_4 h_4 = \dot{m}_1 h_1 + \dot{Q}_l$
EnBE: $\dot{m}_4 s_4 + \dot{S}_{gen,CON} = \dot{m}_1 s_1 + \frac{\dot{Q}_l}{T_b}$
ExBE: $\dot{m}_4 ex_4 = \dot{m}_1 ex_1 + \dot{E}x^{\dot{Q}_l} + \dot{E}x_{d,l}$

In order to calculate the performance of the steam Rankine cycle, the energy and exergy efficiencies should be calculated. The objective of the steam Rankine cycle is to generate power by utilizing the heat. From this aspect, the useful output of the steam Rankine cycle is the power generated by the turbine, and total input is the heat input to the cycle at the steam generator. However, the pump component consumes power, which requires the calculation of the net power as follows:

$$\dot{W}_{net} = \dot{W}_t - \dot{W}_p$$

The net power generation can be considered as the net useful output in performance calculations. According to these, the energy and exergy efficiencies of the steam Rankine cycle can be calculated as follows:

$$\eta_{en,ov} = \frac{\dot{W}_{El,net}}{\dot{Q}_{SG}}$$

and

$$\eta_{ex,ov} = \frac{\dot{W}_{El,net}}{\dot{E}x^{\dot{Q}_{SG}}}$$

where $\dot{E}x_{\dot{Q}_{SG}}$ is the exergy version of the heat input at the steam generation step.

In Fig. 7.16 an ideal and actual steam Rankine cycles are drawn on a temperature-entropy diagram. The main difference between the ideal and actual ones is the entropy generation in the pump and turbine components. The lines between 3 and 4, 1 and 2 are vertical in Fig. 7.16a, where there is no entropy generation in the ideal steam Rankine cycle. On the other hand, the lines between 3 and 4, 1 and 2 are growing in terms of entropy value, according to the entropy generation.

Illustrative Example 7.1 For the illustrative example, two different power plants based on typical commercial power plants which utilize oil and coal, are designed. Oil-fired power plant (illustrated in Fig. 7.17a) consists of a system where the water is pressurized to 139 bar and superheated to 535°C by using the oil. On the other hand, a coal power plant (illustrated in Fig. 7.17b) is designed by considering 600°C of superheated steam and 620°C reheated steam temperatures. The steam generator pressure is considered as 285 bar in the coal-fired system (Table 7.1).

Both systems are designed with regenerative heating as well as preheating stages. According to their process temperatures, the coal power plant is designed with seven preheaters, and the oil power plant is designed with six preheaters. Both of the condensing pressures are considered as 0.05 bar. Reheating pressure is designed as 40 bar in the oil power plant and 60 bar in the coal power plant. The maximum temperature obtained at the outlet of reheater for the coal power plant is 620°C.

The process is highly similar in oil and coal power plants of the illustrative example. Water is pressurized in the pump at state points between 1 and 2. High-pressure water enters the steam generator and gets superheated. Superheated steam enters the high-pressure turbine. In oil power plant, there is only one outlet exists in the high-pressure steam turbine, at state point 5, which is later on diverted in two streams 6 and 7. Stream 6 enters the reheater and 7 is the input to the preheater 6 in the oil power plant. On the other hand, two outlets exit from the high-pressure turbine of the coal power plant.

The balance equations for the oil power plant (as illustrated in Fig. 7.17a) can be given as follows:

The balance equations for the pump component can be written as:

MBE: $\dot{m}_1 = \dot{m}_2$

EBE: $\dot{m}_1 h_1 + \dot{W}_p = \dot{m}_2 h_2$

EnBE: $\dot{m}_1 s_1 + \dot{S}_{gen,p} = \dot{m}_2 s_2$

ExBE: $\dot{m}_1 ex_1 + \dot{W}_p = \dot{m}_2 ex_2 + \dot{E}x_{d,p}$

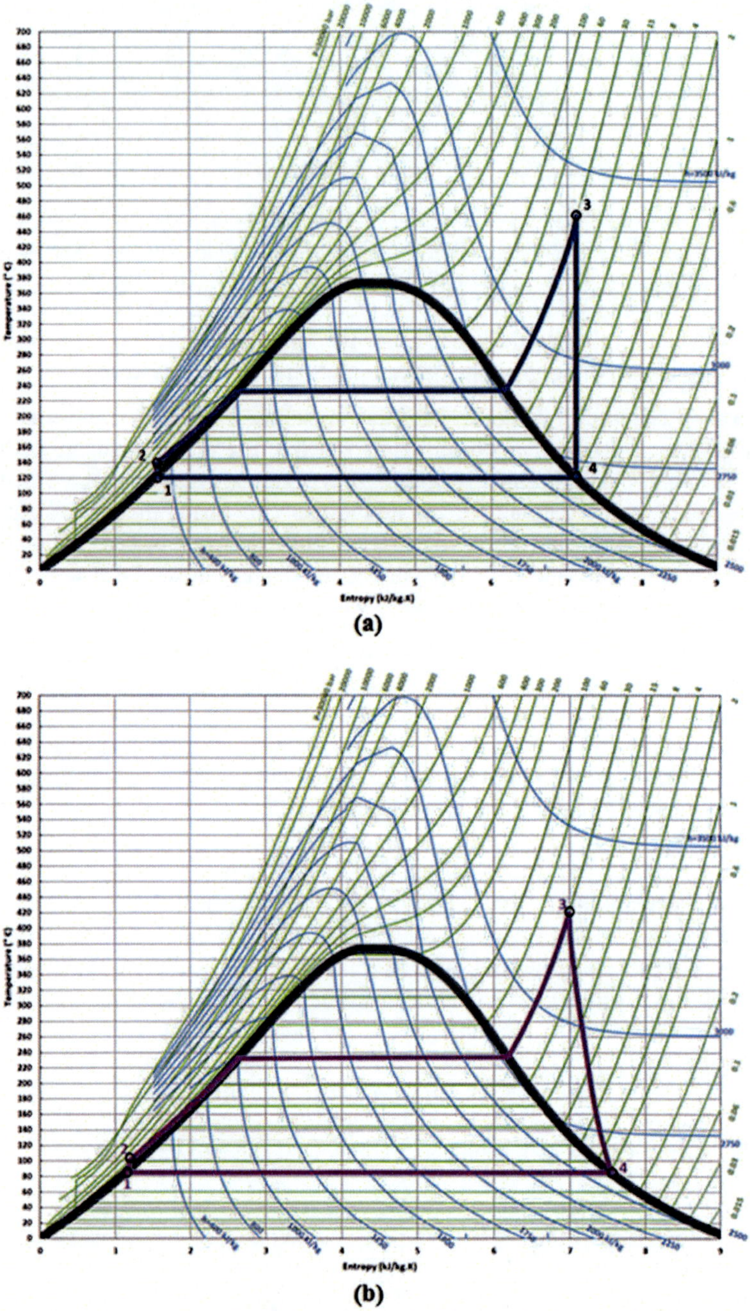

Fig. 7.16 Temperature-specific entropy (T-s) diagram of (**a**) an ideal and (**b**) an actual steam Rankine cycles

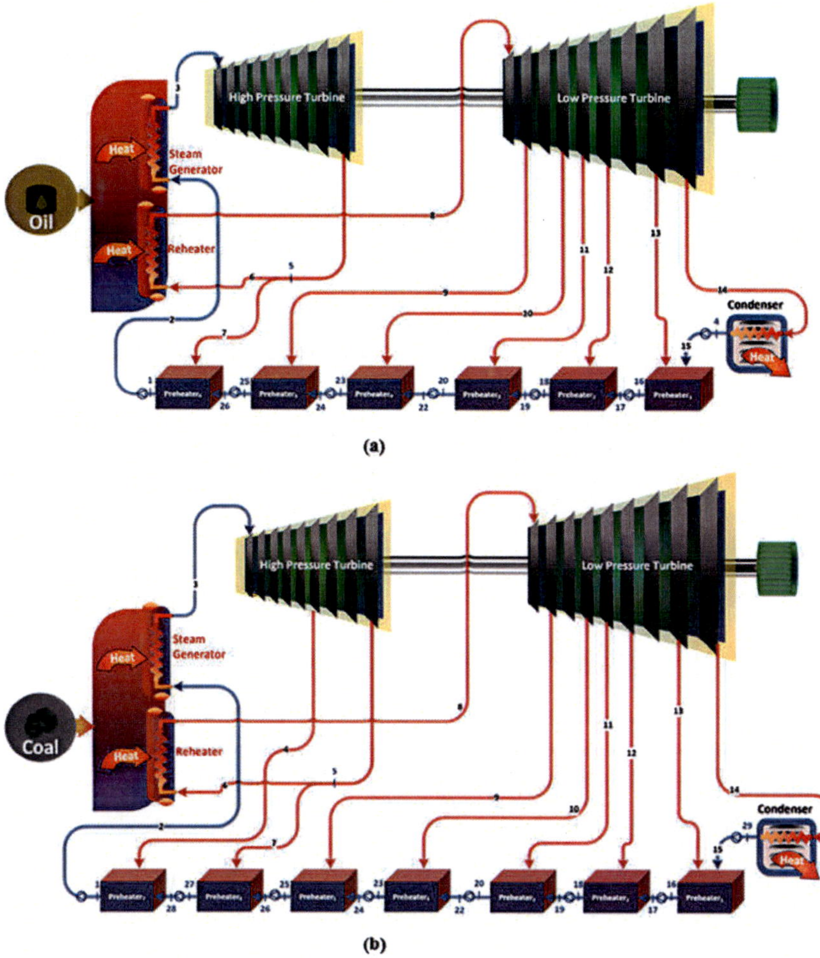

Fig. 7.17 Illustrations of (**a**) an oil-driven regenerative steam Rankine cycle with seven preheaters and (**b**) a coal-driven regenerative steam Rankine cycle with six preheaters

Table 7.1 The design parameters of the power plants with regenerative Rankine cycle

Parameter	Oil power plant	Coal power plant
Steam generator outlet T	535°C	600 °C
Steam generator outlet P	139 bar	285 bar
Reheater outlet T	535°C	620 °C
Reheater outlet P	40 bar	60 bar
Feedwater heaters	6	7
Condenser P	0.05 bar	0.05 bar
Preheaters outlet T	253°C	303 °C

In the second step, heat enters to superheat the steam. The steam generation process can be balanced as follows:

MBE: $\dot{m}_2 = \dot{m}_3$
EBE: $\dot{m}_2 h_2 + \dot{Q}_{SG} = \dot{m}_3 h_3$
EnBE: $\dot{m}_2 s_2 + \frac{\dot{Q}_{SG}}{T_b} + \dot{S}_{gen,SG} = \dot{m}_3 s_3$
ExBE: $\dot{m}_2 ex_2 + \dot{E}x_{\dot{Q}_{SG}} = \dot{m}_3 ex_3 + \dot{E}x_{d,SG}$

In the high-pressure turbine component, the work output along with other parameters according to Fig. 7.17b can be written as:

MBE: $\dot{m}_3 = \dot{m}_5$
EBE: $\dot{m}_3 h_3 = \dot{W}_{HPT} + \dot{m}_5 h_5$
EnBE: $\dot{m}_3 s_3 + \dot{S}_{gen,HPT} = \dot{m}_5 s_5$
ExBE: $\dot{m}_3 ex_3 = \dot{W}_{HPT} + \dot{m}_5 ex_5 + \dot{E}x_{d,HPT}$

For the low-pressure turbine, the work output along with other parameters according to Fig. 7.17b can be written as:

MBE: $\dot{m}_8 = \dot{m}_9 + \dot{m}_{10} + \dot{m}_{11} + \dot{m}_{12} + \dot{m}_{13} + \dot{m}_{14}$
EBE: $\dot{m}_8 h_8 = \dot{W}_{LPT} + \dot{m}_9 h_9 + \dot{m}_{10} h_{10} + \dot{m}_{11} h_{11} + \dot{m}_{12} h_{12} + \dot{m}_{13} h_{13} + \dot{m}_{14} h_{14}$
EnBE: $\dot{m}_8 s_8 + \dot{S}_{gen,LPT} = \dot{m}_9 s_9 + \dot{m}_{10} s_{10} + \dot{m}_{11} s_{11} + \dot{m}_{12} s_{12} + \dot{m}_{13} s_{13} + \dot{m}_{14} s_{14}$
ExBE: $\dot{m}_8 ex_8 = \dot{W}_{LPT} + \dot{m}_9 ex_9 + \dot{m}_{10} ex_{10} + \dot{m}_{11} ex_{11} + \dot{m}_{12} ex_{12} + \dot{m}_{13} ex_{13} + \dot{m}_{14} ex_{14} + \dot{E}x_{d,LPT}$

The heat output from the condenser along with other parameters can be balanced as:

MBE: $\dot{m}_{14} = \dot{m}_4$
EBE: $\dot{m}_{14} h_{14} = \dot{m}_4 h_4 + \dot{Q}_l$
EnBE: $\dot{m}_{14} s_{14} + \dot{S}_{gen,CON} = \dot{m}_4 s_4 + \frac{\dot{Q}_l}{T_b}$
ExBE: $\dot{m}_{14} ex_{14} = \dot{m}_4 ex_4 + \dot{E}x_{\dot{Q}_l} + \dot{E}x_{d,l}$

The balance equations for preheater 1 can be written as:

MBE: $\dot{m}_{15} + \dot{m}_{13} = \dot{m}_{16}$
EBE: $\dot{m}_{15} h_{15} + \dot{m}_{13} h_{13} = \dot{m}_{16} h_{16}$
EnBE: $\dot{m}_{15} s_{15} + \dot{m}_{13} s_{13} + \dot{S}_{gen,PH1} = \dot{m}_{16} s_{16}$
ExBE: $\dot{m}_{15} ex_{15} + \dot{m}_{13} ex_{13} = \dot{m}_{16} ex_{16} + \dot{E}x_{d,PH1}$

The balance equations for preheater 2 can be written as:

MBE: $\dot{m}_{17} + \dot{m}_{12} = \dot{m}_{18}$
EBE: $\dot{m}_{17} h_{17} + \dot{m}_{12} h_{12} = \dot{m}_{18} h_{18}$
EnBE: $\dot{m}_{17} s_{17} + \dot{m}_{12} s_{12} + \dot{S}_{gen,PH2} = \dot{m}_{18} s_{18}$
ExBE: $\dot{m}_{17} ex_{17} + \dot{m}_{12} ex_{12} = \dot{m}_{18} ex_{18} + \dot{E}x_{d,PH2}$

The balance equations for preheater 3 can be written as:

MBE: $\dot{m}_{19} + \dot{m}_{11} = \dot{m}_{20}$
EBE: $\dot{m}_{19}h_{19} + \dot{m}_{11}h_{11} = \dot{m}_{20}h_{20}$
EnBE: $\dot{m}_{19}s_{19} + \dot{m}_{11}s_{11} + \dot{S}_{gen,PH3} = \dot{m}_{20}s_{20}$
ExBE: $\dot{m}_{19}ex_{19} + \dot{m}_{11}ex_{11} = \dot{m}_{20}ex_{20} + \dot{E}x_{d,PH3}$

The balance equations for preheater 4 can be written as:

MBE: $\dot{m}_{22} + \dot{m}_{10} = \dot{m}_{23}$
EBE: $\dot{m}_{22}h_{22} + \dot{m}_{10}h_{10} = \dot{m}_{23}h_{23}$
EnBE: $\dot{m}_{22}s_{22} + \dot{m}_{10}s_{10} + \dot{S}_{gen,PH4} = \dot{m}_{23}s_{23}$
ExBE: $\dot{m}_{22}ex_{22} + \dot{m}_{10}ex_{10} = \dot{m}_{23}ex_{23} + \dot{E}x_{d,PH4}$

The balance equations for preheater 5 can be written as:

MBE: $\dot{m}_{24} + \dot{m}_{9} = \dot{m}_{25}$
EBE: $\dot{m}_{24}h_{24} + \dot{m}_{9}h_{9} = \dot{m}_{25}h_{25}$
EnBE: $\dot{m}_{24}s_{24} + \dot{m}_{9}s_{9} + \dot{S}_{gen,PH5} = \dot{m}_{25}s_{25}$
ExBE: $\dot{m}_{24}ex_{24} + \dot{m}_{9}ex_{9} = \dot{m}_{25}ex_{25} + \dot{E}x_{d,PH5}$

The balance equations for preheater 6 can be written as:

MBE: $\dot{m}_{26} + \dot{m}_{7} = \dot{m}_{1}$
EBE: $\dot{m}_{26}h_{26} + \dot{m}_{7}h_{7} = \dot{m}_{1}h_{1}$
EnBE: $\dot{m}_{26}s_{26} + \dot{m}_{7}s_{7} + \dot{S}_{gen,PH6} = \dot{m}_{1}s_{1}$
ExBE: $\dot{m}_{26}ex_{26} + \dot{m}_{7}ex_{7} = \dot{m}_{1}ex_{1} + \dot{E}x_{d,PH6}$

In order to reheat the steam for further expansion in the low-pressure turbine, reheater component is used, which can be balanced as follows (Table 7.2):

MBE: $\dot{m}_{6} = \dot{m}_{8}$
EBE: $\dot{m}_{6}h_{6} + \dot{Q}_{RH} = \dot{m}_{8}h_{8}$
EnBE: $\dot{m}_{6}s_{6} + \frac{\dot{Q}_{RH}}{T_b} + \dot{S}_{gen,RH} = \dot{m}_{8}s_{8}$
ExBE: $\dot{m}_{6}ex_{6} + \dot{E}x_{\dot{Q}_{RH}} = \dot{m}_{8}ex_{8} + \dot{E}x_{d,RH}$

By using the balance equations, thermophysical properties can be completed as given in Table 7.3.

The balance equations for the coal power plant (as illustrated in Fig. 7.17b) can be given as follows:

The balance equations for the pump component can be written as:

MBE: $\dot{m}_{1} = \dot{m}_{2}$
EBE: $\dot{m}_{1}h_{1} + \dot{W}_{p} = \dot{m}_{2}h_{2}$
EnBE: $\dot{m}_{1}s_{1} + \dot{S}_{gen,p} = \dot{m}_{2}s_{2}$
ExBE: $\dot{m}_{1}ex_{1} + \dot{W}_{p} = \dot{m}_{2}ex_{2} + \dot{E}x_{d,p}$

Table 7.2 The thermophysical properties of water in oil-driven regenerative steam Rankine cycle

State P	ṁ (kg/s)	T (°C)	P (bar)	h (kJ/kg)	s (kJ/kgK)	ex (kJ/kg)
0	0	25	1	105	0.367	0
1	303	253	41	1098	2.817	270
2	303	296	139	1318	3.195	379
3	303	535	139	3422	6.520	1502
4	202	33	0.05	138	0.477	0.7
5	303	356	40	3108	6.608	1162
6	280	356	40	3108	6.608	1162
7	23	356	40	3108	6.608	1162
8	280	535	40	3526	7.193	1407
9	20	449	22	3354	7.238	1222
10	13	334	8.9	3126	7.303	975
11	13	255	4.4	2974	7.355	808
12	11	166	1.8	2804	7.410	621
13	20	88	0.7	2654	7.464	456
14	202	33	0.05	2341	7.676	80
15	202	33	0.05	138	0.477	0.7
16	222	88	0.7	369	1.171	27
18	234	116	1.8	487	1.485	52
20	247	147	4.4	620	1.813	89
23	259	175	8.9	741	2.091	128
25	280	217	22	932	2.494	200

Table 7.3 The thermophysical properties of water in coal-driven regenerative steam Rankine cycle

State P	ṁ (kg/s)	T (°C)	P (bar)	h (kJ/kg)	s (kJ/kgK)	ex (kJ/kg)
0	0	25	1	105	0.367	0
1	397	303	90	1362	3.284	397
2	397	303	285	1618	3.728	522
3	397	600	285	3461	6.274	1614
4	30	411	90	3151	6.337	1285
5	367	355	60	3057	6.357	1185
6	306	355	60	3057	6.357	1185
7	61	355	60	3057	6.357	1185
8	306	620	60	3706	7.223	1579
9	21	412	15	3282	7.309	1129
10	15	293	5.8	3048	7.365	879
11	13	203	2.5	2875	7.415	691
12	17	121	1.0	2718	7.471	517
13	12	60	0.2	2494	7.562	267
14	228	33	0.05	2329	7.637	80
15	228	33	0.05	138	0.477	0.7
16	240	60	0.2	251	0.831	9.3
18	257	100	1.0	417	1.301	37
20	270	127	2.5	535	1.606	65
23	285	158	5.8	665	1.918	103
25	306	198	15	845	2.315	165
27	367	276	60	1214	3.028	324

In the second step, heat enters to superheat the steam. The steam generation process can be balanced as follows:

MBE: $\dot{m}_2 = \dot{m}_3$
EBE: $\dot{m}_2 h_2 + \dot{Q}_{SG} = \dot{m}_3 h_3$
EnBE: $\dot{m}_2 s_2 + \frac{\dot{Q}_{SG}}{T_b} + \dot{S}_{gen,SG} = \dot{m}_3 s_3$
ExBE: $\dot{m}_2 ex_2 + \dot{E}x_{\dot{Q}_{SG}} = \dot{m}_3 ex_3 + \dot{E}x_{d,SG}$

In the high-pressure turbine component, the work output along with other parameters according to Fig. 7.17b can be written as:

MBE: $\dot{m}_3 = \dot{m}_4 + \dot{m}_5$
EBE: $\dot{m}_3 h_3 = \dot{W}_{HPT} + \dot{m}_4 h_4 + \dot{m}_5 h_5$
EnBE: $\dot{m}_3 s_3 + \dot{S}_{gen,HPT} = \dot{m}_4 s_4 + \dot{m}_5 s_5$
ExBE: $\dot{m}_3 ex_3 = \dot{W}_{HPT} + \dot{m}_4 ex_4 + \dot{m}_5 ex_5 + \dot{E}x_{d,HPT}$

For the low-pressure turbine, the work output along with other parameters according to Fig. 7.17b can be written as:

MBE: $\dot{m}_8 = \dot{m}_9 + \dot{m}_{10} + \dot{m}_{11} + \dot{m}_{12} + \dot{m}_{13} + \dot{m}_{14}$
EBE: $\dot{m}_8 h_8 = \dot{W}_{LPT} + \dot{m}_9 h_9 + \dot{m}_{10} h_{10} + \dot{m}_{11} h_{11} + \dot{m}_{12} h_{12} + \dot{m}_{13} h_{13} + \dot{m}_{14} h_{14}$
EnBE: $\dot{m}_8 s_8 + \dot{S}_{gen,LPT} = \dot{m}_9 s_9 + \dot{m}_{10} s_{10} + \dot{m}_{11} s_{11} + \dot{m}_{12} s_{12} + \dot{m}_{13} s_{13} + \dot{m}_{14} s_{14}$
ExBE: $\dot{m}_8 ex_8 = \dot{W}_{LPT} + \dot{m}_9 ex_9 + \dot{m}_{10} ex_{10} + \dot{m}_{11} ex_{11} + \dot{m}_{12} ex_{12} + \dot{m}_{13} ex_{13} + \dot{m}_{14} ex_{14} + \dot{E}x_{d,LPT}$

The heat output from the condenser along with other parameters can be balanced as:

MBE: $\dot{m}_{14} = \dot{m}_{29}$
EBE: $\dot{m}_{14} h_{14} = \dot{m}_{29} h_{29} + \dot{Q}_l$
EnBE: $\dot{m}_{14} s_{14} + \dot{S}_{gen,CON} = \dot{m}_{29} s_{29} + \frac{\dot{Q}_l}{T_b}$
ExBE: $\dot{m}_{14} ex_{14} = \dot{m}_{29} ex_{29} + \dot{E}x_{\dot{Q}_l} + \dot{E}x_{d,l}$

The balance equations for preheater 1 can be expressed as:

MBE: $\dot{m}_{15} + \dot{m}_{13} = \dot{m}_{16}$
EBE: $\dot{m}_{15} h_{15} + \dot{m}_{13} h_{13} = \dot{m}_{16} h_{16}$
EnBE: $\dot{m}_{15} s_{15} + \dot{m}_{13} s_{13} + \dot{S}_{gen,PH1} = \dot{m}_{16} s_{16}$
ExBE: $\dot{m}_{15} ex_{15} + \dot{m}_{13} ex_{13} = \dot{m}_{16} ex_{16} + \dot{E}x_{d,PH1}$

The balance equations for preheater 2 can be expressed as:

MBE: $\dot{m}_{17} + \dot{m}_{12} = \dot{m}_{18}$
EBE: $\dot{m}_{17} h_{17} + \dot{m}_{12} h_{12} = \dot{m}_{18} h_{18}$
EnBE: $\dot{m}_{17} s_{17} + \dot{m}_{12} s_{12} + \dot{S}_{gen,PH2} = \dot{m}_{18} s_{18}$
ExBE: $\dot{m}_{17} ex_{17} + \dot{m}_{12} ex_{12} = \dot{m}_{18} ex_{18} + \dot{E}x_{d,PH2}$

The balance equations for preheater 3 can be expressed as:

MBE: $\dot{m}_{19} + \dot{m}_{11} = \dot{m}_{20}$
EBE: $\dot{m}_{19}h_{19} + \dot{m}_{11}h_{11} = \dot{m}_{20}h_{20}$
EnBE: $\dot{m}_{19}s_{19} + \dot{m}_{11}s_{11} + \dot{S}_{gen,PH3} = \dot{m}_{20}s_{20}$
ExBE: $\dot{m}_{19}ex_{19} + \dot{m}_{11}ex_{11} = \dot{m}_{20}ex_{20} + \dot{E}x_{d,PH3}$

The balance equations for preheater 4 can be expressed as:

MBE: $\dot{m}_{22} + \dot{m}_{10} = \dot{m}_{23}$
EBE: $\dot{m}_{22}h_{22} + \dot{m}_{10}h_{10} = \dot{m}_{23}h_{23}$
EnBE: $\dot{m}_{22}s_{22} + \dot{m}_{10}s_{10} + \dot{S}_{gen,PH4} = \dot{m}_{23}s_{23}$
ExBE: $\dot{m}_{22}ex_{22} + \dot{m}_{10}ex_{10} = \dot{m}_{23}ex_{23} + \dot{E}x_{d,PH4}$

The balance equations for preheater 5 can be expressed as:

MBE: $\dot{m}_{24} + \dot{m}_{9} = \dot{m}_{25}$
EBE: $\dot{m}_{24}h_{24} + \dot{m}_{9}h_{9} = \dot{m}_{25}h_{25}$
EnBE: $\dot{m}_{24}s_{24} + \dot{m}_{9}s_{9} + \dot{S}_{gen,PH5} = \dot{m}_{25}s_{25}$
ExBE: $\dot{m}_{24}ex_{24} + \dot{m}_{9}ex_{9} = \dot{m}_{25}ex_{25} + \dot{E}x_{d,PH5}$

The balance equations for preheater 6 can be expressed as:

MBE: $\dot{m}_{26} + \dot{m}_{7} = \dot{m}_{27}$
EBE: $\dot{m}_{26}h_{26} + \dot{m}_{7}h_{7} = \dot{m}_{27}h_{27}$
EnBE: $\dot{m}_{26}s_{26} + \dot{m}_{7}s_{7} + \dot{S}_{gen,PH6} = \dot{m}_{27}s_{27}$
ExBE: $\dot{m}_{26}ex_{26} + \dot{m}_{7}ex_{7} = \dot{m}_{27}ex_{27} + \dot{E}x_{d,PH6}$

The balance equations for preheater 7 can be written as:

MBE: $\dot{m}_{28} + \dot{m}_{4} = \dot{m}_{1}$
EBE: $\dot{m}_{28}h_{28} + \dot{m}_{4}h_{4} = \dot{m}_{1}h_{1}$
EnBE: $\dot{m}_{28}s_{28} + \dot{m}_{4}s_{4} + \dot{S}_{gen,PH7} = \dot{m}_{1}s_{1}$
ExBE: $\dot{m}_{28}ex_{28} + \dot{m}_{4}ex_{4} = \dot{m}_{1}ex_{1} + \dot{E}x_{d,PH7}$

In order to reheat the steam for further expansion in the low-pressure turbine, reheater component is used, which can be balanced as follows:

MBE: $\dot{m}_{6} = \dot{m}_{8}$
EBE: $\dot{m}_{6}h_{6} + \dot{Q}_{RH} = \dot{m}_{8}h_{8}$
EnBE: $\dot{m}_{6}s_{6} + \frac{\dot{Q}_{RH}}{T_b} + \dot{S}_{gen,RH} = \dot{m}_{8}s_{8}$
ExBE: $\dot{m}_{6}ex_{6} + \dot{E}x_{\dot{Q}_{RH}} = \dot{m}_{8}ex_{8} + \dot{E}x_{d,RH}$

By utilizing the balance equations, thermophysical properties can be completed as given in Table 7.3.

For the performance evaluation of the regenerative steam Rankine cycles, both oil- and coal-fired systems will be calculated in the same way. The net power output

Table 7.4 The results summary of the power plants with regenerative Rankine cycle

Parameter	Oil power plant	Coal power plant
Steam generator flowrate	302.7 kg/s	397.4 kg/s
Reheated steam flowrate	280 kg/s	306 kg/s
Heat input to the steam generator	637.5 MW	731.7 MW
Heat input to the reheater	117 MW	198.6 MW
Total heat input	754.5 MW	930.3 MW
Power generation of the low-pressure turbine	284.1 MW	374.9 MW
Power generation of the high-pressure turbine	95.1 MW	157.6 MW
Power consumption of the pumps	66.7 MW	101.6 MW
Net power generation	312.6 MW	430.9 MW
Energy efficiency	41.42%	46.31%
Exergy efficiency	61.13%	65.23%

need to be calculated as a useful output by removing the power consumption of pump from the power generation of the turbine as follows:

$$\dot{W}_{net} = \dot{W}_t - \dot{W}_p$$

Considering the heat inputs of the regenerative Rankine cycle, the heat (thermal energy) enters the system at both steam generator and reheater components. Therefore, the total heat input can be calculated as:

$$\dot{Q}_{in} = \dot{Q}_{SG} + \dot{Q}_{RH}$$

In order to calculate energy and exergy efficiencies, net power can be considered as the useful output and divided into the total heat input as follows:

$$\eta_{en,ov} = \frac{\dot{W}_{net}}{\dot{Q}_{in}}$$

and

$$\eta_{ex,ov} = \frac{\dot{W}_{net}}{\dot{E}x^{\dot{Q}_{in}}}$$

where $\dot{E}x^{\dot{Q}_{in}}$ is the thermal exergy rate of the heat input at the steam generation step (Table 7.4).

Both of the regenerative cycles are analyzed, and different results are obtained. A brief explanation can be made as the temperature and pressure lead efficiency changes. In the oil power plant, the system to reach a maximum temperature of 535°C and a maximum pressure of 139 bar.

In order to effectively use the waste heat from the turbine, multiple outlets are designed to feed a series of preheaters. In the oil power plant, six preheaters are used to heat 33°C condensed water to 253°C, which is an important leap in terms of heat additions. Moreover, coal power plant is designed to reach higher temperatures and higher pressures, with a maximum temperature of 620°C and a maximum pressure of 285 bar. For the effective preheating process, seven preheaters are used in the coal power plant, which is more than the oil power plant.

In the oil power plant, 637.5 MW of heat entered from the steam generator and a 117 MW of heat entered from the reheater; in total, 754.5 MW of heat input is calculated. The high-pressure turbine generates 95.1 MW of power and the low-pressure turbine generates 284.1 MW of power. The total pumping power is calculated as 66.7 MW. The net power generation of the oil power plant is calculated as 312.6 MW.

When we look at the coal power plant, a 731.7 MW of heat entered from the steam generator and a 198.6 MW of heat input is found for the reheater. In total, 930.3 MW of heat enters the coal power plant. Net power generation is calculated as 430.9 MW, where 157.6 MW is generated by the high-pressure turbine, 374.9 MW of power is generated by the low-pressure turbine, and 101.6 MW of power consumption is calculated for total pumping of the system.

The overall energy and exergy efficiencies for the oil power plant is found as 41.42% and 61.13%. A better overall energy and exergy efficiencies are found for the coal power plant as 46.31% and 65.23%. Figure 7.18 compares the overall energy and exergy efficiencies of both coal and oil-based regenerative steam Rankine cycles.

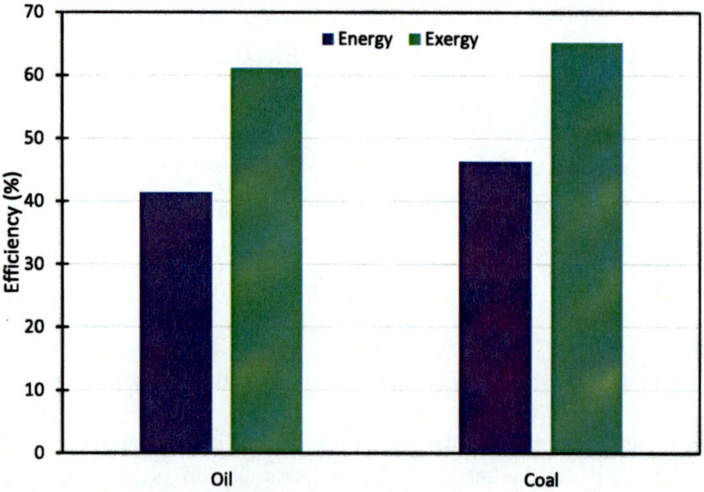

Fig. 7.18 Energy and exergy efficiencies of oil- and coal-driven regenerative steam Rankine cycles

7.3.2 *Organic Rankine Cycle*

Organic Rankine cycles are principally the same process as the steam Rankine cycle, only difference is the working fluid. In steam Rankine cycles, water is used as the thermofluid; on the other hand, organic fluids, such as hydrocarbons or carbon dioxide are used as the thermofluid in organic Rankine cycles.

Waste heat recovery and geothermal are the typical applications for the organic Rankine cycle systems. When we look at commercial systems, there are power generation systems available as a product, typically between 6 kW and 70 MW of power range and 80–400°C of temperature range. R134a, n-pentane, and octamethyltrisiloxane are some of the commercially available fluids for the organic Rankine cycles.

A simple organic Rankine cycle is depicted in Fig. 7.19. There are four steps available in the simple organic Rankine cycle as:

- The pressurization of the organic fluid in the pump component between state points 1 and 2.
- The organic fluid is superheated in the steam generator between state points 2 and 3.
- The superheated organic fluid enters the organic cycle turbine, where it expands and generates power between state points 3 and 4.
- The condensation process occurs between state points 4 and 1 at the condenser component.

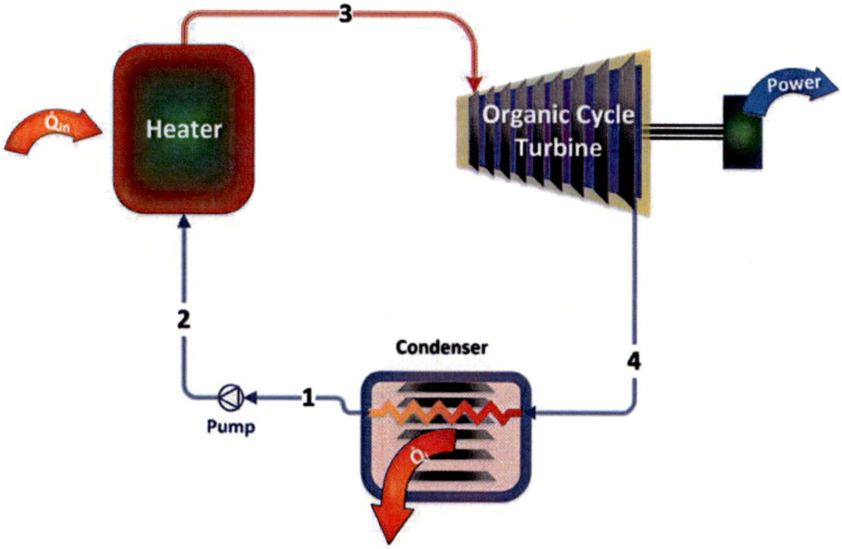

Fig. 7.19 Illustration of a simple organic Rankine cycle

For the performance evaluation of the organic Rankine cycle, heat and work inputs, along with entropy and exergy changes, need to be calculated by using the balance equations.

For the pump component in Fig. 7.19, the balance equations can be expressed as follows:

MBE: $\dot{m}_1 = \dot{m}_2$
EBE: $\dot{m}_1 h_1 + \dot{W}_p = \dot{m}_2 h_2$
EnBE: $\dot{m}_1 s_1 + \dot{S}_{gen,p} = \dot{m}_2 s_2$
ExBE: $\dot{m}_1 ex_1 + \dot{W}_p = \dot{m}_2 ex_2 + \dot{E}x_{d,p}$

The superheating process of the thermofluid in the heater can be balanced as follows:

MBE: $\dot{m}_2 = \dot{m}_3$
EBE: $\dot{m}_2 h_2 + \dot{Q}_H = \dot{m}_3 h_3$
EnBE: $\dot{m}_2 s_2 + \frac{\dot{Q}_H}{T_b} + \dot{S}_{gen,H} = \dot{m}_3 s_3$
ExBE: $\dot{m}_2 ex_2 + \dot{E}x_{\dot{Q}_H} = \dot{m}_3 ex_3 + \dot{E}x_{d,H}$

The power generation with the expansion of the organic thermofluid in the turbine can be expressed as follows:

MBE: $\dot{m}_3 = \dot{m}_4$
EBE: $\dot{m}_3 h_3 = \dot{W}_t + \dot{m}_4 h_4$
EnBE: $\dot{m}_3 s_3 + \dot{S}_{gen,t} = \dot{m}_4 s_4$
ExBE: $\dot{m}_3 ex_3 = \dot{W}_t + \dot{m}_4 ex_4 + \dot{E}x_{d,t}$

The heat output from the condenser along with other parameters can be balanced as:

MBE: $\dot{m}_4 = \dot{m}_1$
EBE: $\dot{m}_4 h_4 = \dot{m}_1 h_1 + \dot{Q}_l$
EnBE: $\dot{m}_4 s_4 + \dot{S}_{gen,CON} = \dot{m}_1 s_1 + \frac{\dot{Q}_l}{T_b}$
ExBE: $\dot{m}_4 ex_4 = \dot{m}_1 ex_1 + \dot{E}x_{\dot{Q}_l} + \dot{E}x_{d,l}$

For the energy and exergy efficiencies of the organic Rankine cycle, net power generation is considered as the useful output and the heat input from the heater is considered as the total input as follows:

$$\eta_{en,ov} = \frac{\dot{W}_{net}}{\dot{Q}_H}$$

and

$$\eta_{ex,ov} = \frac{\dot{W}_{net}}{\dot{E}x_{\dot{Q}_H}}$$

where $\dot{E}x_{\dot{Q}_H}$ is the exergy version of the heat input at the heater.

Table 7.5 The design parameters of the organic Rankine cycle

Parameter	Value
Substance	n-octane
Heater temperature	275.8°C
Heater pressure	2000 kPa
Condensing pressure	35.7 kPa

Table 7.6 The thermophysical properties of the thermofluid in the organic Rankine cycle

State Point	Substance	ṁ (kg/s)	T (°C)	P (kPa)	h (kJ/kg)	s (kJ/kgK)	ex (kJ/kg)
1	n-octane	11	91.8	35.7	157	0.476	15.6
2	n-octane	11	92.7	2000	161	0.4781	18.8
3	n-octane	11	275.8	2000	708	1.674	209.2
4	n-octane	11	159.7	35.7	625	1.723	111.2

Illustrative Example 7.2 An illustrative example is carried out by considering the simple organic Rankine cycle configuration in Fig. 7.19.

As the thermofluid, n-octane is used in the cycle. N-octane is pressurized in the pump component from 35.7 kPa to 2000 kPa. The pressurized n-octane enters the heater component where it is heated to 275.8°C. At state point 3, n-octane enters the organic cycle turbine, in order to expand and generate power. At the exit of the turbine, pressure drops to the 35.7 kPa and temperature drops to the 159.7°C. Design parameters of the organic Rankine cycle is given in Table 7.5.

Thermophysical properties are given for the designed organic Rankine cycle in Table 7.6. Mass flow rate of the cycle is designed as 11 kg/s. According to the designed parameters, 6018 kW of heat enters the heater and superheats n-octane in the heater. If the source temperature is considered as 600 K, the exergy version of the heat input is calculated as 3059.2 kW. The expansion of n-octane in turbine can generate 917.4 kW of power. On the other hand, pressurization of n-octane needs 41.8 kW of power. The energy and exergy efficiencies of the organic Rankine cycle are 14.55% and 28.62%, according to the calculations (Table 7.6).

7.4 Gas Cycles

Gas cycles are basically the power generation cycles where thermofluid is always in the gaseous form. Especially for mobility, gas cycles are the dominant cycle. Otto and Diesel cycles are the main cycles for transportation. Carnot, Stirling, and Ericson cycles are the theoretical cycles which are considered totally reversible. Similar to the vapor cycles, gas cycles also consist of four steps.

The general process of conventional gas cycles begins with the utilization of fossil fuels. Typically, fossil fuels with oxidizing agent carry out an exothermic reaction, where heat or hot gases are used to create movement, as shown in Fig. 7.20.

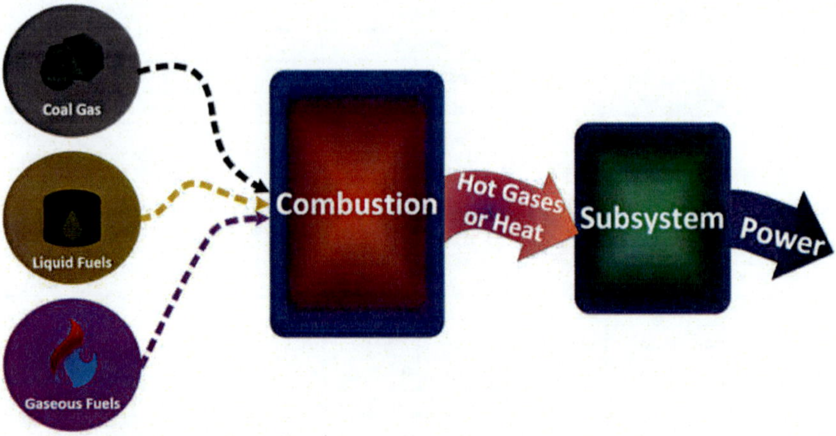

Fig. 7.20 General pathway of the gas cycles from source to useful output

In open systems, fuel and air enter the gas cycle and then the exhaustion process occurs where the combustion products released to the open ambient. On the other hand, in closed systems, only heat and work pass through the gas cycle boundary, and no mass transfer occurs in or out the cycle.

Combustion is the process of utilizing gaseous or liquid fuels with an oxidizing agent in order to generate heat or hot gases for further processes. According to its occurrence, there are open, closed, and adiabatic/non-adiabatic combustion systems, as shown in Fig. 7.21. According to the heat transfer from and to the combustion chamber, the process can be categorized as adiabatic or non-adiabatic. In adiabatic combustion, there is no heat transfer occurrs in or out of the system boundary. All the heat generated from the system is transferred into the product of the chemical reaction. In the non-adiabatic processes, heat is transferred from and to the system boundary. Apart from heat transfer, mass transfer also classifies the combustion system. Typically, fuel and air enter the system and combustion products leave the system, in the open systems. No mass transfer occurs in the closed systems, only heat can be transferred.

In order to express the mass balance of the systems, the mass balance equation of the open system (Fig. 7.21a) and the adiabatic system (Fig. 7.21c) can be introduced accordingly as follows:

$$\sum \dot{n}_R = \sum \dot{n}_P$$

where $\dot{n}$ denotes the mole flow rate of the substances in kmol/s. For the closed system (Fig. 7.21b), there is no mass flow, however, the initial and final balances can be written as follows:

$$\sum n_R = \sum n_P$$

where n denotes the mole of the substances in kmol.

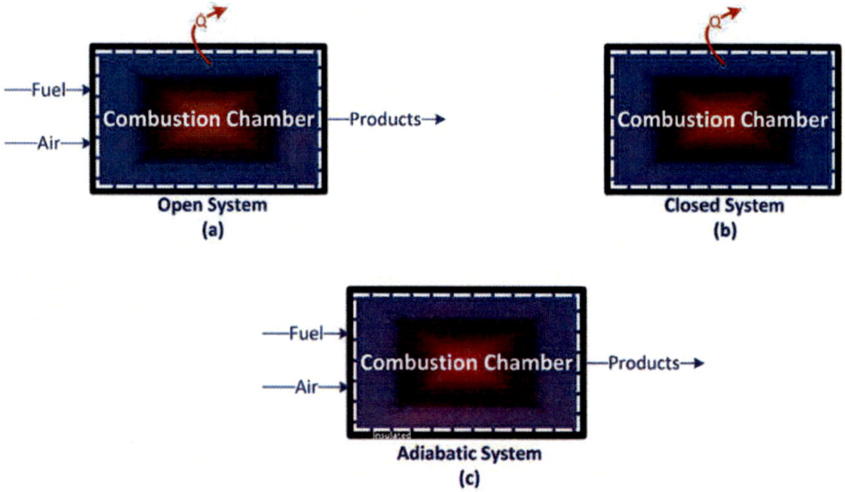

Fig. 7.21 Classification of the combustion systems

The energy balance equation of the open system in Fig. 7.21a can be written as follows:

$$H_R = H_P + \dot{Q}$$

where

$$H_R = \sum \dot{n}_R \left(\overline{h_f{}^\circ} + \overline{h} - \overline{h^\circ} \right)_R$$

and

$$H_P = \sum \dot{n}_P \left(\overline{h_f{}^\circ} + \overline{h} - \overline{h^\circ} \right)_P$$

where $\overline{h_f{}^\circ}$ denotes the enthalpy of formation, $\overline{h}$ denotes the enthalpy value obtained at the specific temperature of the substance, and $\overline{h^\circ}$ denotes the enthalpy value at the reference temperature. The energy balance equation for the closed system in Fig. 7.21b is then expressed as follows:

$$U_R = U_P + Q$$

where

$$U_R = \sum n_R \left(\overline{h_f{}^\circ} + \overline{h} - \overline{h^\circ} - P\overline{v} \right)_R$$

and

$$U_{\mathrm{P}} = \sum n_{\mathrm{P}} \left(\overline{h_{\mathrm{f}}{}^{\circ}} + \overline{h} - \overline{h^{\circ}} - P\overline{v} \right)_{\mathrm{P}}$$

In the adiabatic system in Fig. 7.21c, there is no heat transfer occurs in and out the system. According to that, the energy balance equation is:

$$H_{\mathrm{R}} = H_{\mathrm{P}}$$

where

$$H_{\mathrm{R}} = \sum \dot{n}_{\mathrm{R}} \left(\overline{h_{\mathrm{f}}{}^{\circ}} + \overline{h} - \overline{h^{\circ}} \right)_{\mathrm{R}}$$

and

$$H_{\mathrm{P}} = \sum \dot{n}_{\mathrm{P}} \left(\overline{h_{\mathrm{f}}{}^{\circ}} + \overline{h} - \overline{h^{\circ}} \right)_{\mathrm{P}}$$

The entropy balance equation for an open system as shown in Fig. 7.21a can be written as follows:

$$\dot{S}_{\mathrm{R}} + \dot{S}_{\mathrm{gen}} = \dot{S}_{\mathrm{P}} + \frac{\dot{Q}}{T_{\mathrm{comb}}}$$

where

$$\dot{S}_{\mathrm{R}} = \sum (\dot{n}\overline{s})_{\mathrm{R}}$$

and

$$\dot{S}_{\mathrm{R}} = \sum (\dot{n}\overline{s})_{\mathrm{P}}$$

where $\overline{s}$ stands for the absolute entropy for a specific temperature (T) and pressure (P), that can be further expressed as follows:

$$\overline{s}(T, P) = \overline{s^{\circ}}(T_0, P) - R_{\mathrm{u}} \ln\left(\frac{P}{P_0}\right)$$

where $\overline{s^{\circ}}$ denotes the reference absolute entropy at the reference temperature and pressure, R_{u} denotes the universal gas constant, and P_0 refers to the atmospheric pressure, T_0 refers to the reference temperature.

The entropy balance equation for the closed system in Fig. 7.21b is:

$$S_{\mathrm{R}} + S_{\mathrm{gen}} = S_{\mathrm{P}} + \frac{Q}{T_{\mathrm{comb}}}$$

where

$$S_R = \sum (n\bar{s})_R$$

and

$$S_R = \sum (n\bar{s})_P$$

And the entropy balance equation for the adiabatic system (Fig. 7.21c) is then obtained as follows:

$$\dot{S}_R + \dot{S}_{gen} = \dot{S}_P$$

For the open system as illustrated in Fig. 7.21a, the exergy balance equation is:

$$\dot{Ex}_R = \dot{Ex}_P + \dot{Ex}^Q + \dot{Ex}_d$$

where the exergy destruction ($\dot{Ex}_d$) is written through the entropy generation and the reference temperature as:

$$\dot{Ex}_d = T_0\dot{S}_{gen}$$

Moreover, the exergy flows of reactants ($\dot{Ex}_R$) and products ($\dot{Ex}_P$) can be expressed as:

$$\dot{Ex}_R = \sum \dot{n}_R \left(\overline{h_f}^\circ + \bar{h} - \bar{h}^\circ - T_0\bar{s}\right)_R$$

and

$$\dot{Ex}_P = \sum \dot{n}_P \left(\overline{h_f}^\circ + \bar{h} - \bar{h}^\circ - T_0\bar{s}\right)_P$$

For the closed system as illustrated in Fig. 7.19b, the exergy balance equation can be denoted as:

$$Ex_R = Ex_P + Ex^Q + Ex_d$$

where the exergy destruction (Ex_d) can be expressed from entropy generation and the reference temperature as:

$$Ex_d = T_0 S_{gen}$$

Moreover, the exergy of reactants (Ex_R) and products (Ex_P) can be expressed as:

$$Ex_R = \sum n_R \left(\overline{h_f}^\circ + \bar{h} - \overline{h^\circ} - P\bar{v} - T_0\bar{s}\right)_R$$

and

$$Ex_p = \sum n_P \left(\overline{h_f}^\circ + \bar{h} - \overline{h^\circ} - P\bar{v} - T_0\bar{s}\right)_P$$

For the adiabatic system (as clearly seen in Fig. 7.21c), the exergy balance equation is then expressed as:

$$\dot{E}x_R = \dot{E}x_P + \dot{E}x_d$$

It is important to highlight that, air-fuel ratio is one of the critical parameters of the utilization of fossil fuels in traditional energy systems. According to the air fuel ratio (AFR), a burning process can occur with lower than required oxidizing agent, higher than required oxidizing agent, or the equal amount of oxidizing to the requirement. The complete combustion refers to the process where there is just enough amount of the oxidizing agent to completely burn the fuel. If the amount of oxidizing agent higher than then the air required, the process occurs with excess air. On the other hand, if the process happens with the less amount of air required., the process is called incomplete combustion. AFR can be calculated as follows:

$$\text{AFR} = \frac{m_{\text{air}}}{m_{\text{fuel}}}$$

The heating values are another parameters that is important for the combustion process. There are lower and higher heating values, where the higher heating value (HHV) is calculated by including the heat of vaporization of water, and the lower heating value (LHV) is calculated without it as follows:

$$\text{HHV}\left(\frac{\text{kJ}}{\text{kg}}\right) = \text{LHV}\left(\frac{\text{kJ}}{\text{kg}}\right) + x_{\text{H}_2\text{O}} h_{fg,\text{H}_2\text{O}}$$

where the HHV at reference conditions is defined as:

$$\text{HHV} = \sum n_P \left(\overline{h_f}^\circ\right)_P - \sum n_R \left(\overline{h_f}^\circ\right)_R$$

Here, HHV includes the enthalpy of water/steam mixture, on the other hand, LHV does not include the enthalpy of water/steam mixture.

7.4.1 Carnot Cycle

Sadi Carnot proposed two different cycles, to generate power or to upgrade/downgrade the heat for heat pump/refrigeration process. Carnot cycle is a conceptual model where the process is reversible, which is the upper limit for practical applications. The process for the heat engine in Carnot cycle is depicted in Fig. 7.22. A Carnot cycle utilizes high temperature heat source in order to generate power and low temperature heat, when it works as a heat engine. Since the process is reversible, efficiency of Carnot heat engine in Fig. 7.22 can be expressed as follows:

$$\eta_{\text{rev}} = 1 - \frac{T_L}{T_H}$$

where T_L and T_H are the heat sink and heat source temperatures in Kelvin.

Efficiency calculation for the Carnot heat engine is practical to understand the limits of the heat engine. Carnot cycle considers processes where the ideal gas is compressed with an adiabatic and isentropic compressor, later on heat exchanger transferred the heat from heat source to the ideal gas, after that, the ideal gas is expanded by the turbine with an adiabatic and isentropic process. When the energy

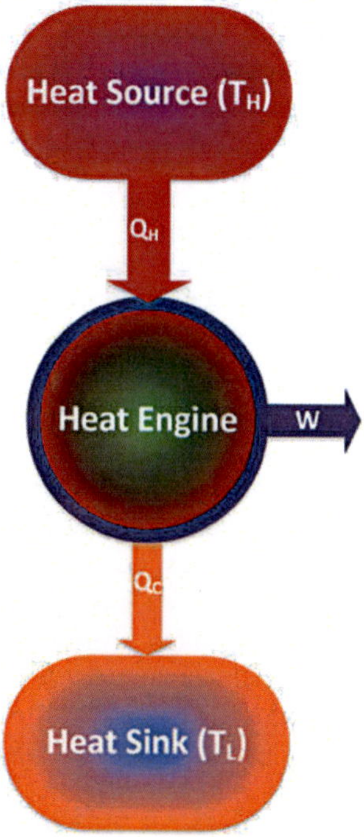

Fig. 7.22 Schematic diagram of a heat engine inputs and outputs

balance of a reversible process is considered, the energy efficiency of a Carnot cycle can be written as:

$$\eta_{\text{en}} = \frac{W_{\text{net}}}{Q_H}$$

where

$$W_{\text{net}} = Q_H - Q_C$$

from there, it can be recalculated as:

$$\eta_{\text{en}} = 1 - \frac{Q_C}{Q_H}$$

where Q_H and Q_C are the heat input from the hot side and heat output to the cold side.

7.4.2 Brayton Cycle

Among the conventional methods, one of the most efficient ways to turn liquid or gaseous fuels into electricity is the gas turbine. Brayton cycle is typically considered for large-scale applications such as power plants.

On the other hand, the Otto and Diesel cycles are more suitable for smaller scale, mobile applications. Natural gas and oil are suitable for gas turbines as well as Brayton cycles. However, in order to use coal, coal gas can be generated from coal when the coal is heated strongly in the absence of air, such as the coal gasification reaction.

Compared to the Rankine cycle, Brayton cycles or gas turbine cycles offer better efficiency and shorter startup times, typically. Especially the Brayton-Rankine combined cycles can pass more than 55% efficiencies since the exhaust heat can be recovered more than 600 K temperatures, which is very suitable for Rankine cycle. The working fluids for the Brayton cycle are the combustion gases, which is the mixture of air and gaseous or liquid fuel. Typically, excess air is used with the Brayton cycle to prevent unburned fuels.

A simple configuration of the Brayton cycle consists of four steps as shown in Fig. 7.23. In the closed-type Brayton cycle (Fig. 7.23a), heat input from the combustion enters the heat exchanger to heat the thermofluid which is typically dry air or helium. On the other hand, fuel and compressed air enters the combustion chamber and combustion occurs in the open-type Brayton cycle (Fig. 7.23b). In the open-type Brayton cycles, air inlet (state point 1) and gas exit (state point 4) need to be equal to the atmospheric pressure since these points are open to the atmosphere. On the other hand, the closed-type Brayton cycles enable the expand gases to the

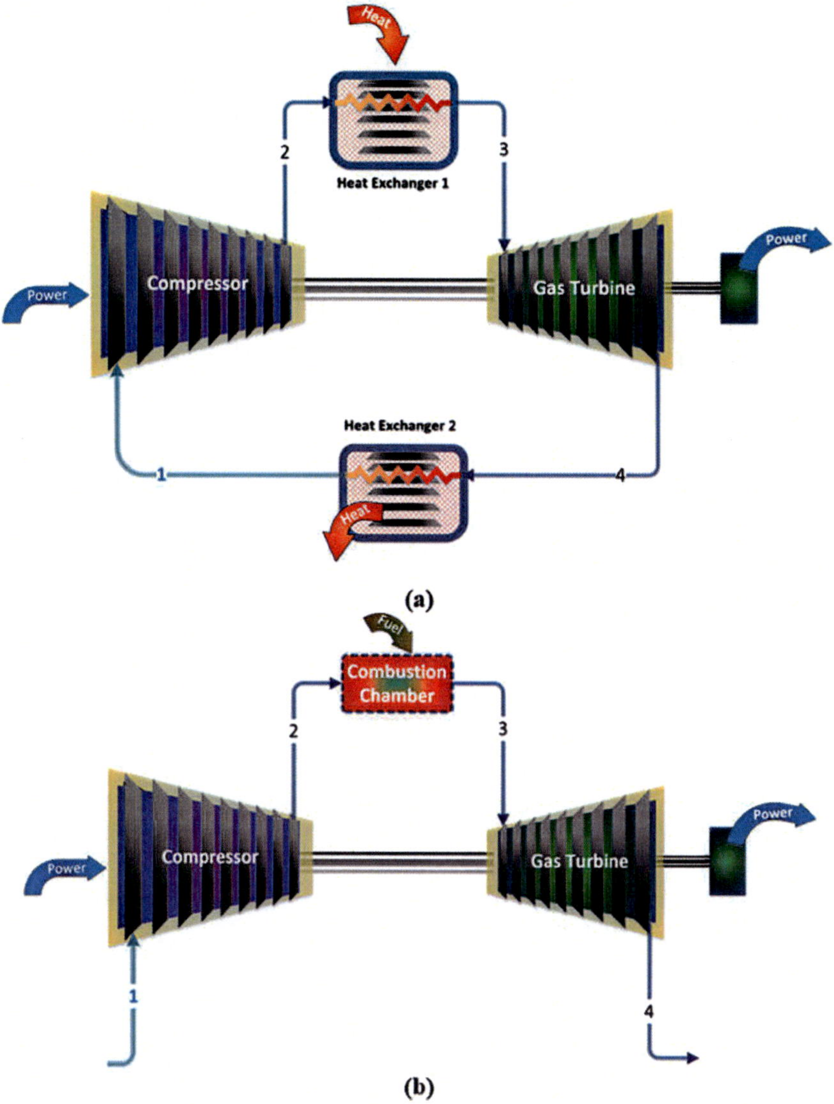

Fig. 7.23 Schematics of (**a**) a closed type and (**b**) an open type of Brayton cycles

vacuum pressures, which brings an enhanced power generation. A temperature-specific entropy diagram is depicted in Fig. 7.24.

Balance equations of both closed and open types of Brayton cycles are the same. As mentioned earlier, the working fluid is compressed according to the compression ratio. For the compressor, the balance equations can be written as:

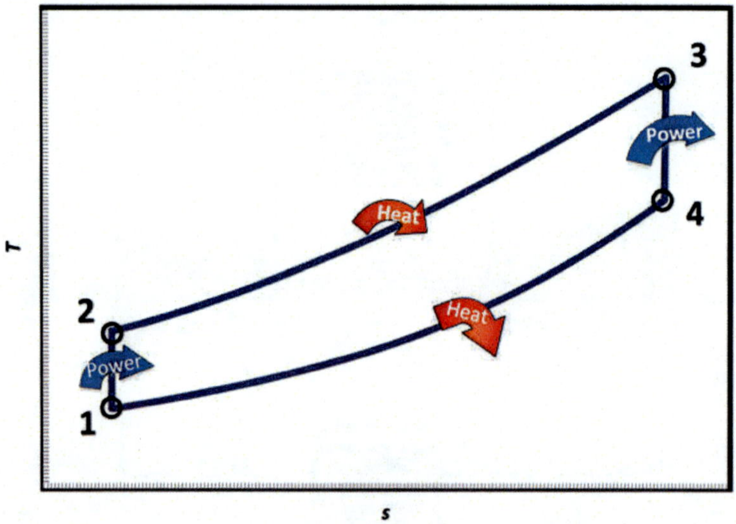

Fig. 7.24 Representation of an ideal Brayton cycle in the temperature-specific entropy (T-s) diagram

MBE: $\dot{m}_1 = \dot{m}_2$
EBE: $\dot{m}_1 h_1 + \dot{W}_c = \dot{m}_2 h_2$
EnBE: $\dot{m}_1 s_1 + \dot{S}_{gen,c} = \dot{m}_2 s_2$
ExBE: $\dot{m}_1 ex_1 + \dot{W}_c = \dot{m}_2 ex_2 + \dot{E}x_{d,c}$

In the second step, either heat from combustion chamber heated the working fluid or the combustion occurred. In order to simplify the equations, two assumptions are made: (i) the fuel mass is not considered in the equations due to its very low mass content compared to the air, (ii) the heat from combustion process is considered as the heat input to the component. The balance equations for the combustion chamber/heat exchanger are:

MBE: $\dot{m}_2 = \dot{m}_3$
EBE: $\dot{m}_2 h_2 + \dot{Q}_{com} = \dot{m}_3 h_3$
EnBE: $\dot{m}_2 s_2 + \frac{\dot{Q}_{com}}{T_b} + \dot{S}_{gen,com} = \dot{m}_3 s_3$
ExBE: $\dot{m}_2 ex_2 + \dot{E}x^{\dot{Q}_{com}} = \dot{m}_3 ex_3 + \dot{E}x_{d,com}$

In the gas turbine, the working fluid is expanded, and power is generated, which can be balanced as:

MBE: $\dot{m}_3 = \dot{m}_4$
EBE: $\dot{m}_3 h_3 = \dot{W}_t + \dot{m}_4 h_4$
EnBE: $\dot{m}_3 s_3 + \dot{S}_{gen,t} = \dot{m}_4 s_4$
ExBE: $\dot{m}_3 ex_3 = \dot{W}_t + \dot{m}_4 ex_4 + \dot{E}x_{d,t}$

This step is only available in the closed-type Brayton cycle (Fig. 7.23a), since the gases at turbine outlet are exhausted to the atmosphere in the open-type Brayton cycle. The balance equations for cooling the working fluid in the heat exchanger of the closed-type Brayton cycle are:

MBE: $\dot{m}_4 = \dot{m}_1$

EBE: $\dot{m}_4 h_4 = \dot{m}_1 h_1 + \dot{Q}_{HX2}$

EnBE: $\dot{m}_4 s_4 + \dot{S}_{gen,HX2} = \dot{m}_1 s_1 + \frac{\dot{Q}_{HX2}}{T_b}$

ExBE: $\dot{m}_4 ex_4 = \dot{m}_1 ex_1 + \dot{E}x^{\dot{Q}_{HX2}} + \dot{E}x_{d,HX2}$

To evaluate the performance of the Brayton cycle, the energy and exergy efficiencies can be calculated. The useful output from the Brayton cycle is the power generated by the turbine. However, a portion of the power is used in the compressor. The net power output can be calculated as:

$$\dot{W}_{net} = \dot{W}_t - \dot{W}_c$$

The net power generation can be considered as the net useful output in performance calculations. According to these, the energy and exergy efficiencies of the Brayton cycle can be calculated as follows:

$$\eta_{en,ov} = \frac{\dot{W}_{net}}{\dot{Q}_{com}}$$

and

$$\eta_{ex,ov} = \frac{\dot{W}_{net}}{\dot{E}x^{\dot{Q}_{com}}}$$

where $\dot{E}x^{\dot{Q}_{com}}$ is the exergy version of the heat input from the combustion process.

Illustrative Example 7.3 An illustrative example is carried out by considering an open-type Brayton cycle depicted in Fig. 7.23b. A compression ratio of 10 is used, where the data for state points are used from [4].

As the working fluid, air is used in the cycle since and open-type cycle is considered. The air is pressurized in the compressor from atmospheric pressure with the compression ratio of 10. The pressurized air is heated in the heat exchanger and reached at 960.85°C and 1013.25 kPa. The hot and pressurized air enters the turbine and expands there to generate power. The exhaust gases are released to the atmosphere at state point 4. The thermophysical properties of the open type of Brayton cycle are given in Table 7.7.

Table 7.7 The thermophysical properties of the air in an open type of Brayton cycle

State point	Substance	ṁ (kg/s)	T (°C)	P (kPa)	h (kJ/kg)	s (kJ/kgK)	ex (kJ/kg)
1	Air	292.8	25	101.325	298.6	5.69	0
2	Air	292.8	297.85	1013.25	576.9	5.69	278.3
3	Air	292.8	960.85	1013.25	1318	6.55	764.5
4	Air	292.8	411.85	101.325	698.1	6.55	144.8

The compressor is used 81.5 MW of power in order to pressurize the air. On the other hand, gas turbine generates 181.5 MW of power. When we look at the net power output as follows:

$$\dot{W}_{net} = \dot{W}_t - \dot{W}_c$$

The net power generation is found as 100 MW. The heat input from the combustion process is found as 217 MW. When the source temperature is taken as 1350 K, the exergy of heat input is found as 169.5 MW. Therefore, 46.09% of energy and 58.98% of exergy efficiencies are found for the proposed open-Brayton cycle, according to the calculations.

7.4.3 Otto and Diesel Cycles

Otto and Diesel cycles are suitable cycles for smaller scale applications compared to the Brayton cycle. In transportation, otto and diesel cycles are the dominant cycles. More than 80% of the vehicles are powered by Otto or Diesel cycles. Engine-driven power generators and internal combustion engines are suitable to employ Otto and Diesel cycles.

The Otto cycle is typically used via spark-ignition engines. There are typically four steps in the Otto cycle (as shown in Fig. 7.25) as follows:

- In the process between state points 1 and 2, a compression occurs, which would be isentropic in the ideal case. This process consumes power to compress the gas, however, no heat transfer occurs which makes it an adiabatic process.
- Isochoric heat input occurs in the second process, between state points 2 and 3. During this process, volume is constant, while the pressure and temperature are increased. The heat input in this process can be considered as the total energy input to the system, which is important for the efficiency calculations.
- The working fluid is expanded which creates a movement to change the position of the pistons in this process. In the ideal Otto cycle, the expansion process occurs isentropically.
- Between state points 4 and 1, an isochoric heat rejection process occurs. The volume remains constant, but the temperature and pressure decrease.

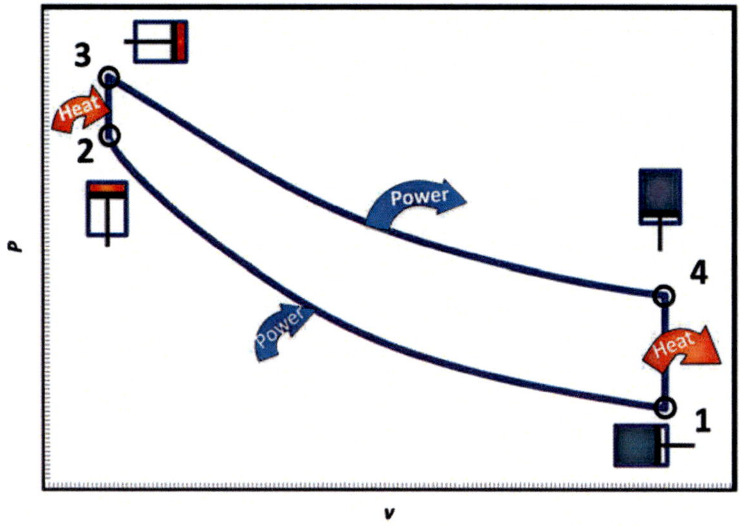

Fig. 7.25 The pressure-specific volume (P-v) diagram of an ideal Otto cycle

For the performance assessment of the Otto cycle, the net power can be calculated to removing the work consumed by the compressor from the work generated by the turbine as follows:

$$\dot{W}_{net} = \dot{W}_t - \dot{W}_c$$

The energy and exergy efficiencies of the Otto cycle can be calculated by dividing the net power output to the heat input as follows:

$$\eta_{en,ov} = \frac{\dot{W}_{net}}{\dot{Q}_{in}}$$

and

$$\eta_{ex,ov} = \frac{\dot{W}_{net}}{\dot{E}x^{\dot{Q}_{in}}}$$

where $\dot{E}x^{\dot{Q}_{in}}$ is the exergy version of the heat input to the system.

The Diesel cycle is typically used for larger scale engine-generator groups compared to the Otto cycle. In the Diesel cycle, auto-ignition is obtained by compressing the air, where spark is not needed. In contrast to the Otto engines, the Diesel engines are considered as compression-ignition engines. At high pressures, air reaches high temperature, once the fuel enters the cylinder, ignition occurs. Compared to the Otto cycle, Diesel cycle has a slower combustion process due to

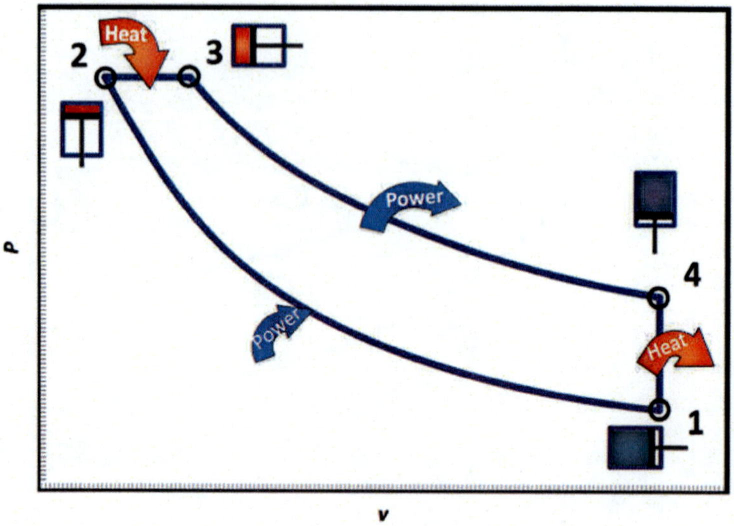

Fig. 7.26 The pressure-specific volume (P-v) diagram of the diesel cycle

the lack of spark. The four steps of the simple Diesel cycle (as shown in Fig. 7.26) can be written as follows:

- The first process is the compression of the working fluid. In the ideal case, this process occurs isentropically. The power is consumed by the compressor, on the other hand, the process is adiabatic.
- Between state points 2 and 3, heat enters the system boundary. During this process, the pressure is maintained constant, but the temperature and the volume are increased.
- In the third process, between state points 3 and 4, air-fuel mixture expands. In the ideal diesel cycle, this process occurs isentropically. Once the volume reaches to the volume at the firs state point, the expansion stops.
- Between state points 4 and 1, heat is rejected at the constant volume.

To evaluate the performance of the Diesel cycle, net power is the useful output which is calculated by removing the work consumed by the compressor from the work generated by the turbine as follows:

$$\dot{W}_{net} = \dot{W}_t - \dot{W}_c$$

The energy and exergy efficiencies of the Diesel cycle can be calculated as follows:

$$\eta_{en,ov} = \frac{\dot{W}_{net}}{\dot{Q}_{in}}$$

and

$$\eta_{ex,ov} = \frac{\dot{W}_{net}}{\dot{E}x^{\dot{Q}_{in}}}$$

where $\dot{E}x^{\dot{Q}_{in}}$ is the exergy version of the heat input to the system.

7.5 Combined Cycles

A combined cycle typically refers to the combination of a vapor cycle and a gas cycle. Exhaust gases from the gas cycles consist of high potential of further utilization. Most of the gas cycles exhaust more than 600 K exhaust gases from the gas turbine. A vapor cycle, such as a Rankine cycle, is very suitable to exploit the excess and turn it into a useful energy. The gas cycle–based conventional power plants can add a waste heat recovery system to recover the heat and typically generate steam for a Rankine cycle. Waste heat recovery systems exploit the rejected low-pressure hot flue gasses from the gas turbine and drive another power cycle. It is critical to design the pressures and temperatures of the integrated cycle in order to maximize the useful outputs.

The two sections where heat is lost from source to service are depicted in Fig. 7.27. When we consider a conventional power generation cycle, there are typically two sections from source to service, where the heat is lost. The first section where heat gets lost is the utilization process of the heat from the source. The combustion process, or burning fossil-based fuels, exploits the chemical energy from the fuel and generates heat. Heat is typically transferred into the cycle, which can be a Rankine cycle or a

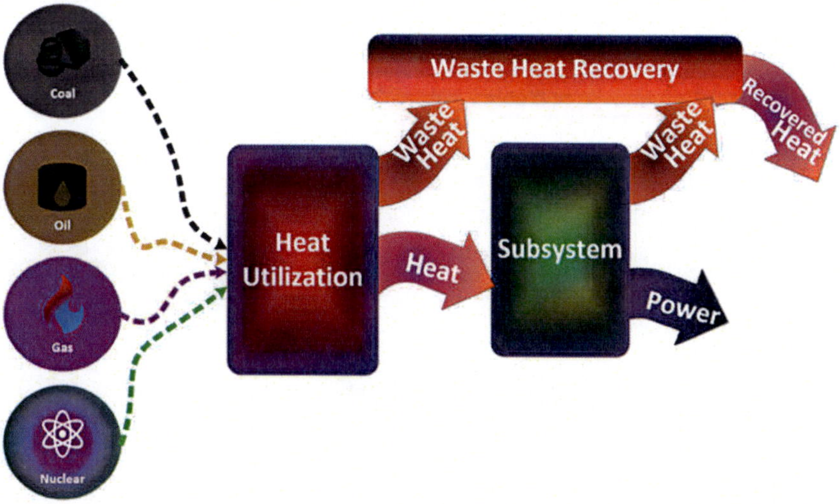

Fig. 7.27 Heat recovery options for the combined cycles

Brayton cycle. The chemical product from the burning process typically consists high temperature of gases, which is exhausted. Apart from that, the cycle that generates power from the heat consists of a step where the working fluid is getting cooled, which is the second section where heat is lost from source to service.

7.5.1 Brayton-Rankine Combined Cycle

The most typical example of the combined cycles is the Brayton-Rankine combined cycle (as shown in Fig. 7.28). Both open- and closed-type of Brayton cycles can be combined with a Rankine cycle. For the open-type Brayton cycle, the expanded gas mixture in the gas turbine, is exhausted at the atmospheric pressure in order to generate power. The exhausted gases can be flown through a heat exchanger, where Rankine cycle's steam generation step conncted. After the heat recovery process, the

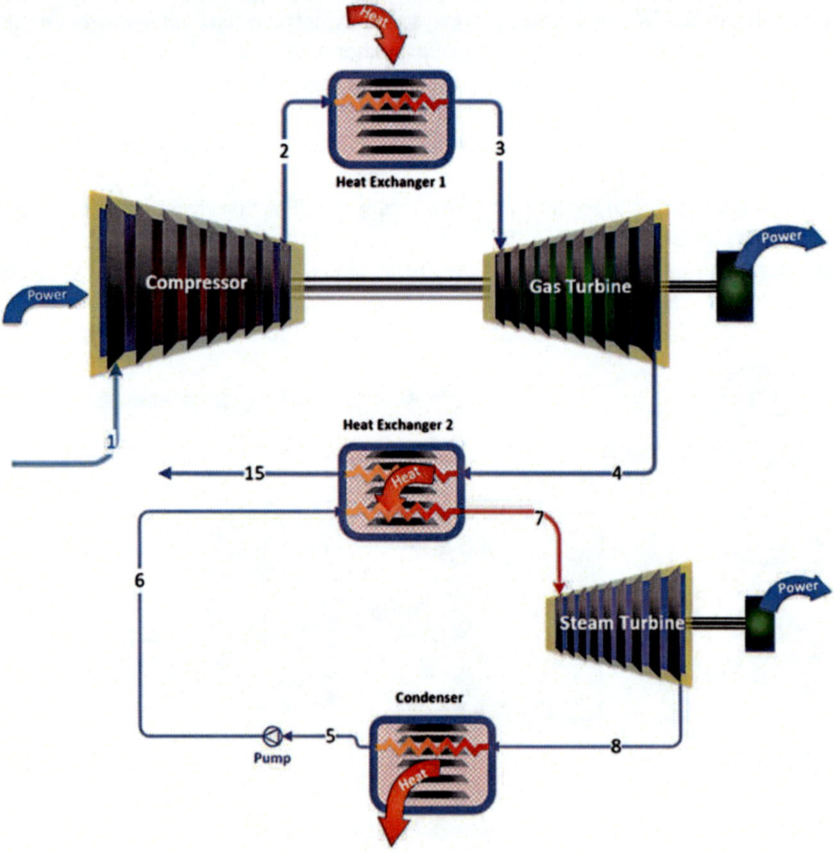

Fig. 7.28 Illustration of a Brayton-Rankine combined cycle

exhaust gases of Brayton cycle can be released to the atmosphere. On the other hand, the gas can be turned within the Brayton cycle and can be combined with the Rankine cycle as shown in Figure 7.28, in the closed-type Brayton-Rankine combined system.

A simple Brayton-Rankine combined cycle is given in Fig. 7.28. The first four step is belonged to the Brayton cycle, where the air is compressed to the desired pressure at the first step, or the working fluid can be used as helium if the process is a closed-type Brayton cycle. At the second step, there are two considerations can be made, either heat from a combustion chamber, or directly a combustion chamber can be used in the process between the state points 2 and 3. In this example, a closed type of system with a heat exchanger at the heat input is used. The heated gases enter to the gas turbine and expand there to generate power. In the closed-type Brayton cycles, the expansion can be done until vacuum pressures, however, the open-type Brayton cycles cannot be expanded lower than atmospheric pressure. In heat exchanger 2, heat is recovered from the expanded gases and transferred into the secondary cycle. In the secondary cycle, the pressurized water from the process between state points 5 and 6 enters the heat exchanger, and the pressurized water is superheated with the heat from the expanded gases. The superheated steam enters the steam turbine, and a second expansion happens in the steam turbine. The expanded steam enters the condenser to release the further heat contained in the vapor and condense the working fluid. The condensing process usually happens under the vacuum pressure conditions. In order to calculate the performance, the balance equations are given below.

The balance equations of both closed and open types of Brayton cycles connected combined cycles are the same. For the compressor, the balance equations can be written as:

MBE: $\dot{m}_1 = \dot{m}_2$
EBE: $\dot{m}_1 h_1 + \dot{W}_c = \dot{m}_2 h_2$
EnBE: $\dot{m}_1 s_1 + \dot{S}_{gen,c} = \dot{m}_2 s_2$
ExBE: $\dot{m}_1 ex_1 + \dot{W}_c = \dot{m}_2 ex_2 + \dot{E}x_{d,c}$

The balance equations for the heat exchanger 1 are:

MBE: $\dot{m}_2 = \dot{m}_3$
EBE: $\dot{m}_2 h_2 + \dot{Q}_{HX1} = \dot{m}_3 h_3$
EnBE: $\dot{m}_2 s_2 + \frac{\dot{Q}_{HX1}}{T_b} + \dot{S}_{gen,HX1} = \dot{m}_3 s_3$
ExBE: $\dot{m}_2 ex_2 + \dot{E}x^{\dot{Q}_{HX1}} = \dot{m}_3 ex_3 + \dot{E}x_{d,HX1}$

In the gas turbine, the working fluid is expanded, and power is generated, which can be formulated as follows:

MBE: $\dot{m}_3 = \dot{m}_4$
EBE: $\dot{m}_3 h_3 = \dot{W}_{gt} + \dot{m}_4 h_4$
EnBE: $\dot{m}_3 s_3 + \dot{S}_{gen,gt} = \dot{m}_4 s_4$
ExBE: $\dot{m}_3 ex_3 = \dot{W}_{gt} + \dot{m}_4 ex_4 + \dot{E}x_{d,gt}$

The balance equations for the heat recovery process at the second heat exchanger are:

MBE: $\dot{m}_4 = \dot{m}_{15}$, $\dot{m}_6 = \dot{m}_7$
EBE: $\dot{m}_4 h_4 + \dot{m}_6 h_6 = \dot{m}_1 h_1 + \dot{m}_7 h_7$
EnBE: $\dot{m}_4 s_4 + \dot{m}_6 s_6 + \dot{S}_{gen,HX2} = \dot{m}_{15} s_{15} + \dot{m}_7 s_7$
ExBE: $\dot{m}_4 ex_4 + \dot{m}_6 ex_6 = \dot{m}_{15} ex_{15} + \dot{m}_7 ex_7 + \dot{E}x_{d,HX2}$

In the steam turbine, the working fluid is expanded, and power is generated, which can be written as follows:

MBE: $\dot{m}_7 = \dot{m}_8$
EBE: $\dot{m}_7 h_7 = \dot{W}_{st} + \dot{m}_8 h_8$
EnBE: $\dot{m}_7 s_7 + \dot{S}_{gen,st} = \dot{m}_8 s_8$
ExBE: $\dot{m}_7 ex_7 = \dot{W}_{st} + \dot{m}_8 ex_8 + \dot{E}x_{d,st}$

The balance equations of the pump can be written as:

MBE: $\dot{m}_8 = \dot{m}_6$
EBE: $\dot{m}_8 h_8 + \dot{W}_p = \dot{m}_6 h_6$
EnBE: $\dot{m}_8 s_8 + \dot{S}_{gen,p} = \dot{m}_6 s_6$
ExBE: $\dot{m}_8 ex_8 + \dot{W}_p = \dot{m}_6 ex_6 + \dot{E}x_{d,p}$

In order to evaluate the performance, there are two power generation and two power consumption steps. The net power generation can be calculated by adding the power generations of the steam and gas turbines and removing the pump and compressor consumptions as follows:

$$\dot{W}_{net} = \dot{W}_{gt} + \dot{W}_{st} - \dot{W}_c - \dot{W}_p$$

The net power generation can be considered as the net useful output in performance calculations. Although at the second heat exchanger, the Rankine cycle is heated, however, to the overall system boundary, there is only one heat addition which is at the heat exchanger 1. According to these, the energy and exergy efficiencies of the Brayton cycle can be calculated as follows:

$$\eta_{en,ov} = \frac{\dot{W}_{net}}{\dot{Q}_{HX1}}$$

and

$$\eta_{ex,ov} = \frac{\dot{W}_{net}}{\dot{E}x^{\dot{Q}_{HX1}}}$$

where $\dot{E}x^{\dot{Q}_{HX1}}$ is the exergy version of the heat input from the combustion process.

Illustrative Example 7.4 An illustrative example is designed for a Brayton-Rankine combined cycle configuration as illustrated in Fig. 7.28.

An open type of Brayton cycle in the illustrative example 7.3 is considered for the primary cycle in this example. In addition to that, a Rankine cycle is added to the system. There are two critical parameters to design the given system. The maximum temperature of the secondary cycle cannot achieve the exhaust temperature of the primary cycle. Moreover, heat transfer rate need to be arranged accordingly from the exhaust gas to the secondary cycle. Between the state points 5 and 6, the water is pressurized via pump component. The pressurized water recovers the heat from exhaust air and gets heated up until 383.1°C. The superheated water expands in the steam turbine and generates power from the secondary cycle. The thermophysical properties of the open type of Brayton cycle combined with a steam Rankine cycle are given in Table 7.8.

In the primary loop, the compressor is used 81.5 MW of power in order to pressurize the air. On the other hand, gas turbine generates 181.5 MW of power. In the secondary loop, the steam turbine generated 9.6 MW of power while the pump component consumed 200 kW of power. When we look at the net power output as follows:

$$\dot{W}_{net} = \dot{W}_{gt} + \dot{W}_{st} - \dot{W}_{c} - \dot{W}_{p}$$

The net power generation is found as 109.4 MW. The total heat input to the system boundary is the heat input from the heat exchanger 1, which is found as 217 MW. When the source temperature is taken as 1373 K, the exergy of heat input is found as 170.4 MW. Therefore, 46.09% of energy and 58.71% of exergy efficiencies are found for the proposed open-Brayton cycle, according to the calculations. When we look at the combined cycle, the energy and exergy efficiencies are found as 50.41% and 64.20%, as shown in Fig. 7.29. Which improved the source utilization and prevented to production of waste in the combined cycle, compared to the only Brayton cycle.

Table 7.8 Thermophysical properties of the thermofluid in the organic Rankine cycle

State point	Substance	$\dot{m}$ (kg/s)	T (°C)	P (kPa)	h (kJ/kg)	s (kJ/kgK)	ex (kJ/kg)
1	Air	292.8	25	101.3	298.6	5.69	0
2	Air	292.8	297.85	1013.2	576.9	5.69	278.3
3	Air	292.8	960.85	1013.2	1318	6.55	764.5
4	Air	292.8	411.85	101.3	698.1	6.55	144.8
5	Water	29.8	180	1000	762.5	2.13	132
6	Water	29.8	180.6	7700	769.2	2.13	138.7
7	Water	29.8	383.1	7700	3098	6.32	1218.3
8	Water	29.8	180	1000	2777	6.56	825.7
15	Air	292.8	185	101.3	460	6.13	32

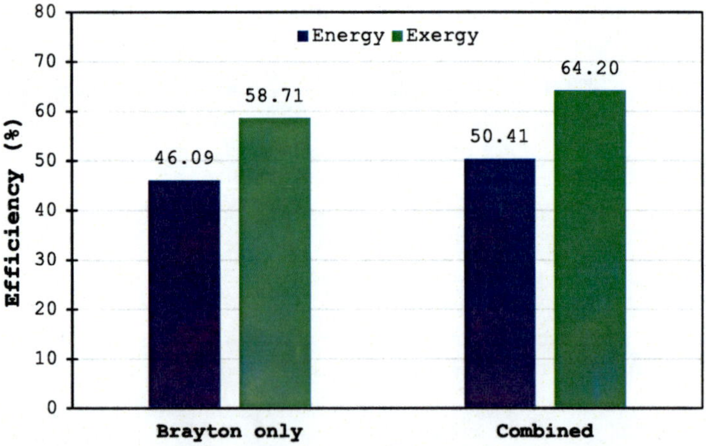

Fig. 7.29 The energy and exergy efficiencies of Brayton only and combined cycles

7.5.2 Combined Heat and Power

During the traditional energy cycles, the heat is rejected during the heat utilization process as well as in the subsystem, from source to service. Waste heat is harmful to the environment in fact it consists of energy that can be converted into useful outputs. A heat recovery system can be developed, and the waste heat can be recovered. In conventional combined heat and power plants, waste heat from the process is recovered and used for space heating or domestic heating applications.

Combined heat and power applications are commercially mature technologies that are widely accepted in the energy industry. Already available power plants can adapt this technology and add a heat recovery component to drive the heating systems by utilizing the waste heat rather than the direct use of fossil fuels for heating. The majority of space heating applications are driven by fossil fuels. From this aspect, replacing fossil fuels with waste heat would contribute significantly to reducing greenhouse gas emissions.

Illustrative Example 7.5 An illustrative example is designed for a combined heat and power system, where the power generation part is used from the illustrative example 7.4.

Traditional fuels are utilized to obtain heat as the first step of the process. A Brayton cycle is used to generate power, where cooling process is also occurred to release a portion of heat. A Rankine cycle is employed in order to exploit the waste heat from the Brayton cycle process. Moreover, the Rankine cycle has as condensing step where the heat is rejected from the thermofluid. In this step, a district heating system is added, which is illustrated in the energy flow (Fig. 7.30). The overall integrated energy system for heat and power generation is illustrated in Fig. 7.31.

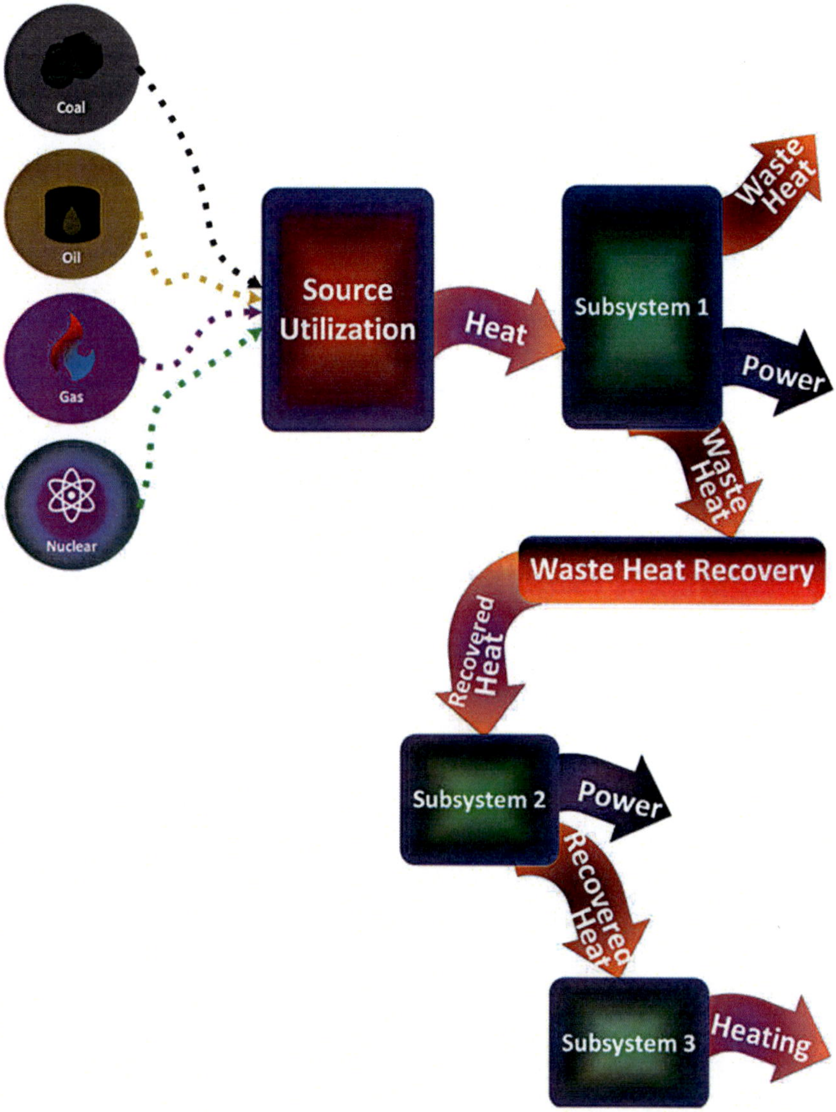

Fig. 7.30 The energy flow from sources to services of a combined heat and power system

Table 7.9 is given for the combined heat and power cycle. State points 1–4 are for the Brayton cycle where air is used as the thermofluid. Air is heated up to 960.85°C at 1013.25 kPa. A gas turbine is used for the expansion of air to 411.85°C and 101.3 kPa. The process of heat recovery from the Brayton cycle occurs between the state points 4 and 15, where heat is exploited by the Rankine cycle.

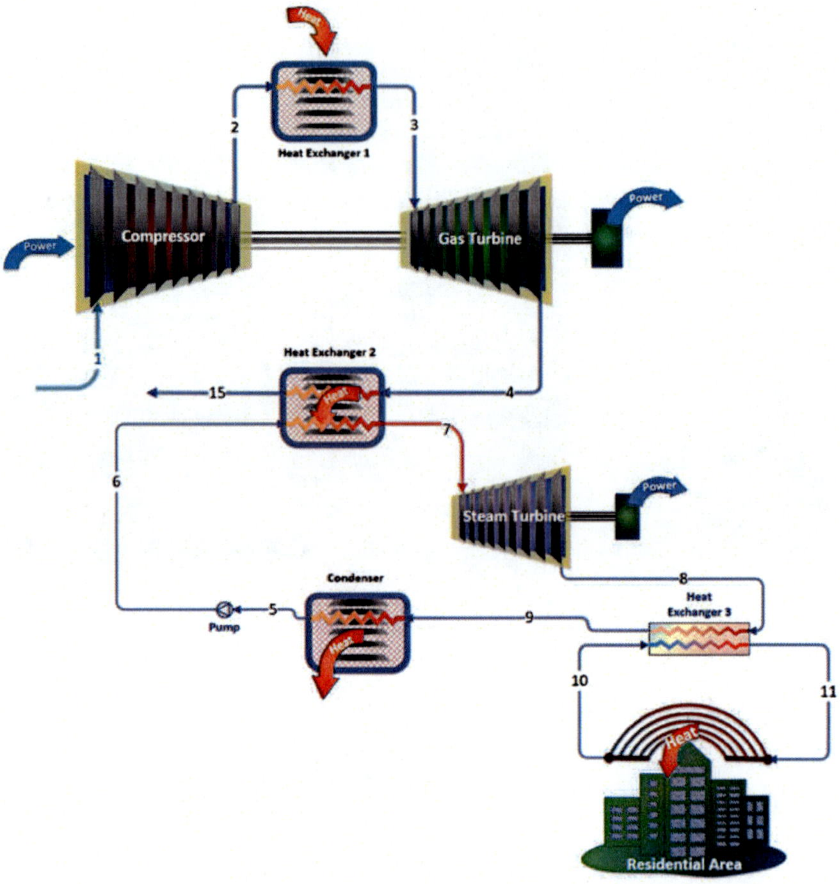

Fig. 7.31 The illustration of a combined heat and power system

Table 7.9 Thermophysical properties of the thermofluid in the combined heat and power system

State point	Substance	ṁ (kg/s)	T (°C)	P (kPa)	h (kJ/kg)	s (kJ/kgK)	ex (kJ/kg)
1	Air	292.8	25	101.325	298.6	5.69	0
2	Air	292.8	297.85	1013.25	576.9	5.69	278.3
3	Air	292.8	960.85	1013.25	1318	6.55	764.5
4	Air	292.8	411.85	101.325	698.1	6.55	144.8
5	Water	29.8	180	1000	762.5	2.13	132
6	Water	29.8	180.6	7700	769.2	2.13	138.7
7	Water	29.8	383.1	7700	3098	6.32	1218.3
8	Water	29.8	180	1000	2777	6.56	825.7
9	Water	29.8	180	1000	1448	3.651	332
10	Water	450	70	200	293.2	0.955	14.9
11	Water	450	85	200	356.1	1.135	24.7
15	Air	292.8	185	101.3	460	6.13	32

In the Rankine cycle, the water as the thermofluid, is heated to 383.1°C at 7700 kPa. A steam turbine is used for the expansion of the water. Expansion is occurred to 180°C and 1000 kPa, which is used for the heat recovery process for the space district heating. A district heating system is used the rejected heat from the condensing process of the Rankine cycle and turned it into useful heat for residential heating application. Space heating process used 85°C of water which is supplied to the residential area. Hot water returned at 70°C to the system.

When we look at the efficiencies of the system, total input is the same with only Brayton and Brayton-Rankine cycle examples in illustrative example 7.4. On the other hand, the useful output is multiplied with the additional output from the district heating system as useful heat. By adding these, the energy and exergy efficiencies of the combined cycle system can be calculated as follows:

$$\eta_{en,ov} = \frac{\dot{W}_{net} + \dot{Q}_{RH}}{\dot{Q}_{HX1}}$$

and

$$\eta_{ex,ov} = \frac{\dot{W}_{net} + \dot{E}x^{\dot{Q}_{RH}}}{\dot{E}x^{\dot{Q}_{HX1}}}$$

From the Brayton cycle, the system generated 100 MW of power and release 69.6 MW of heat. The Rankine cycle used the heat in order to generate power with steam turbine. The steam turbine generated 9.56 MW of power. The released heat from the condenser is exploited by the district heating system where 28.3 MW of useful heat generated.

As shown in Fig. 7.32, the efficiencies are increased as the source utilization is improved with additional useful outputs from the same source. A Brayton cycle used

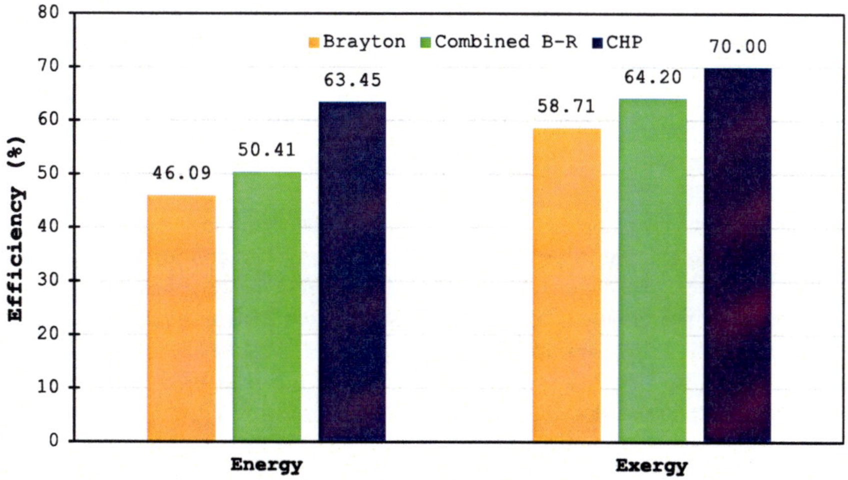

Fig. 7.32 The energy and exergy efficiencies of Brayton only and combined cycles

conventional fuels to generate power, which has a 46.09% energy and 58.71% exergy efficiencies. The rejected heat from the Brayton cycle is used by the Rankine cycle as a combined Brayton-Rankine cycle, where the energy and exergy efficiencies are found as 50.41% and 64.20%. A district heating system is added as a combined heat and power system, where energy and exergy efficiencies are calculated as 63.45% and 70.00%.

7.5.3 Combined Cooling, Heat and Power

Similar to the previous combined systems, the waste heat can be further utilized for additional useful output generation. The ultimate goal is to improve the utilization of the source by multiplying the useful outputs. From this aspect, waste heat can be utilized in numerous ways. As a commercially available method, cooling is one of the alternatives to exploit the waste heat to turn it into a useful output. The further details are explained with an illustrative example.

Illustrative Example 7.6 An illustrative example is designed for a combined cooling, heat and power generation system, where the power generation part as well as the heat generation part is used from the previous illustrative example 7.4 and 7.5. Additionally, a lithium-bromide refrigeration cycle is added where heat is used to generate cooling effect.

The overall process and the layout are depicted in Fig. 7.33. In Fig. 7.33a, the process is illustrated from source to service. The traditional systems utilize the fuel in order to generate heat. Air is pressurized, heated and expanded in order to generate power in the Brayton cycle. The waste heat is recovered from the Brayton cycle in order to drive Rankine cycle. The rejected heat from the condenser of Rankine cycle is exploited by the heating and cooling systems in order to provide heat and cooling effect for the residential area. The thermophysical properties of the combined cooling, heat and power system are given in Table 7.10.

As shown in Table 7.10, the Brayton cycle, Rankine cycle and the district heating systems are kept at same conditions with illustrative example 7.5. The lithium-bromide absorption chiller generated chilled water which supplied at 5°C and returned at 10°C as shown at state points 12 and 13.

The efficiency calculation is changed due to the additional useful output as cooling. The total input to the system is the same, which is considered as the heat entered from heat exchanger 1. The generated work from the system is calculated by adding the net power outputs from the and Brayton and Rankine cycles. In addition to the net work output, residential heating system is considered as another useful output. The third output is the residential cooling obtained by using a lithium-bromide absorption chiller.

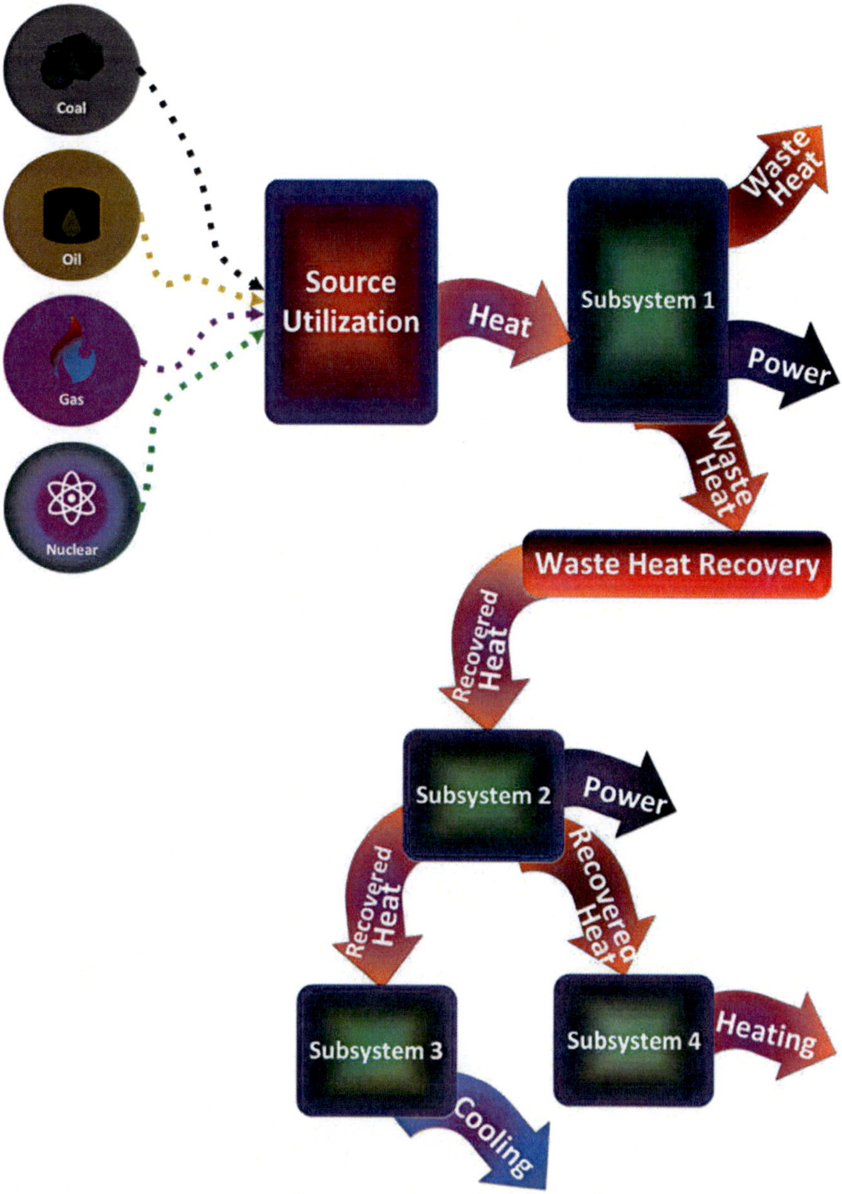

(a)

Fig. 7.33 The illustration of a combined cooling, heat and power generation system

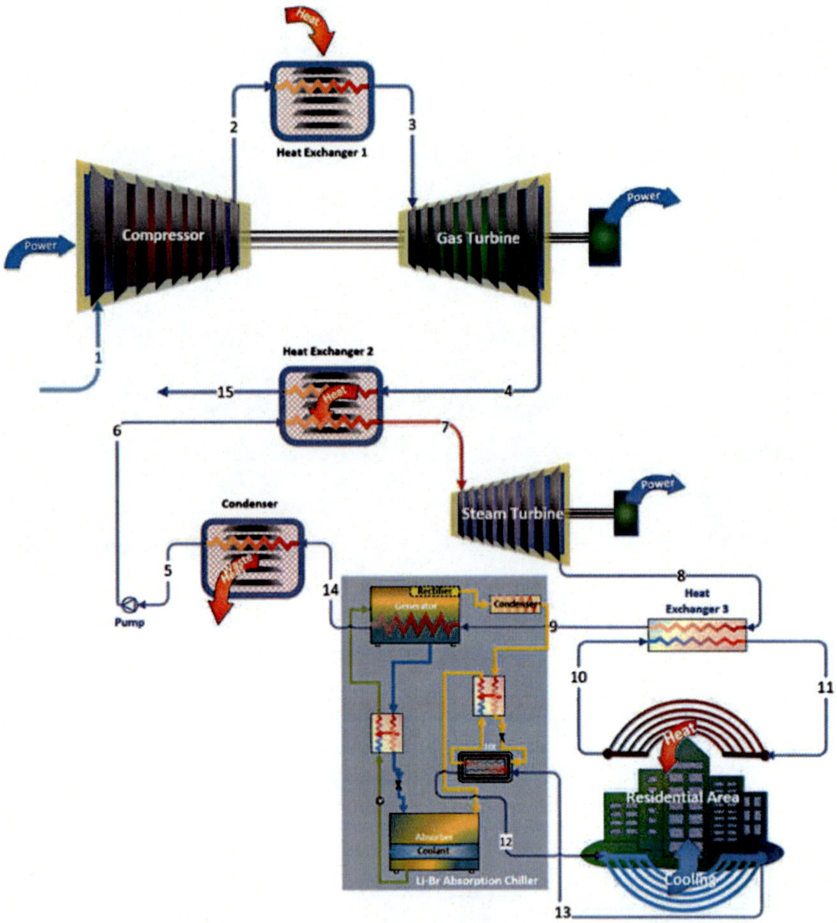

(b)

Fig. 7.33 (continued)

$$\eta_{en,ov} = \frac{\dot{W}_{net} + \dot{Q}_{RH} + \dot{Q}_{RC}}{\dot{Q}_{HX1}}$$

and

$$\eta_{ex,ov} = \frac{\dot{W}_{net} + \dot{E}x^{\dot{Q}_{RH}} + \dot{E}x^{\dot{Q}_{RC}}}{\dot{E}x^{\dot{Q}_{HX1}}}$$

In total, 116 MW of net power is generated from the combined Brayton-Rankine cycle. Moreover, the district heating system supplied 22.6 MW of heat to the residential area. The lithium-bromide absorption chiller used the waste heat to generate chilled water at 5°C, which is used for residential cooling purposes.

Table 7.10 Thermophysical properties of thermofluids in combined cooling, heat and power system

State point	Substance	ṁ (kg/s)	T (°C)	P (kPa)	h (kJ/kg)	s (kJ/kgK)	ex (kJ/kg)
1	Air	292.8	25	101.325	298.6	5.69	0
2	Air	292.8	297.85	1013.25	576.9	5.69	278.3
3	Air	292.8	960.85	1013.25	1318	6.55	764.5
4	Air	292.8	411.85	101.325	698.1	6.55	144.8
5	Water	29.8	180	1000	762.5	2.13	132
6	Water	29.8	180.6	7700	769.2	2.13	138.7
7	Water	29.8	383.1	7700	3098	6.32	1218.3
8	Water	29.8	180	1000	2777	6.56	825.7
9	Water	29.8	180	1000	1448	3.651	374.4
10	Water	450	70	200	293.2	0.955	14.9
11	Water	450	85	200	356.1	1.135	24.7
12	Water	345	5	200	21.2	0.076	2.2
13	Water	345	10	200	42.2	0.151	1.1
14	Water	29.8	180	1000	1098	2.879	332
15	Air	292.8	185	101.3	460	6.13	32

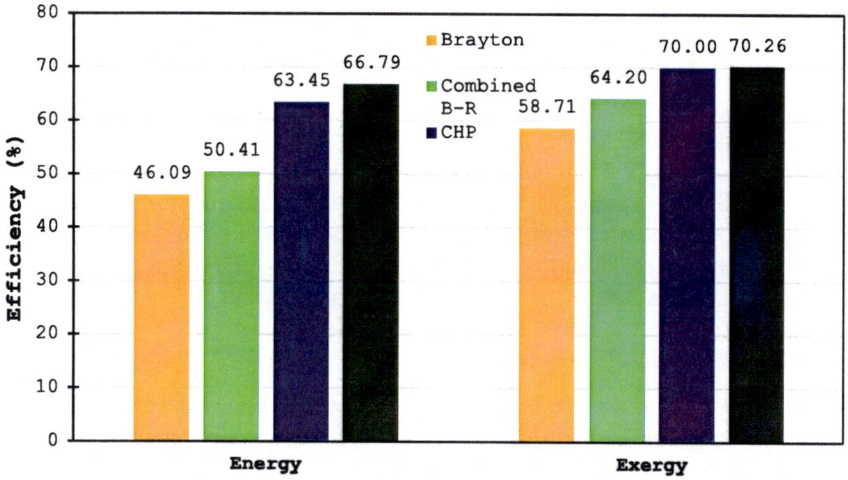

Fig. 7.34 The energy and exergy efficiencies of Brayton only and combined cycles

As shown in Fig. 7.34, the efficiencies are increased due to the better source utilization. The energy and exergy efficiencies of the Brayton cycle without integration was 46.09% and 58.71%. The combined Brayton-Rankine cycle efficiencies were improved, where the energy and exergy efficiencies were calculated as 50.41%

and 64.20%. A district heating system was added as a combined heat and power system, where energy and exergy efficiencies were found as 63.45% and 70.00%. Finally, the combined cooling, heating and power system has 66.79% of energy and 70.26% of exergy efficiencies.

7.5.4 Traditional Energy Systems Combined with Desalination

The same concept for other combined cycles, desalination can be combined with traditional energy systems in order to generate fresh water. There are different options and pathways to generate fresh water by using or combining conventional energy systems. One of the options is to use heat directly from the source via heat-driven desalination methods. Another option is to use power from the traditional power generation systems and utilize the generated power in desalination systems.

There are more than twelve methods for desalination, where heat and/or power are utilized. Heat-driven desalination methods are mainly, multi-stage flashing, multi-effect distillation, thermal vapor compression, adsorption desalination, membrane distillation, humidification dehumidification, and freezing. Heat-driven methods can be used by utilizing direct heat or waste heat from conventional energy sources. On the other hand, reverse-osmosis, forward-osmosis, electro dialysis, and mechanical vapor compression systems can be combined with traditional power generation systems. Hybrid desalination methods can be used via traditional energy systems by both utilizing heat and power.

Globally, among the currently operating desalination plants, the dominant desalination technology is reverse osmosis. Reverse osmosis is a power-driven technology, which is responsible for more than 65% of global desalination. Two heat-driven methods, multi-stage flashing and multi-effect distillation, follows the reverse osmosis. Multi-stage flashing is a heat-driven method, which shares a 21% among global desalination methods. Multi-effect distillation is the third method that responsible for 7% of global desalination (Fig. 7.35).

Illustrative Example 7.7 An illustrative example is designed for a traditional energy system combined with desalination. When we look at the traditional energy systems combined with desalination, some of the desalination methods can be highlighted as commercially mature methods. These are reverse osmosis, multi-effect distillation and multi-stage flashing. In order to combine these with conventional energy systems, pathways are depicted in Fig. 7.36.

A Brayton cycle is considered which is used in illustrative example 7.3. The heat rejection from the Brayton cycle is used in three different options for desalination. The rejected heat from the Brayton cycle is directly utilized by heat-driven desalination method. The multi-stage flashing and the multi-effect distillation systems used directly the rejected heat from heat exchanger 2.

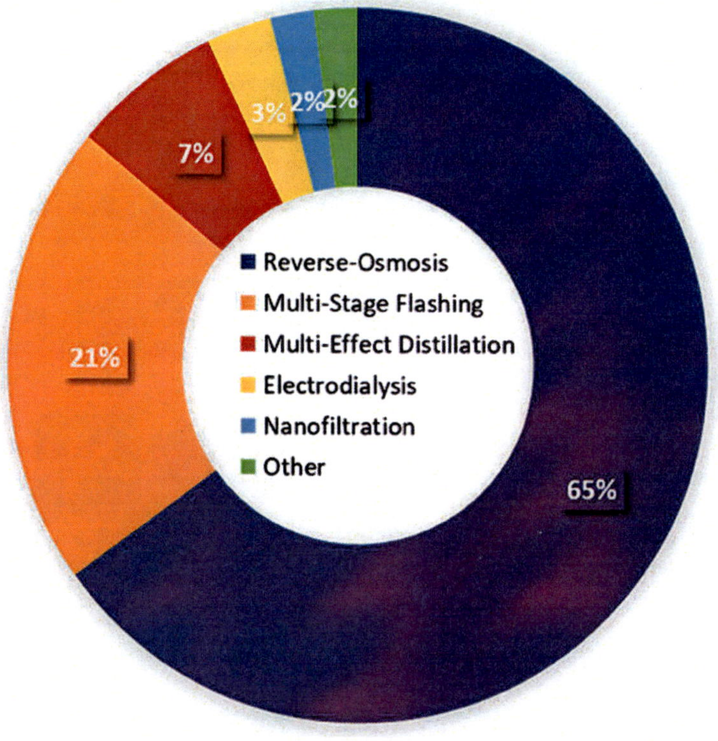

Fig. 7.35 The global installed desalination plants by technology (data from Ref. [5])

A reverse osmosis system is used, however, the rejected heat from the Brayton cycle is used in a Rankine cycle to generate power first, later on power is utilized by the reverse osmosis system to generate fresh water. Each of the systems depicted in Fig. 7.36b–d.

The thermal energy inputs to the systems studied in this example are the rejected heat from the Brayton cycle, which is 116.9 MW of heat. When we look at the generated fresh water, reverse osmosis with Rankine cycle system generated 919.24 kg/s fresh water, which is the most effective method in terms of freshwater generation. Multi-effect distillation system utilized the waste heat and generated 524.9 kg/s of fresh water. Multi-stage flashing generated fresh water at 402.93 kg/s rate. Figure 7.37 shows freshwater generation rates from the proposed desalination systems combined with traditional energy systems.

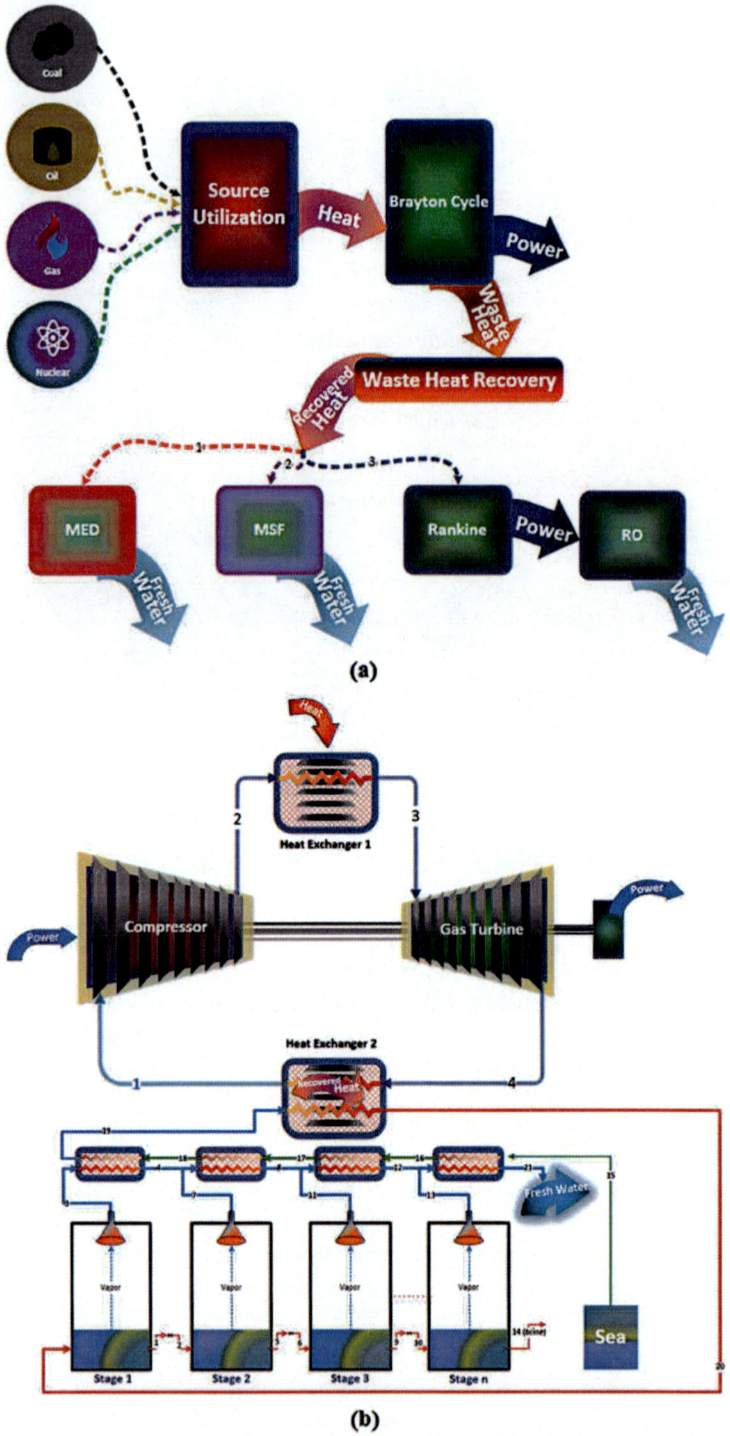

Fig. 7.36 The traditional energy systems combined with desalination (**a**) pathways, (**b**) with multi-stage flashing, (**c**) multi-effect distillation, and (**d**) reverse osmosis

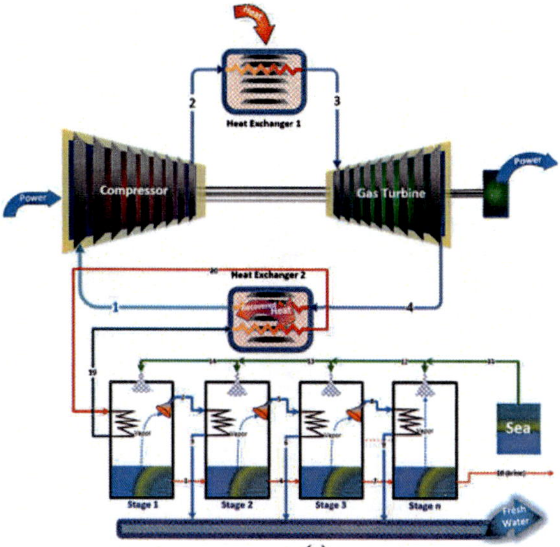

(c)

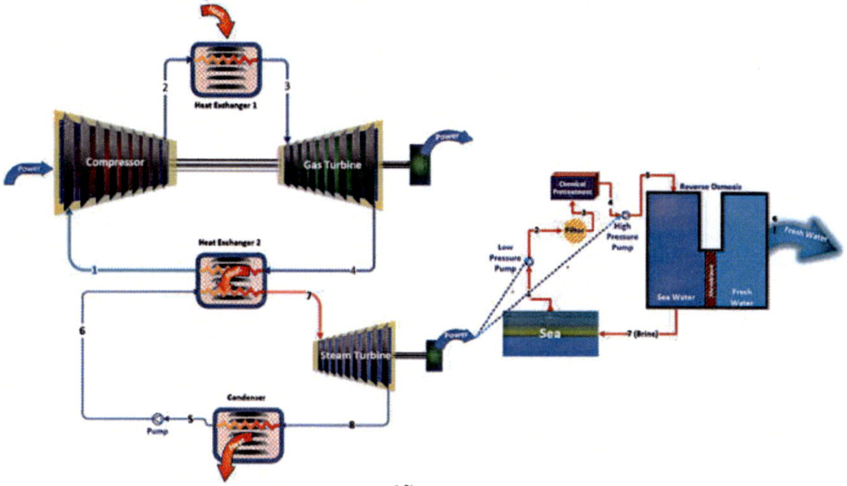

(d)

Fig. 7.36 (continued)

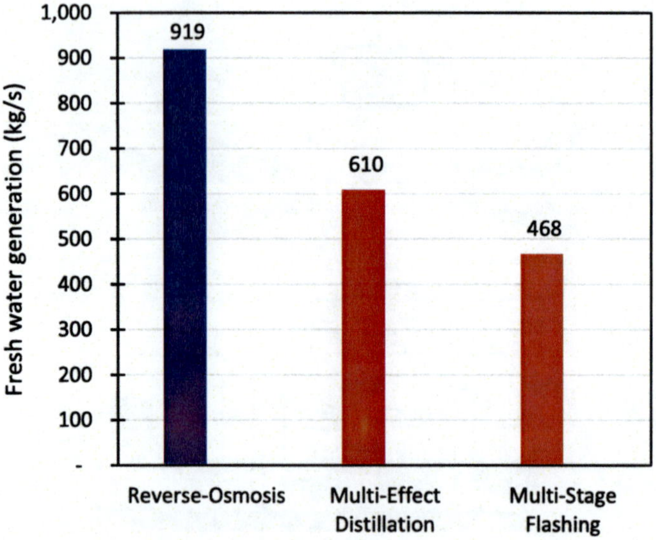

Fig. 7.37 The freshwater generations from reverse osmosis, multi-effect distillation and multi-stage flashing options

7.6 Closing Remarks

The current chapter aims to cover traditional energy systems. Traditional energy systems, typically conventional systems which are turning thermal energy into movement or power. Historically, water and air are very critical as the most common heat transfer mediums to turn heat into power. Since the first century, steam has been studied for its expansion to create movement. Air is also another medium that is critical to create movement with its expansion when heated. Currently, air and water-based systems are responsible for more than 80% of power generation. The heated air and water are expanded, and turbines turn that expansion movement into turning power. However, the current systems are relying on fossil fuels. As shown in the maps at the beginning of the chapter, the world is highly dependent on fossil fuels. Which possess one of the most critical challenges to the society, since the fossil fuels are changing and warming the globe, which is a serious threat to the life on Earth. Jurisdictions are agreed and took measures to replace the entire fossil fuel–based energy systems with renewable alternatives. From this aspect, it is critical to develop knowledge on what will be changed, in order to propose solid potential solutions. Thermodynamics is the critical science of energy and exergy to understand both traditional and modern energy systems. In this chapter, conventional energy systems and their variations are extensively analyzed within the thermodynamic aspects of energy and exergy in order to contribute to the knowledge in the area. Renewable alternatives to traditional energy systems are discussed in previous chapters. The following chapter will discuss how to further improve the performance and feasibility of the alternative energy systems, in an integrated manner.

References

1. Ritchie H, Rosado P (2022) Fossil fuels. Our world in data. Retrieved from: https://ourworldindata.org/fossil-fuels (Online Resource)
2. United Nations (2015) Paris agreement – UNFCCC
3. Dincer I (2021) Thermodynamics: a smart approach. Wiley, Chichester
4. Dincer I, Zamfirescu C (2014) Advanced power generation systems. Elsevier, Amsterdam
5. Aende A, Gardy J, Hassanpour A (2020) Seawater desalination: a review of forward osmosis technique, its challenges, and future prospects. Processes 8:901. https://doi.org/10.3390/PR8080901

Chapter 8
Integrated Energy Systems

Nomenclature

A	area (m^2)
DCMD	direct-contact membrane distillation
EBE	energy balance equation
EnBE	entropy balance equation
ExBE	exergy balance equation
ex	specific exergy (kJ/kg)
$\dot{E}x$	exergy rate (kW)
g	gravitational acceleration (m/s^2)
GHI	global horizontal irradiation (W/m^2)
h	specific enthalpy (kJ/kg)
$\bar{h}$	absolute enthalpy (kJ/kmol)
HP	heat pump
LHV	lower heating value (kJ/kg)
$\dot{m}$	mass flow rate (kg/s)
MBE	mass balance equation
MED	multi-effect distillation
MSF	multi-stage flashing
$\dot{N}$	molar flow rate (mol/s)
ORC	organic Rankine cycle
P	pressure (kPa)
PEM	proton exchange membrane
PV	photovoltaic
Q	heat (kJ)
$\dot{Q}$	heat rate (kW)
S	total entropy (kJ /K)
s	specific entropy (kJ/kg K)
$\bar{s}$	absolute entropy (kJ/kmol K)

I. Dincer, M. Temiz, *Renewable Energy Options for Power Generation and Desalination*, https://doi.org/10.1007/978-3-031-53437-9_8

$\dot{S}$	entropy rate (kW/K)
T	temperature (°C)
V	overpotential (V)
w	work
$\dot{W}$	power (kW)
WOS	Web of Science
X	mass ratio of component

Subscripts

c	complete, condenser
comb	combustion
f	fresh
GT	gas turbine
in	injection
P	product
PH	process heat
R	reactant
RC	residential cooling
RH	residential heating
re-in	re-injection
s	source, saline
SCFR	sodium cooled fast reactor
ST	steam turbine
w	water

8.1 Introduction

Designing energy systems effectively and efficiently has always been a critical subject in this century for many applications in various sectors. There has been a sort of traditional practice of designing systems in a simple manner with no or limited complexities. In this regard, conventional energy systems are designed as single-input and single-output systems. However, single-generation processes typically generate waste and excess energy. In order to utilize the waste and excess energy, subsystems can be integrated to multiply useful outputs. When we look at the single-generation energy systems, it is systematically clear that a significant amount of energy will be wasted. Integration of the energy systems has been done, even with traditional energy systems, such as with combined heat and power or combined cooling, heat, and power systems, in order to multiply the useful outputs by utilizing the same input. The possibilities are only limited by the principles of thermodynamics and the imagination of the designer. Useful outputs are unlimited: electric power, heat, cooling, fresh water, clean air, clean food, dry food, renewable fuel, food, and dry air are only a few examples that can be mentioned.

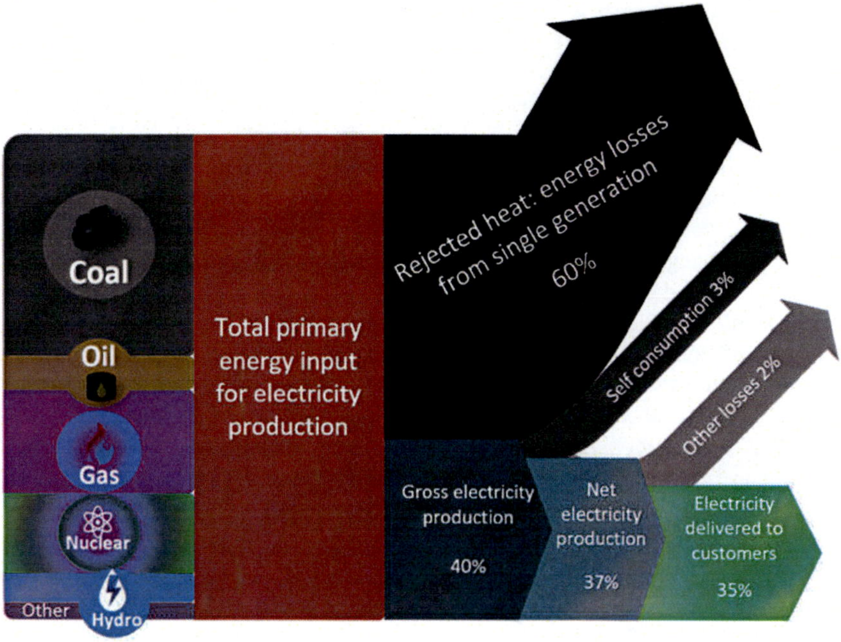

Fig. 8.1 An illustration of global power generation and losses

It is clearer to understand the scale of energy losses when we look at one of the most important energy utilization processes, power generation, on a global scale. Figure 8.1 illustrates the energy flow of global power generation and related losses. More than 60% of the energy is rejected from the currently available power generation systems as losses. Only 35% is used by the customers. The objective behind the system integration lies in this graph. A majority of the losses are actually recoverable, depending on the system components and their operating conditions. Integrated energy systems essentially have multiple subsystems to utilize in the best possible way to turn the input energy(ies) into useful outputs in an effective and efficient manner. They are also expected to recover and utilize any variety of waste or excess energy. When we specifically look at the global power generation process, 60% of energy is rejected as a low-temperature heat. This heat can be easily recovered by integrating a district heating system. Moreover, low-temperature power generation cycles such as the organic Rankine cycle or Kalina cycle can be integrated to generate more power. Other than these, an absorption chiller can be added in order to turn the waste heat into cooling. There are numerous combinations to integrate a waste energy recovery system to minimize the production of waste.

Among the subsystems to recover waste or excess energy, desalination possesses critical importance in order to secure one of the vital sources for life on Earth. When we look at the currently available literature about integrated energy systems with desalination, a bibliometric map can be designed or developed to depict the current status. The bibliometric analysis method is used to identify historical and current developments in a particular field of study. It focuses on the outputs of academic literature databases to produce new data and offer a broad overview of substantial academic literature in a specific area of expertise. In this regard, a comprehensive systematic literature review of research and review articles and book chapters from 2015 to 2023 are performed to evaluate and visualize research hotspots, trends, and frontiers in integrated energy systems and desalination worldwide. The Web of Science (WoS) database is used to collect publications about integrated energy systems and desalination, and the VOSviewer software is utilized for the science mapping of collected data, as shown in Fig. 8.2. The search parameters are as follows: editions = "Science Citation Index Expanded" and "Sciences Citation Index," database = "Web of Science Core Collection". Moreover, combining the topic keywords (integrated energy systems and desalination) is used to achieve the related papers to guarantee the validity of the retrieved data.

Later on, all information was collected, and any duplicate papers were removed after obtaining the documents under each topic keyword combination, and a total of

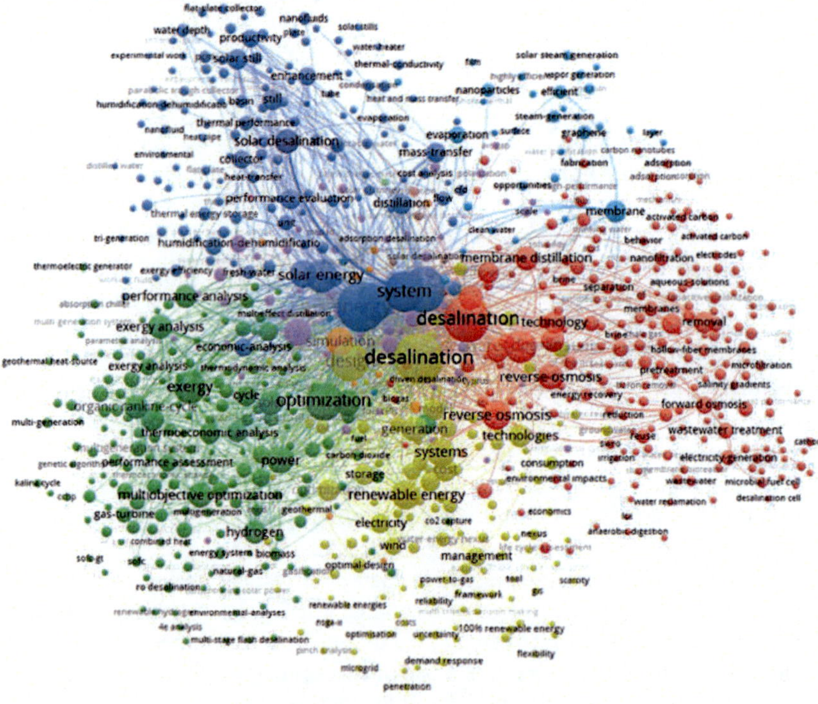

Fig. 8.2 Bibliometric map of co-occurrence network of integrated energy systems and desalination keywords

2334 papers were obtained. Figure 8.2 shows the results of the co-occurrence analysis of keywords using VOSviewer and nodes in this co-occurrence network regarding a keyword. The frequency of the keyword increases with node size. The distance between keywords might reveal information about their relations. The higher the distance between two nodes, the lower the relation between these keywords. As can be observed from Fig. 8.2, the keywords were divided into seven clusters. The yellow and red clusters comprise the largest quantity of keywords, such as "desalination," "renewable energy," "removal," "membrane distillation," and "reverse osmosis". In the green cluster, which comprises energy issues like "exergy," "power," "hydrogen," "optimization," and so on. On the other hand, the blue cluster especially focuses on the performance and operational parameters considered in integrated energy systems. For instance, the keywords "thermal performance" and "nanofluids".

In Fig. 8.3, the bibliometric mapping illustrates the time distribution of terms co-occurrence in 2015–2023. The color of the node approaching yellow signifies that this keyword is still in the beginning stages of research. On the other hand, if the node color is dark blue, it demonstrates that this keyword is deeply investigated. For instance, researchers mostly concentrated on water desalination with membrane processes, such as osmosis and nanofiltration, during the time period from 2015 to 2023. Currently, researchers are focusing on multi-generation systems, various cycles (organic and steam ranking cycles), subsystems, and exergy approaches, which provide sustainable and promising enhancement for energy production in the future.

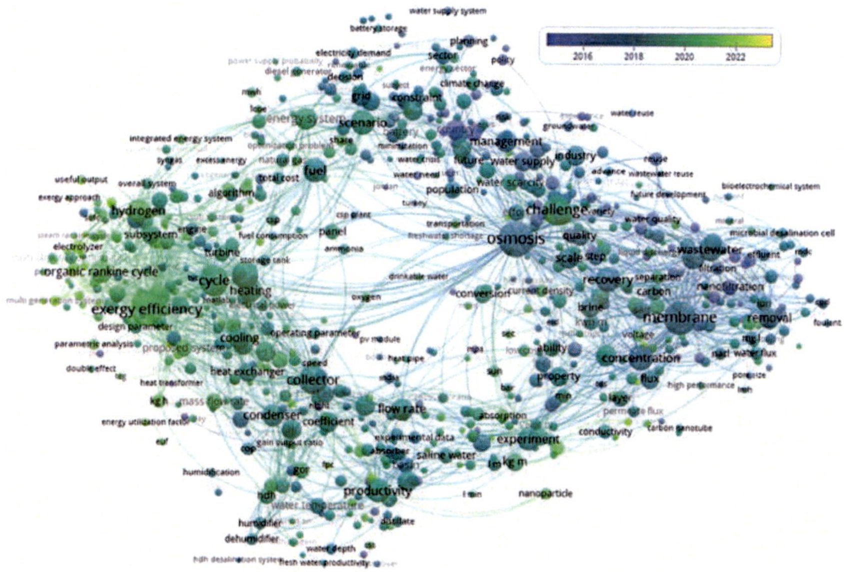

Fig. 8.3 Timeline map of integrated energy systems and desalination keywords from 2015 to 2023

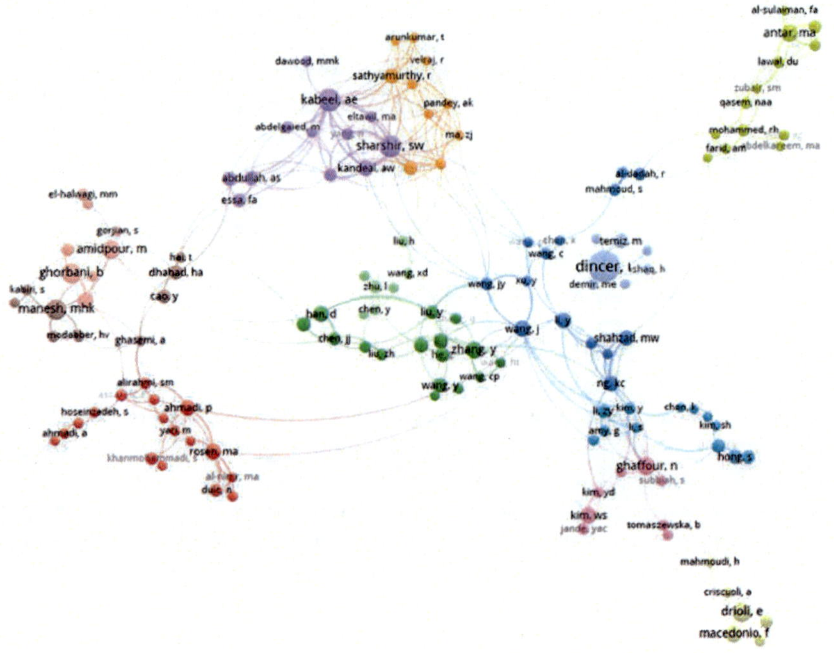

Fig. 8.4 Collaboration network and top authors in integrated energy systems research

Furthermore, the researchers who dominated the studies on integrated energy systems and desalination are also summarized in Fig. 8.4. It should be noted that a number of papers may have international scope due to international partnerships between scholars and institutions. In general, based on the data from the WoS database, 11 different research groups have contributed to the publications of integrated energy systems studies. The two biggest clusters obtained with their corresponding authors are as follows: light blue cluster (corresponding author: Dincer I.) and purple cluster (referring to two international authors). These corresponding authors provided significant connections with other authors and considerably strong collaboration with their co-authors, having a rather strong cooperation relationship, indicating their considerable research collaboration activities. Overall, among them, Dincer, I. portrayed as the most productive researcher with the highest number of publications about integrated energy systems, considering energy and exergy approaches.

As extensively discussed in [1], the development of integrated energy systems starts with primitive systems, where proof of concept systems are used to generate useful output in a certain specific time. To develop a single-generation system requires ensuring the continuous operation of the system by using a certain source to generate single output. Cogeneration systems generate two useful outputs, either as a modification of a single-generation system with energy recovery or as an initially designed system for cogeneration. Trigeneration systems generate three

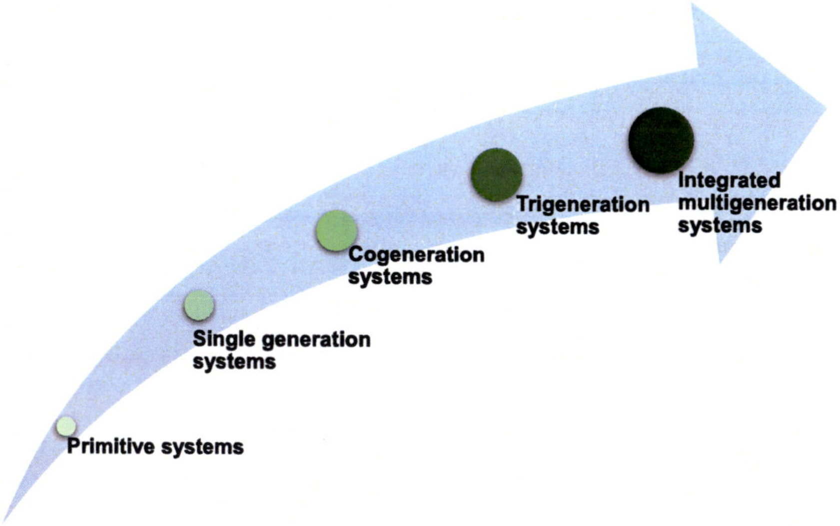

Fig. 8.5 Developmental steps of energy systems into integrated multigenerational systems

useful outputs; these also can be designed initially to generate three useful outputs or can be modified versions of a single or cogeneration system. Figure 8.5 summarizes the development steps of the integrated energy systems. The integrated energy systems, from this aspect, can be the ultimate goal to maximize source utilization, especially when the source is limited.

8.2 The Importance of the Integrated Energy Systems

Integration surrounds our daily life, everywhere from business to personal life, such as integrated plans, integrated action, integrated services, integrated users, and many more. Specifically in the food industry, it is very popular to release a product with multiple ingredients. Moreover, it is almost impossible to find packaged food with a single ingredient. In information and communication technologies, there are no devices consisting of single component. Computers have mainly five hardware components, for input, processing, storage, output, and communication purposes. Considering the biotechnology area, none of the products consists of a single input. From vaccines to cold medicines, typically, they consist of more than one component. However, energy sector is behind this trend as its current status. Globally, the energy sector requires integrated energy systems in order to generate more useful outputs from a single system and a single source, such as power, heat, cooling, hydrogen, fresh water, other renewable fuels, etc. Although traditional energy systems were designed in a way to utilize a single input for a single output, there is a precious opportunity for the energy sector when switching into renewables, switching into the integrated energy systems as well. Therefore, designing and building integrated energy systems have to be a primary goal.

The integration of energy systems can be done by combining the number of components, subsystems, units, and plats while using thermodynamic principles in order to form a larger and more effective system. The ultimate target behind the integration of energy systems is to achieve better sustainability, which will be brought by achieving:

- Better efficiency
- Better source utilization
- Better cost-effectiveness
- Better environmental effect
- Better energy security
- Better water security
- Better management
- Better design

and many others, as shown in Fig. 8.6. It needs to be addressed whether the achievement of all of these benefits is possible or not. When we consider the thermodynamic principles from energy and exergy aspects, these goals and benefits can be addressed and achieved with integrated energy systems. Current and future generations will need useful commodities such as electric power, heat, cooling, fresh water, fresh air, food, waste disposal, refrigeration, carbon dioxide capture, renewable fuels, and many others. It is possible to generate many of these when integrated systems applications are used in the energy industry. There are already integrated

Fig. 8.6 Six domains of the integrated energy systems

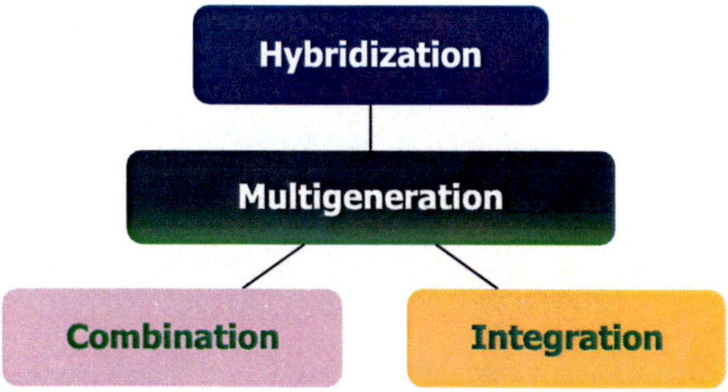

Fig. 8.7 Key pillars of multigeneration

systems, used in different industries and areas. Having a similar concept will enable the energy industry to achieve multiple benefits with multiple useful outputs from a single source of energy.

In order to achieve sustainability as the ultimate target, multigeneration is the key. To achieve the multigeneration, there are three pillars can be mentioned as: hybridization, combination, and integration, as shown in Fig. 8.7. Hybridization is using different types of energy sources in the same system, such as nuclear and renewables or fossil fuels and renewables. This is an alternative way to achieve short-term sustainability goals while transitioning the net-zero economy. Combination is to combine similar processes in order to improve source utilization, such as combining the Brayton cycle and the Rankine cycle in order to better utilize the high-grade heat. Integration, on the other hand, uses different subsystems by utilizing the available sources in order to generate more than one of the useful outputs.

When we look at the system integration concept from thermodynamic aspects, it is exploiting the unused or wasted energy, typically released to the environment in a thermal energy form, and converting it into something useful for the communities and people in the area. One of the key points of view is to ensure that minimizing the production of waste. The currently available energy mixture in the grid is heavily dominated by fossil fuels. The fossil fuel utilization and currently available energy infrastructure are presented in the traditional energy systems section. As mentioned previously, the fossil fuel deployment is a less sustainable practice from many aspects, such as efficiency, source utilization, cost-effectiveness, environmental impact, energy security, water security, etc. When fossil fuel deployment by traditional energy systems is then replaced with smart energy systems where renewable energy sources are deployed in an integrated manner, the above-mentioned aspects can be achieved.

It is important to highlight that different sources have different natures, which possess the importance of better design. The currently available energy systems heavily depend on the heat generation by cracking the chemical bonds, in other words, by burning fossil fuels. Later on, the generated heat from fossil fuels is used in thermodynamic cycles to generate useful outputs. On the other hand, renewable sources are available in many forms, and heat generation is not always the best approach to proceed with the generation of useful output. The concept of exergy has an important role in this challenge. Exergy can be defined as the measurement of the quality of energy; in other words, it shows how useful is energy in a certain environmental condition. By considering both energy and exergy concepts, the successful integration of energy systems can be formulated. The performance assessment, efficiencies, sustainability, and environmental impact are the important aspects that need to be addressed when the integrated energy systems are comparatively assessed.

In traditional energy practices, integrated energy systems are already available, such as combined heat and power systems, combined cooling, heat and power systems, combined power plants, etc. These are presented with details in the traditional energy systems section. It is shown there the increased useful output can help to improve the efficiencies. Heat has a key role in system integrations. The grade of heat is the temperature of heat, which is important in order to run the certain processes to produce certain useful outputs. For example, steam Rankine cycles typically require more than 180°C grade heat, even higher-grade heat for feasible operations. When we look at the thermochemical hydrogen production cycles, higher level heat grades are needed, such as more than 450°C. Freshwater generation, on the other hand, requires lower-grade heat, such as 60°C. The typical practice to design integrated energy systems is that to determine suitable applications for the available heat grade, then to proceed to place the subsystems. The order of the subsystems can typically start with the process that requires a higher grade of heat and end with the process that requires a lower grade of heat. In other words, processes that require a higher grade of heat typically will be closer to the source, and the processes that require lower grade heat will recover the heat from the processes that require higher grade heat.

Efficiency is an important performance assessment in order to understand how the source is utilized to generate useful outputs. Typically, it is the division of useful outputs to the consumption of primary energy input. Single and multigeneration system efficiencies can be written as follows:

$$\eta_{\text{single generation}} = \frac{\text{Output}_1}{\text{Input}}$$

$$\eta_{\text{cogeneration}} = \frac{\text{Output}_1 + \text{Output}_2}{\text{Input}}$$

$$\eta_{\text{trigeneration}} = \frac{\text{Output}_1 + \text{Output}_2 + \text{Output}_3}{\text{Input}}$$

$$\eta_{\text{quadgeneration}} = \frac{\text{Output}_1 + \text{Output}_2 + \text{Output}_3 + \text{Output}_4}{\text{Input}}$$

and

$$\eta_{\text{n}-\text{generation}} = \frac{\text{Output}_1 + \text{Output}_2 + \text{Output}_3 + \text{Output}_4 + \ldots + \text{Output}_n}{\text{Input}}$$

where the number of useful outputs determines the system type.

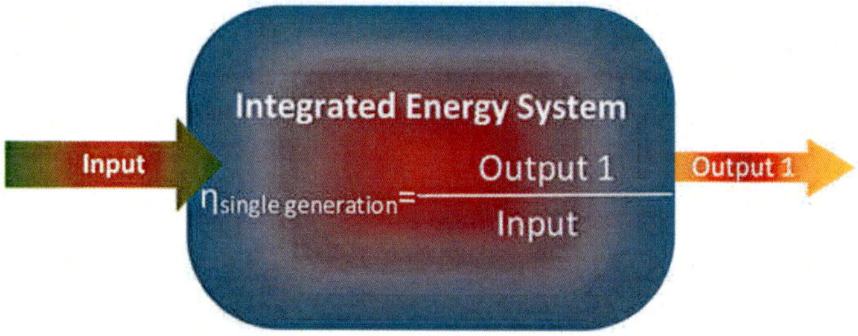

Fig. 8.8 Illustration of an energy system for single-generation

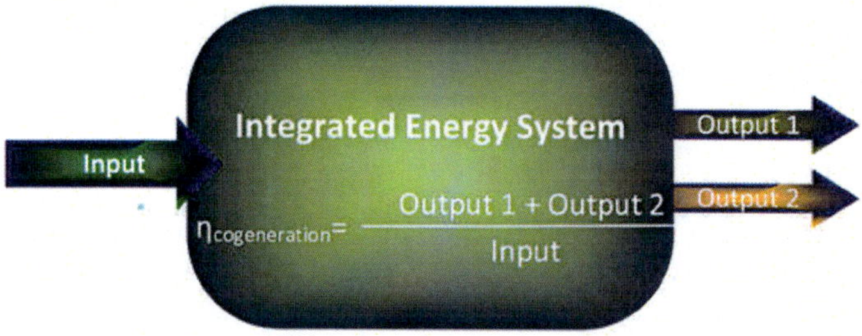

Fig. 8.9 Illustration of an integrated energy system for cogeneration

A single-generation type of energy system is illustrated in Fig. 8.8. A single energy input is utilized in order to generate a single output. This is typical in traditional energy systems. The single-generation process from single input typically contains unused energy, which is released in the form of thermal energy to the environment. From this aspect, single-generation system is a less sustainable method with higher production of wastes. Mathematically, it has only one output in the numerator part for the one denominator.

A cogeneration type of integrated energy system is illustrated in Fig. 8.9. Cogeneration types of systems are both available in traditional energy systems and smart/ modern energy systems categories. A cogeneration-type integrated energy system utilizes the input in order to generate the first output. The waste or excess energy from the process of generating the first output is exploited in order to generate the second output. Compared to the single-generation systems, cogeneration-type integrated energy systems can be mentioned as more sustainable. However, there is still room to reduce inefficiencies by utilizing the wastes with more integrated subsystems in order to achieve trigeneration and more generation.

A trigeneration system is illustrated in Fig. 8.10. By exploiting the wastes from process one and process two, a third process can be carried out to generate the third output by utilizing the same input. There are also examples of the trigeneration systems where fossil fuels or renewables are utilized to generate three useful outputs. Trigeneration systems typically tend to have more efficiencies compared to

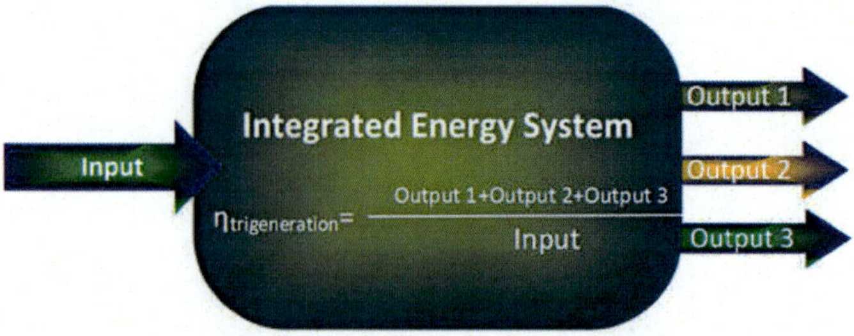

Fig. 8.10 Illustration of an integrated energy system for trigeneration

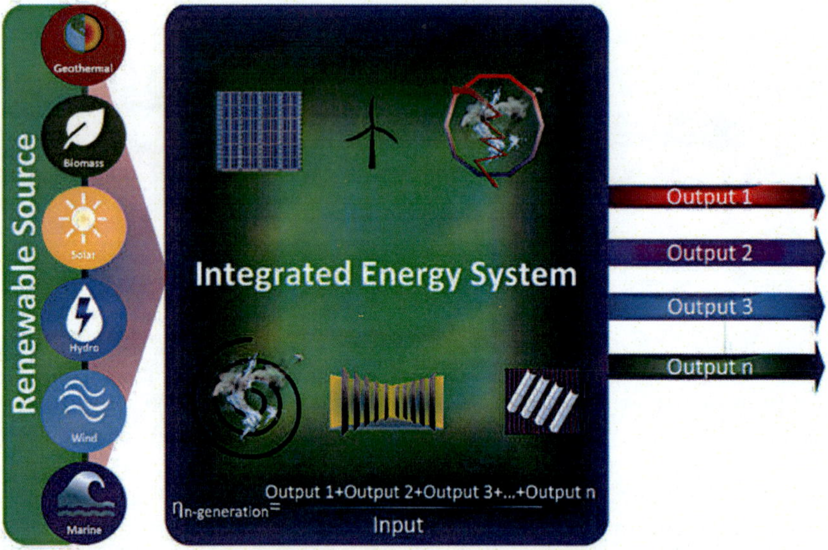

Fig. 8.11 Illustration of the integrated energy systems by using renewable energy sources

cogeneration and single-generation systems, however, still consist of inefficiencies compared to quadro-generation and more generation systems.

The potential of integrated energy systems is only limited to the imagination of the designer. The number of useful outputs can be multiplied by any number. The limits of the thermodynamic laws can be mentioned as the limits of the integrated energy systems. The ultimate goal is to use renewable sources to generate multiple useful outputs in order to achieve ultimate sustainability, as shown in Fig. 8.11.

In order to design integrated energy systems to generate multiple useful outputs, the steps are presented in Fig. 8.12. The source is very critical in order to design the integrated energy systems. Every location is unique; therefore, sources will be in different forms, different quality, and different quantity. Identification of the locally

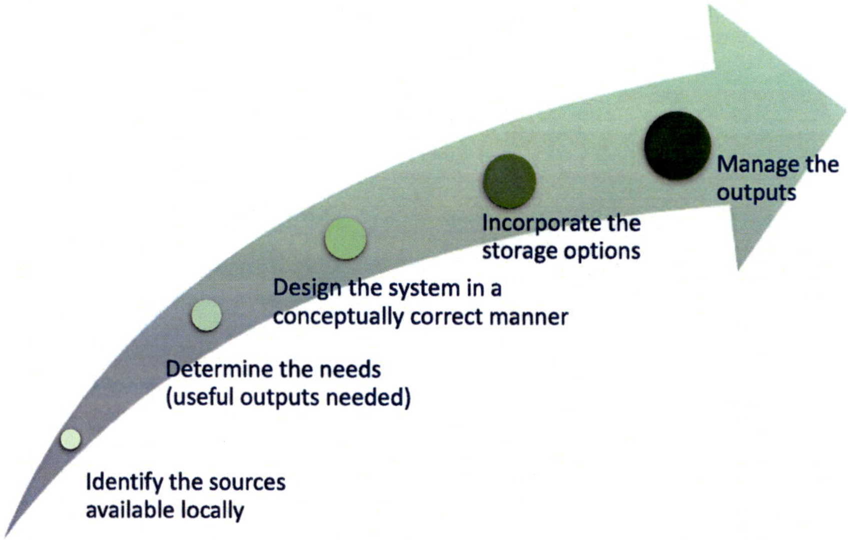

Fig. 8.12 Step-by-step guide to design integrated energy systems

available sources with their details is the first step. Then the needs of the communities or sectors in that location should be addressed in order to determine what needs to be generated. Further, the integrated system needs to be designed by considering the thermodynamic principles. Storage is also key in order to achieve the continuous energy supply and self-sufficiency for better sustainability. Managing, directing, diverting, and transporting the outputs in an effective manner is the last step to design a successful integrated energy system.

Further to note that sustainability is crucial in order to ensure the sources will be available for the needs of future generations. It is key to better utilize the sources while generating multiple useful outputs in an integrated fashion. Figure 8.13 illustrates how generating more outputs from the primary energy input contributes to the sustainability.

8.3 Case Studies

Ten different case studies from the literature are presented in this section. The main objective is to design integrated energy systems in a way to minimize the production of wastes and to maximize the useful outputs. Using renewables as the energy input is crucial in order to minimize environmental impact to achieve the ultimate sustainability. Integrated energy systems are key from this aspect, since renewable

Fig. 8.13 Illustration of sustainability improved by multigeneration

energy source availability highly varies from location to location. It is important to determine what is available at the location and what is needed by the people who lives in that location. These case studies are specifically designed by considering what is available and how it can be improved by integrating subsystems to minimize wastes and maximize the benefit.

8.3.1 Case Study 1

El-Emam and Dincer [2] proposed a reverse osmosis system integrated with a Pelton turbine in order to recover energy waste during the freshwater generation process, as shown in Fig. 8.14. The reverse osmosis system was designed to generate 7586 m^3 of fresh water per day. A low-pressure pump pressurizes the seawater from the sea source at atmospheric conditions, to 6.5 bar and directs the pressurized seawater to a cartridge filter. The suspended particles are filtered in the cartridge filter. A bypass is used after the cartridge filter in order to add salinity to the freshwater product, up until 500 ppm of salinity. The main line after the cartridge filter enters the chemical pretreatment unit where water is treated by adding specific chemicals. A high-pressure pump is used in order to pressurize the water after chemical pretreatment. Reverse osmosis modules are used after the high-pressure pumps, where water is entered at 60 bar and rejected at 51 bar. The fresh water was left at 1.8 bars and entered a mixing chamber where filtered seawater was mixed with the fresh water in order to keep the salinity content up until 500 ppm. Polyamide-based membranes are considered in the reverse osmosis unit, in their study. The design parameters of the seawater reverse osmosis plant are given in Table 8.1.

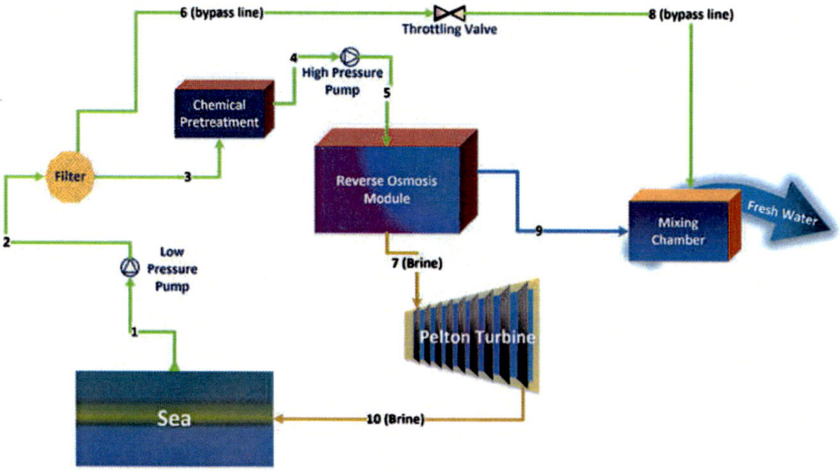

Fig. 8.14 The integrated system layout of case study 1

Table 8.1 The design parameters of the seawater reverse osmosis system

Parameter	Value
Freshwater production	7586 m^3/day
Salinity of the fresh water	<500 ppm of salt
Seawater base salinity	35,000 ppm
Seawater temperature	25°C
Seawater pressure	1 bar
Isentropic efficiency of the high-pressure pump	0.9
Isentropic efficiency of the low-pressure pump	0.87
Isentropic efficiency of the Pelton turbine	0.79
The load factor of the system	0.9
The recovery ratio of the membrane	0.55
The replacement factor of the membrane	0.1
The salt rejection ratio of the membrane	0.99
The levelized cost of the chemical treatment process	\$0.018/m^3 of water
The levelized cost of the cartridge filters	\$0.01\$/m^3
The coefficient of saline permeability	$2.03 \times 10^{-5} m^3/m^2 s$ kPa
The coefficient of water permeability	$2.05 \times 10^{-6} m^3/m^2 s$ kPa
Reference temperature	25°C
Reference pressure	101.35 kPa

Their proposed reverse osmosis system consists of a Pelton turbine, which generates from the rejected water. Francis turbines are also a common application for waste energy recovery from the brine flow. For a high head, high-velocity applications, Pelton turbines are seen more favorable. Rather than generating electric power, these turbines can be connected with a shaft to high- or low-pressure pumps in order to directly transfer the turning power.

Their proposed system simply separates salt and water contents to generate fresh water as the product, as well as, power from the Pelton turbine. In order to define the mass flow rate of the fresh water, it can be defined as the multiplication of recovery ratio and the seawater input. In this case, 55% of the seawater is converted into the product water. They used the first and second laws of thermodynamics, in order to analyze their system. Energy and exergy aspects are used for the first and second laws of analysis. The exergy efficiency of the desalination process is calculated by considering the minimum work required to separate salt and water, and the actual work used by the system as follows:

$$\eta_{ex} = \frac{\dot{W}_{minimum}}{\dot{W}_{actual}}$$

which can be reformulated as the exergy difference of the inlet and exit waters to the net work used by the system as:

$$\eta_{ex} = \frac{\dot{E}x_{fresh\ water\ output} - \dot{E}x_{sea\ water\ input}}{\dot{W}_{net}}$$

where $\dot{E}x_{fresh\ water\ output}$ denotes the exergy rate of the water product, $\dot{E}x_{sea\ water\ input}$ expresses the exergy rate of seawater input, and the net work calculated as removing the generation by the Pelton turbine from the consumption by the pumps as:

$$\dot{W}_{net} = \dot{W}_{pumps} - \dot{W}_{turbine}$$

Moreover, an economic analysis was carried out in order to understand the economic performance of the system. In order to determine the cost of their proposed system, four main components are considered:

- Seawater intake and pretreatment system with the cartridge filters and the chemical pretreatment
- High-pressure pump
- Reverse osmosis permeators
- Pelton turbine

which are the major parts of the system. A total capital investment parameter is calculated, which is the total cost with the direct and indirect cost components. Direct cost items include:

- The cost of the equipment
- Development and preparation of the site
- The cost of the installation

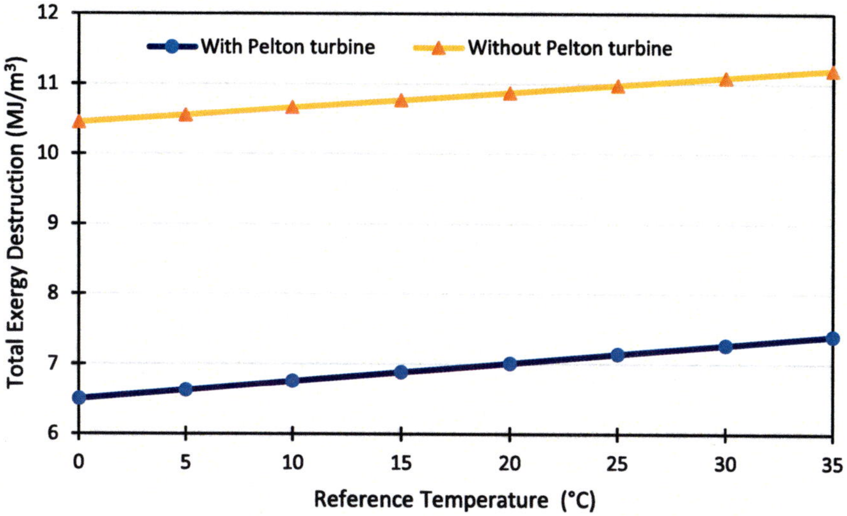

Fig. 8.15 The change in exergy destructions by reference temperature (data from Ref. [2])

When we look at the indirect items in the costs, these are listed as:

- The cost of engineering and supervision
- Building, construction, and contingency

Moreover, they considered an effective discount rate of 8%, a nominal escalation rate of 5%, a lifetime of their proposed system as 20 years, and annual operating hours as 8760 hours.

According to their analysis, they found the levelized cost of fresh water as $2.451/$m^3$ and the exergy efficiency as 5.82%. They showed that the temperature of the seawater has a very slight effect on the cost of water, where 35°C gives a $2.441/$m^3$ of the levelized cost of fresh water.

The reverse osmosis module is the main part of exergy destructions, which is responsible for 67.8% of the total exergy destructions within the system. The high-pressure turbine follows the reverse osmosis module for exergy destructions, which is responsible for 17.16% of total exergy destructions. The work recovery ratio by the Pelton turbine varies between 35% and 75%. In studying the exergy destruction reduction, the case with the Pelton turbine reduces 35.5% of the total exergy destruction as shown in Fig. 8.15. A higher recovery ratio from the Pelton turbine increases the efficiency and decreases the levelized cost of fresh water.

8.3.2 Case Study 2

In case study 2, a solar, wind, and geothermal-driven integrated energy system, developed by Ghosh and Dincer [3], is presented. They developed a multiple input multiple output system in order to generate power, heat, cooling, fresh water, and dry

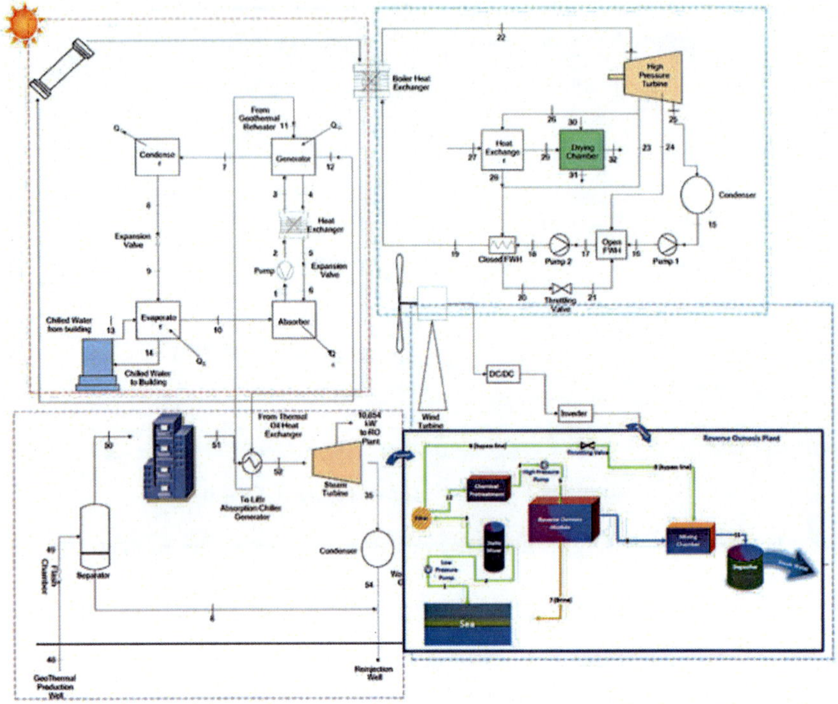

Fig. 8.16 The integrated system layout of case study 2 (modified from Ref. [3])

product. Their system utilizes solar, geothermal, and wind energies with four main subsystems:

- A Rankine cycle
- A geothermal water cycle
- A water desalination cycle
- A vapor absorption cycle

as shown in Fig. 8.16.

In the absorption cycle, heat is utilized in order to generate a cooling effect. The Li-Br is used as the thermofluid in the role of absorbent. Water is also used as thermofluid in the role of primary fluid. The evaporation of the water brings the cooling effect to the system. The heat recovered from the solar collector via thermal oil is utilized within the absorption cycle to generate the cooling effect as one of the useful outputs.

Moreover, a Rankine cycle is employed in order to generate power by utilizing the heat. Two main components are considered while analyzing the Rankine cycle as the heat collection part from solar energy, which is also called the solar receiver and the solar collector; and the power generation part, a heat engine. A heat exchanger–type boiler is used in order to exchange the heat between the working fluid of the solar cycle and the water of the Rankine cycle. A single-stage steam turbine is used to convert the expansion of the superheated steam into mechanical power. Two-stage pumps are used in order to pressurize the water. Open and closed feed water heaters are also employed in order to preheat the water before the heat exchanger–type boiler. At the solar collector, a working fluid at 125°C of temperature enters and collects solar radiation to get heated up to 350°C. In the heat exchanger–type boiler, steam enters at 185°C and leaves at 300°C, where the heat is exchanged with the working fluid from solar collector. The water is condensed at 100 °C when the ambient temperature is 25°C. In the condenser, a drying unit is connected in order to utilize the waste heat for drying applications of agricultural products.

For the desalination process, a reverse osmosis system is considered within the integrated system where power is taken from the geothermal water cycle and the wind turbine. Seawater with waste minerals and other particles is filtered after the static mixer. There are two lines are separated after the filter, one line is directed toward the mixing chamber for the improvement of salinity content at the final product to 230 ppm. A high-pressure turbine is used to further pressurize the seawater and direct it to the reverse osmosis membranes. For the end user, 87.36 kg/s of fresh water is obtained from the reverse osmosis subsystem in case study 2.

They carried out the first and second laws of thermodynamics analysis by considering the energy and exergy aspects. For the ambient conditions, 101.3 kPa of pressure and 298 K of temperature are considered. In order to find the first and second laws of efficiencies, useful outputs are divided into total inputs, in terms of energy and exergy. The overall energy and exergy efficiencies are calculated for the integrated energy system as follows:

$$\eta_{en} = \frac{\dot{W}_{net} + \dot{Q}_{Heating} + \dot{Q}_{Drying} + \dot{Q}_{Cooling} + (\dot{m}_{fw}h_{fw} - \dot{m}_{sw}h_{sw})}{\dot{E}_{geo} + \dot{E}_{solar} + \dot{E}_{wind}}$$

and

$$\eta_{ex} = \frac{\dot{W}_{net} + \dot{E}x^{\dot{Q}_{Heating}} + \dot{E}x^{\dot{Q}_{Drying}} + \dot{E}x^{\dot{Q}_{Cooling}} + (\dot{m}_{fw}ex_{fw} - \dot{m}_{sw}ex_{sw})}{\dot{E}x_{geo} + \dot{E}x_{solar} + \dot{E}x_{wind}}$$

In the geothermal water cycle, underground hot water at 500 kPa is used and flashed in the flash chamber and then separated into gas and liquid at the separator. The vapor is used as a space heater, then it is heated via the solar collector cycle's heat and used in the low-pressure steam turbine to generate more power.

The five useful outputs from the integrated system can be summarized as the work output via the turbines, cooling via absorption refrigeration cycle, drying of agricultural products via hot air, fresh water via reverse osmosis seawater desalination, and space heating via the geothermal water cycle.

When we look at the performance of subunits, geothermal cycle has 48.29% of energy and 15.45% of exergy efficiencies, absorption cycle has 71.2% of energy and 16.1% of exergy efficiencies, solar-driven Rankine cycle has 29.65% of energy and

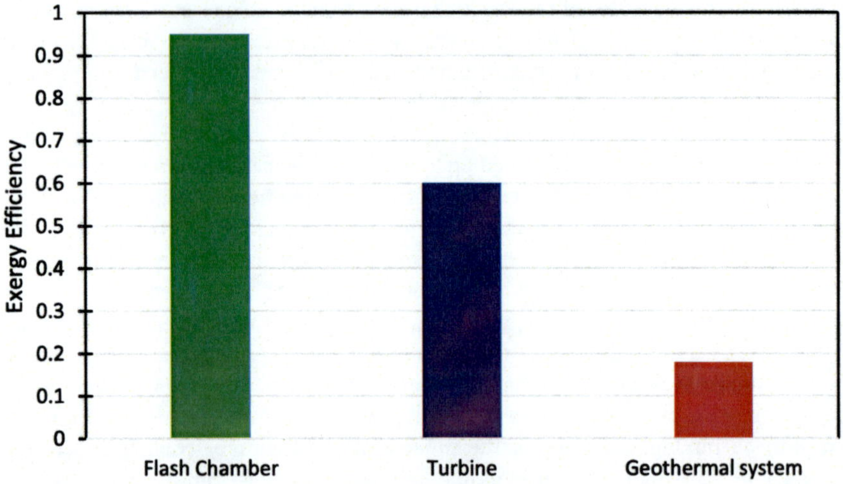

Fig. 8.17 The exergy efficiencies of the subunits and plants (data from Ref. [3])

67.24% of exergy efficiencies, and reverse osmosis desalination cycle has 1.51% of energy and exergy efficiencies. Figure 8.17 shows the subunit and plant exergy efficiencies The overall energy and exergy efficiencies of the integrated energy system in case study 2 are found to be 36.66% and 25.07%.

8.3.3 Case Study 3

A solar-based integrated system for electric power, heat, fresh water, and hydrogen production was developed by Ozlu and Dincer [4]. Their integrated energy system is shown in Fig. 8.18. The solar energy is collected by parabolic trough–type solar collectors. A thermal storage is coupled with the parabolic solar collectors, as a backup solution when the solar radiation is unavailable. A heat exchanger–type boiler is used within the two-stage Rankine cycle, remaining of the remaining heat from the solar collector is used in a domestic hot water system. A portion of the power generated by the system is used with a proton exchange membrane electrolyzer to generate hydrogen as a useful output. Moreover, the condenser of the two-stage steam Rankine cycle is connected with an ammonia-water Rankine cycle, to further utilize the heat for power generation. The condenser part of the ammonia-water Rankine cycle is connected with a multi-stage flashing desalination system in order to generate fresh water.

The overall energy and exergy efficiencies of the system are based on the useful outputs and total inputs in terms of energy and exergy. The work is generated by four

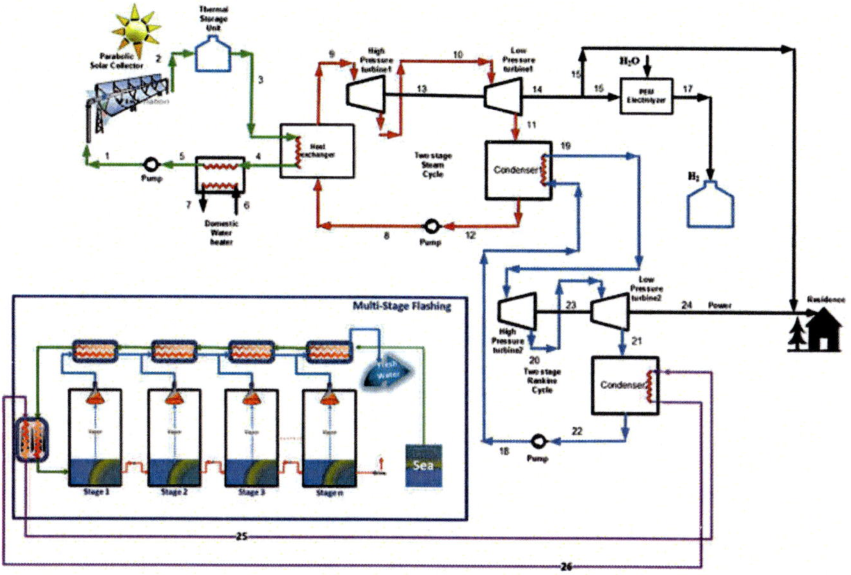

Fig. 8.18 The integrated system layout of case study 3 (modified from Ref. [4])

different components and consumed by different components. The net work output is calculated as follows:

$$\dot{W}_{net} = \dot{W}_{LPT,RC1} + \dot{W}_{HPT,RC1} + \dot{W}_{LPT,RC2} + \dot{W}_{HPT,RC2} - \dot{W}_{consumption}$$

Moreover, heat, hydrogen, and fresh water are also considered as useful outputs. In order to calculate the overall energy and exergy efficiencies, the following can be used:

$$\eta_{en} = \frac{\dot{W}_{net} + \dot{Q}_{Heat} + (\dot{m}_{fw}h_{fw} - \dot{m}_{sw}h_{sw})}{\dot{E}_{solar}}$$

and

$$\eta_{ex} = \frac{\dot{W}_{net} + \dot{E}x^{\dot{Q}_{Heat}} + (\dot{m}_{fw}ex_{fw} - \dot{m}_{sw}ex_{sw})}{\dot{E}x_{solar}}$$

Exergy destructions are also important to examine in order to understand where the exergy is destructed more and where to integrate more systems to minimize the exergy destructions.

The main goal of the current integrated system is to utilize solar energy from high-temperature processes to low-temperature processes to generate electric power, heat, fresh water, and hydrogen. They modeled the system by using thermodynamic

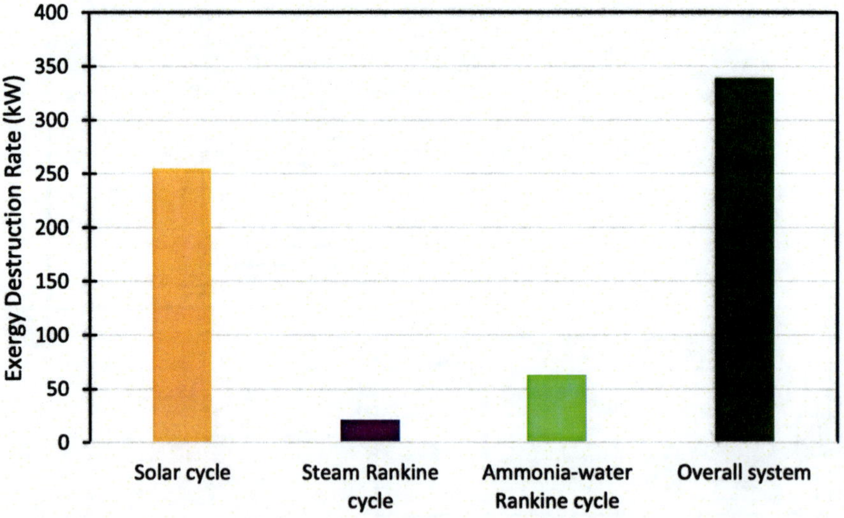

Fig. 8.19 The exergy destructions of subsystems and the overall system (data from Ref. [4])

principles as energy and exergy approaches. In order to understand the performance of the system, energy and exergy efficiencies are calculated. Moreover, they optimized the parameters to obtain better efficiencies and better environmental performances.

They considered different changing parameters throughout the system. Solar radiation is one of the most important parameters since it is the single input to the system. It is considered between 0 and 1000 W/m^2 if solar intensity. The ambient temperature was between 10 and 35°C; the pressure ratio for the ammonia-water Rankine cycle was between 20 and 150; the steam Rankine cycle pressure ratio was between 20 and 150; the mass flow rate of the solar cycle, ammonia-water Rankine cycle, and steam Rankine cycle were between 0.5 and 5 kg/s.

The exergy destructions from the subsystems and the overall system is given in Fig. 8.19. They obtained the optimum energy and exergy efficiencies for the overall integrated system as 36% and 44%. When the solar collector area is 24 m^2, the total turbine output reaches its maximum point at 116 kW. The freshwater generation rate is 0.04 kg/s. They mentioned that the produced commodities can be enough for a residential area with 38 apartments.

8.3.4 *Case Study 4*

Demir and Dincer [5] developed a hybridized system by integrating a heliostat-based solar energy system with a natural gas–based combustion system, which is used for case study 4. The overall integrated system is shown in Fig. 8.20. A thermoelectric

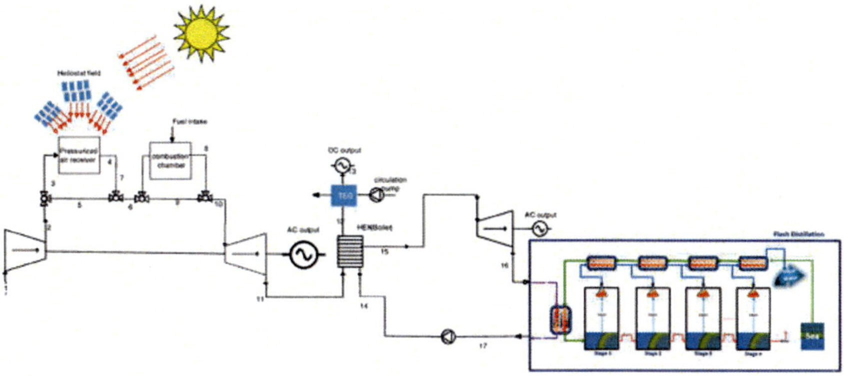

Fig. 8.20 The integrated system layout of case study 4 (modified from Ref. [5])

generator is used where air is compressed first, then heated via solar energy or natural gas combustion due to the availability, and then expanded in the gas turbine. Natural gas–based system is cut in once the solar radiation intensity is below 900 W/m^2, in order to compensate it. With the product gases from the first cycle, a heat exchanger–type boiler is used with a steam Rankine cycle in order to generate power. The condenser part is connected with a flash distillation system.

When we look at the thermophysical properties of the integrated system, the air is compressed to 1410 kPa before the combustion process. The air is heated up to 940.8 K with a solar thermal system and natural gas–based combustion chamber where combustion gasses reached to 1187 K. The exhaust heat at state point 11 in Fig. 8.20 is used to drive heat exchanger–type boiler as well as the steam Rankine cycle. The steam reached 630 K and 800 kPa at heat exchanger–type boiler exit and expanded at the steam turbine to generate power. The steam expanded to 303.2 K and 100 kPa, which is used to heat flash distillation cycle to generate fresh water.

They analyzed the system with the first and second laws of thermodynamics from energy and exergy aspects. They assumed that the ideal gas principles is applied for air, potential and kinetic energy changes are neglected, only the chemical reaction is carried out in the combustion chamber. The compressor pressurized the air with a 14.1 pressure ratio, the ambient conditions are considered as 100 kPa and 295 K. Efficiencies are considered as 87% for the turbine, 85% for the compressor, 95% for the combustor chamber, and 95% for the electromechanical system. Air mass flow rate to the compressor is 42 kg/s and salinity of the water is 42000 ppm at the sea.

There are two cycles that generate power by utilizing thermal energy: the first cycle is the Brayton cycle where a gas turbine generates power with the expansion of air, and the second cycle is the Rankine cycle where the expansion of the water is used to generate power. The net work output can be calculated by removing the compressor work, circulation pump work, vacuum pump work, and the Rankine

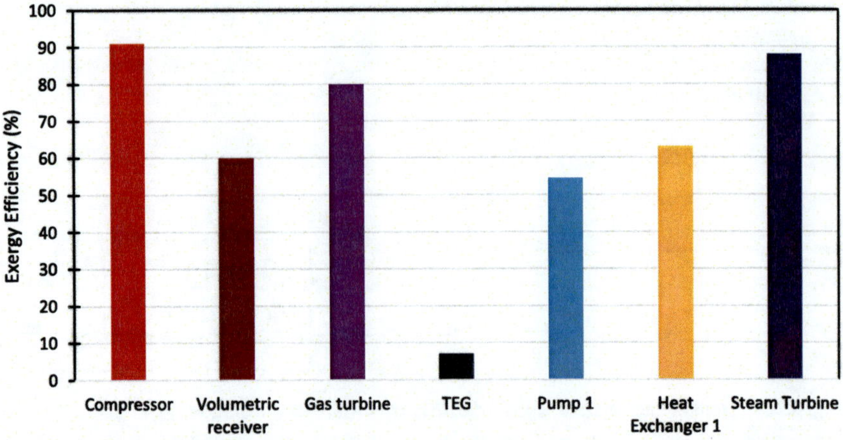

Fig. 8.21 The exergy efficiencies of subsystems and the overall system (data from Ref. [5])

cycle's pump work from the generated work from the gas turbine and the steam turbine as follows:

$$\dot{W}_{net} = \dot{W}_{GT} + \dot{W}_{ST} - \dot{W}_{compressor} - \dot{W}_{circulation\ pump} - \dot{W}_{pump\ 1} - \dot{W}_{vacuum\ pump}$$

This system particularly generates power and freshwater by utilizing solar radiation and natural gas fuel. The energy and exergy efficiencies of the overall system can be calculated as:

$$\eta_{en} = \frac{\dot{W}_{net} + (\dot{m}_{fw}h_{fw} - \dot{m}_{sw}h_{sw})}{\dot{E}_{solar} + \dot{E}_{natural\ gas}}$$

and

$$\eta_{ex} = \frac{\dot{W}_{net} + (\dot{m}_{fw}ex_{fw} - \dot{m}_{sw}ex_{sw})}{\dot{E}x_{solar} + \dot{E}x_{natural\ gas}}$$

The overall energy and exergy efficiencies of the integrated system are 54.9% and 44.5% when the solar contribution is 50%. Figure 8.21 shows the exergy efficiencies of the components and subsystems within the integrated system. The electric power generation capacity of the overall system is 21.8 MW. For the flash distillation unit, the freshwater production rate is 3.36 kg/s. In a daily cycle, the proposed system with multi-stage flash unit can generate 280,775 kg fresh water per day, which is enough for 1000 people. The capacities are enough to meet the needs of small towns.

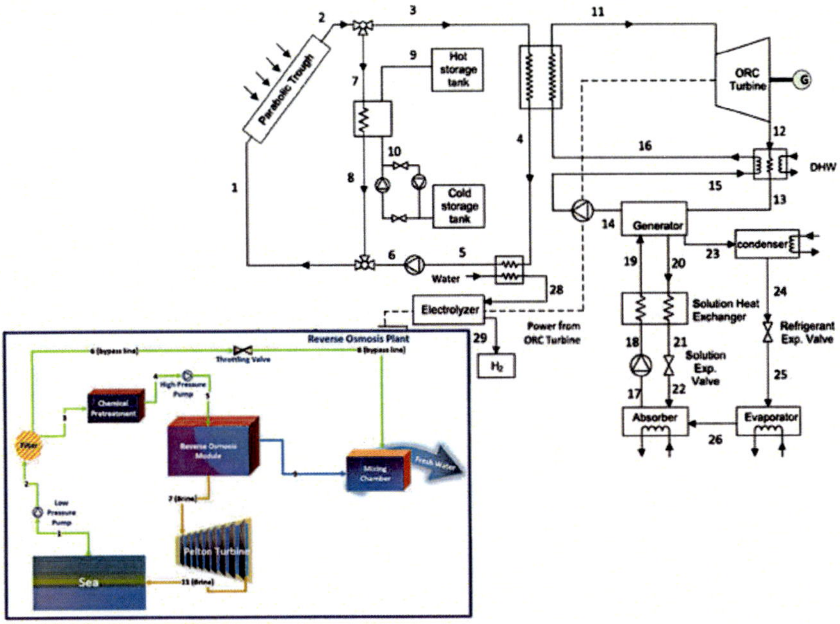

Fig. 8.22 The integrated system layout of case study 5 (modified from Ref. [6])

8.3.5 Case Study 5

A novel solar-driven integrated system was developed by El-Emam and Dincer [6], which is given here in case study 5. Their proposed system is a solar-based integrated energy system with parabolic trough–type of solar collector in order to generate power, hydrogen, cooling, fresh water, and domestic hot water; briefly, a single-input system for five outputs. They used an organic Rankine cycle, absorption refrigeration cycle, and reverse osmosis desalination system in their integrated energy system, which is shown in Fig. 8.22.

They carried out analysis based on the first and second laws of thermodynamics from energy and exergy aspects. The integrated energy system generates power, hydrogen, cooling, domestic hot water and fresh water by using solar energy. According to that, the overall energy and exergy efficiencies can be calculated as follows:

$$\eta_{en} = \frac{\dot{W}_{net} + \dot{Q}_{Cooling} + \dot{m}_{H_2} LHV_{H_2} + \dot{m}_{fw} h_{fw}}{\dot{E}_{solar}}$$

and

$$\eta_{ex} = \frac{\dot{W}_{net} + \dot{E}x^{\dot{Q}_{Cooling}} + \dot{m}_{H_2} ex_{H_2} + \dot{m}_{fw} ex_{fw}}{\dot{E}x_{solar}}$$

Note that the parabolic trough–type solar collectors concentrate the solar radiation on the solar receiver and heat the working fluid. A solar thermal energy storage system with two tanks is coupled with the concentrating solar system. They used Therminol 66 as the operating thermofluid within the solar receiver, and a commercial nitrate salt, Hitec XL, as the thermofluid within the thermal energy storage tanks, which is a non-toxic and low-cost thermofluid with the composition of 48wt% of $Ca(NO_3)_2$, 45wt% KNO_3, and 7wt% $NaNO_3$. They used n-octane as the thermofluid within the organic Rankine cycle, which has a critical temperature of 296°C. For the desalination system, they used a reverse osmosis system with a Pelton turbine, which is a very similar desalination system in case study 1.

The integrated system basically utilizes thermal energy from the concentrating solar plant and uses it in n-octane-based organic Rankine cycle. The remaining heat is used in an absorption refrigeration cycle to generate cooling. The generated electricity from the organic Rankine cycle is used in the reverse osmosis desalination system to generate fresh water. Pelton turbine recovers the portion of the power which makes a more efficient freshwater generation process compared to the typical reverse osmosis processes.

Table 8.2 summarizes the parameters used to design their proposed system. In addition to the thermodynamic analysis, they also carried out an economic analysis. The overall exergy efficiency of the system is found between 21.7% and 39%. According to the parametric studies, organic Rankine cycle turbine inlet temperature can lead to a decrease in the overall energy efficiency and to increase in the overall exergy efficiency as shown in Fig. 8.23.

8.3.6 *Case Study 6*

Case study 6 is based on a solar energy–based integrated system proposed by Siddiqui and Dincer [7]. Their solar-based system generates power and hydrogen along with fresh water. Solar energy is collected by a solar tower–type concentrating system, where thermofluid is used to transfer the heat to the heat exchanger–type boiler. A two-stage steam Rankine cycle and an organic Rankine cycle are used in order to utilize thermal energy to generate power. The generated power is used to drive reverse osmosis-based desalination and proton exchange membrane-based electrolysis systems to generate fresh water and hydrogen.

Their proposed system is given in Fig. 8.24. The heat from the solar tower concentrating solar thermal system is directly used to run the two-stage steam Rankine cycle. In the heat exchanger type, boiler steam is superheated to 423.5°C

Table 8.2 The design parameters of the integrated system in case study 5

Parameter	Value
Heat transfer fluid for the parabolic trough collector	Therminol-66
Collector width	5.8 m
Collector length	12.3 m
Solar intensity range	500–850 W/m^2
Storage medium	Hitec XL
Organic Rankine cycle thermofluid	n-octane
High pressure of the organic Rankine cycle	22 bar
Isentropic efficiency of the turbine	85%
Evaporator effectiveness	88%
Turbine exit pressure	35 kPa
Pinch temperature of heat exchangers	6°C
PEM electrolyzer operating temperature	80°C
PEM electrolyzer operating current density	5000 A/m^2
Activation energy of the anode side	76 kJ/mol
Activation energy of the cathode side	18 kJ/mol
Absorption chiller solution circulation ratio	11.05
Effectiveness of the heat exchanger	0.7
Isentropic efficiency of the pump	85%
Weak solution concentration	0.565
Strong solution concentration	0.62
Freshwater salinity	450 ppm
Seawater salinity	35,000 ppm

at 9000 kPa and entered the high-pressure steam turbine to expand. Later on, steam is superheated to 489.2°C at 2000 kPa and entered the low-pressure steam turbine at the second stage. The expanded steam entered the heat exchanger 2 for the condensation. At heat exchanger 2, it exchanges its heat to the isobutane which is the working fluid of the organic Rankine cycle. The isobutane is heated up and then expanded in the organic Rankine cycle turbine to generate mechanical power. The generated power is diverted into the reverse osmosis subsystem and the proton exchange membrane electrolysis subsystem.

The overall energy and exergy efficiencies are calculated by calculating the ratio of the useful outputs to the total inputs in terms of energy and exergy. The overall energy and exergy efficiencies are expressed as:

$$\eta_{en} = \frac{\dot{W}_{net} + \dot{m}_{H_2} LHV_{H_2} + \dot{m}_{fw} h_{fw}}{\dot{E}_{solar}}$$

and

$$\eta_{ex} = \frac{\dot{W}_{net} + \dot{m}_{H_2} ex_{H_2} + \dot{m}_{fw} ex_{fw}}{\dot{E}x_{solar}}$$

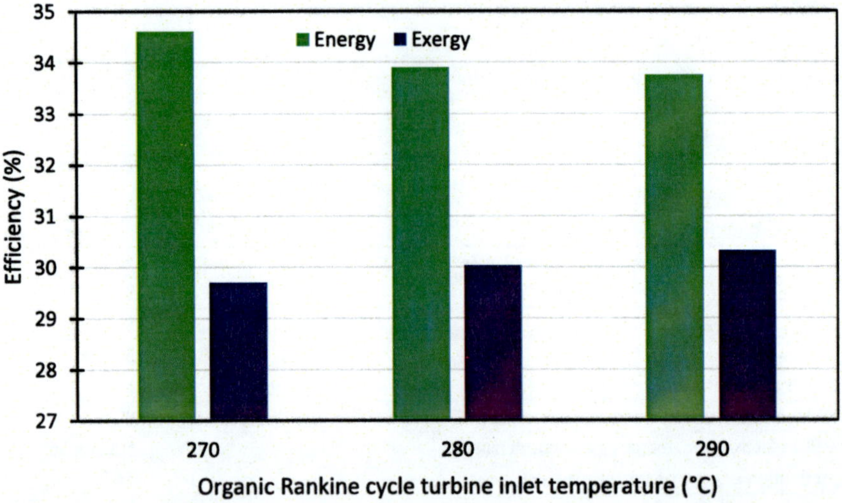

Fig. 8.23 The effect of the changing inlet temperature to the organic Rankine cycle turbine, on the overall energy and exergy efficiencies of the integrated energy system in case study 5 (data from Ref. [6])

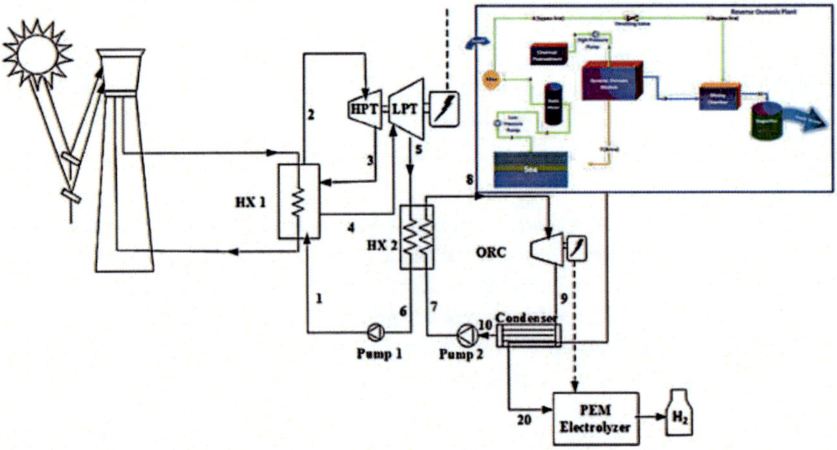

Fig. 8.24 The integrated system layout of case study 6 (modified from Ref. [7])

In the proton exchange membrane electrolysis, they considered the atmospheric pressures at 40°C operating temperature. The activation energies are 76 kJ/mol for the anode side and 18 kJ/mol for the cathode side. In the reverse osmosis plant, the seawater salinity is considered as 35,000 ppm, and freshwater salinity are considered as 450 ppm. The recovery ratio of membrane is considered as 60% and the salt rejection ratio is considered as 90%.

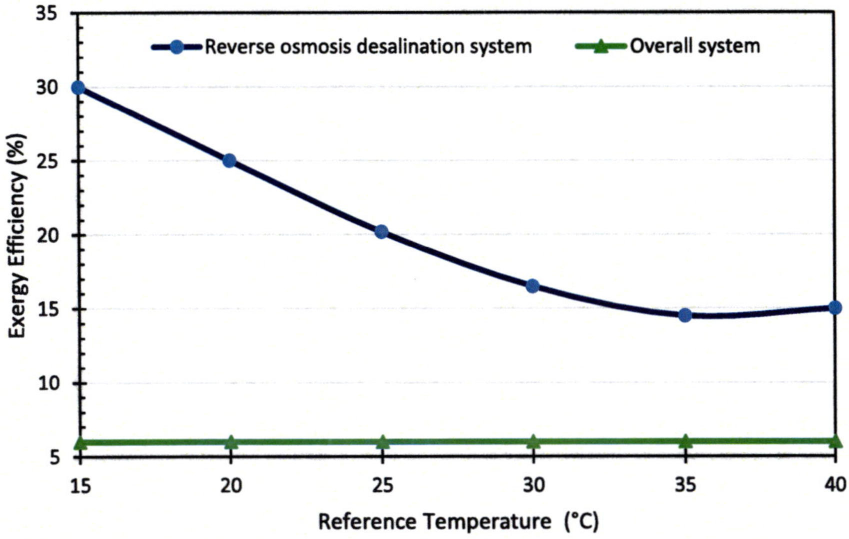

Fig. 8.25 The effect of reference temperature on the overall system and reverse osmosis desalination system efficiencies (data from Ref. [7])

They carried out a case study by considering the meteorological conditions for Jubail in Saudi Arabia. The overall energy and exergy efficiencies are found at 23.2% and 6.2%. The effect of ambient temperature on the exergy efficiency of the overall system and the reverse osmosis process is shown in Fig. 8.25. In particular to the reverse osmosis subsystem, the energy and exergy efficiencies at the optimum conditions are found at 60.3% and 30%. The hydrogen production rate is found between 14.6 kg/h and 270.5kg/h. The freshwater production rate is given as 1500 kg/s at the maximum point during the steady state conditions.

8.3.7 Case Study 7

Gevez and Dincer [8] proposed a new integrated system based on solar and geothermal energies to generate five useful outputs, namely hydrogen, electric power, hot water, heat, and fresh water. Briefly, their proposed system utilizes two inputs to generate five outputs, which is used for case study 7.

Their proposed system collects solar energy via solar towers and uses heat from the underground via hydrothermal sources. In order to generate hydrogen, they used a magnesium-chlorine cycle, which is a thermochemical cycle used for hydrogen production. For the separation of salt and water, they suggested a multi-effect desalination system as illustrated in Fig. 8.26.

They analyzed the system with energy and exergy points of views of thermodynamics, where they balanced mass, energy, entropy, and exergy for each component

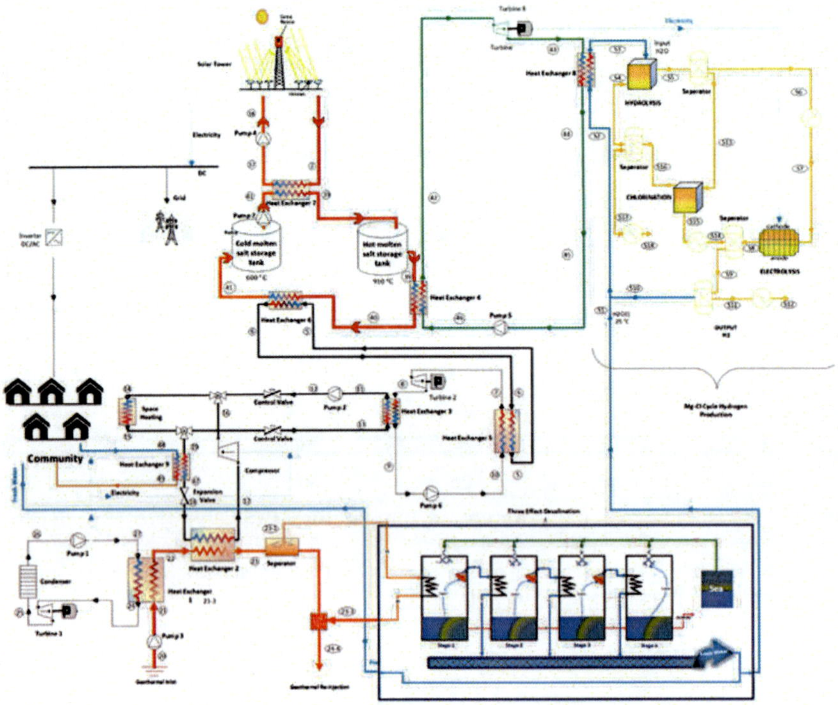

Fig. 8.26 The integrated system layout of case study 7 (modified from Ref. [8])

in order to find the thermophysical properties of each state point. The overall energy and exergy efficiencies of their proposed system are calculated as:

$$\eta_{en,overall} = \frac{\dot{W}_{net} + \dot{Q}_{SH} + \dot{m}_{H_2}LHV_{H_2} + \dot{m}_{fw}h_{fw} + \dot{Q}_{HW}}{\dot{Q}_{solar} + \dot{Q}_{geothermal} + \dot{Q}_{SW}}$$

and

$$\eta_{ex,overall} = \frac{\dot{W}_{net} + \dot{E}x^{Q_{SH}} + \dot{m}_{H_2}ex_{H_2} + \dot{m}_{fw}ex_{fw} + \dot{Q}^{Ex_{HW}}}{\dot{E}x^{Q_{solar}} + \dot{E}x^{Q_{geothermal}} + \dot{Q}^{Ex_{SW}}}$$

The power and heat inputs and outputs to and from each component are summarized in Table 8.3. In addition to these values, the multi-effect desalination system can generate fresh water in the range between 57.7 and 30.7 kg/s. On the other hand, their proposed system can generate 8.7 kg of hydrogen in a typical hour.

Table 8.3 The inputs and outputs of the system components in case study 7 (data from Ref. [8])

Parameter	Value
Outlet of Turbine-1 (kW)	7197
Outlet of Turbine-2 (kW)	38049
Outlet of Turbine-4 (kW)	60187
Inlet of Compressor (kW)	11714
Inlet of Pump-1 (kW)	5395
Inlet of Pump-2 (kW)	22822
Inlet of Pump-3 (kW)	35803
Inlet of Pump-4 (kW)	8262
Electrolysis-step (Mg-Cl Cycle)	261.7
Net power for overall system (kW)	8519
Space heating (kW)	25873
Power from solar receiver (kW)	93323
Fresh water (kg per second)	48.02
Produced hydrogen at Mg-Cl cycle (kg per hour)	8.7

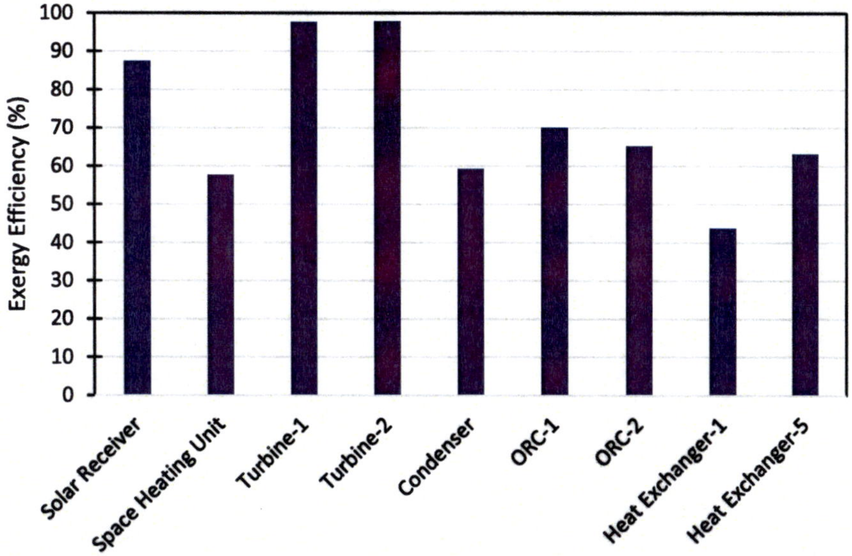

Fig. 8.27 The exergy efficiencies of subsystems, units, and components (data from Ref. [8])

The overall energy and exergy efficiencies of their proposed system are 45.45% and 52.32%, respectively, during the steady state operating conditions. The exergy efficiencies of the different subsystems, units, and components are given in Fig. 8.27. The total exergy destruction rate is 163.9 MW for the overall integrated energy system.

8.3.8 *Case Study 8*

Farsi and Dincer [9] developed and evaluated an integrated energy system with a hybrid desalination system where thermal-based multi-effect distillation desalination and direct-contact membrane-distillation systems are used. The geothermal energy is used by utilizing the hot underground water. A single flash binary steam and organic Rankine cycle power plant is considered at the first contact point with the hot underground water. The generated electricity is used in proton exchange membrane electrolyzer and direct-contact membrane distillation. The excess heat from the Rankine cycles is utilized within the multi-effect distillation desalination system.

By utilizing geothermal energy as single input, the integrated system generates four useful outputs as power, fresh water, cooling, and hydrogen. For the cooling system, the excess heat is used in an ammonia-water absorption refrigeration cycle. They modeled the integrated energy system by using thermodynamic principles, within the aspects of the first and second laws of thermodynamics.

They used the first and second laws of thermodynamics in order to evaluate their proposed system by using the balance equations to balance the mass, energy, entropy, and exergy for each component. They analyzed the system with energy and exergy points of views of thermodynamics, where they balanced mass, energy, entropy, and exergy for each component in order to find the thermophysical properties of each state point. The overall energy and exergy efficiencies of their proposed system are calculated as:

$$\eta_{\mathrm{en,overall}} = \frac{\dot{W}_{\mathrm{net}} + \dot{Q}_{\mathrm{Cool}} + \dot{m}_{\mathrm{H_2}}\mathrm{LHV}_{\mathrm{H_2}} + \dot{m}_{\mathrm{fw,MED}}h_{\mathrm{fg,fw,MED}} + \dot{m}_{\mathrm{fw,DCMD}}h_{\mathrm{fg,fw,DCMD}}}{\dot{m}_1 h_1 - \dot{m}_8 h_8}$$

and

$$\eta_{\mathrm{ex,overall}} = \frac{\dot{W}_{\mathrm{net}} + \dot{E}x^{\dot{Q}_{\mathrm{Cool}}} + \dot{m}_{\mathrm{H_2}}ex_{\mathrm{H_2}} + \dot{m}_{\mathrm{fw,MED}}ex_{\mathrm{fg,fw,MED}} + \dot{m}_{\mathrm{fw,DCMD}}ex_{\mathrm{fg,fw,DCMD}}}{\dot{m}_1 ex_1 - \dot{m}_8 ex_8}$$

where they considered the geothermal energy with reinjection as the total input and the net work output, cooling effect, and fresh water from direct contact membrane desalination and multi-effect distillation desalination subsystems, and the proton exchange membrane electrolyzer as the useful outputs.

The underground fluid is obtained from the production well at 301.2°C at 8734 kPa. A single flashing is applied, and the pressure is reduced to 3000 kPa at 233.90° C. The steam is directed into the steam turbine and expanded until 159.3°C at 606.8 kPa. A heat exchanger–type boiler is used to condense the liquid while transferring its heat to an organic Rankine cycle. A volumetric expander is used with the ammonia working fluid. Ammonia is heated to 129°C at 3800 kPa and expanded in the volumetric expander to 45.22°C and 900 kPa. The expanded ammonia is

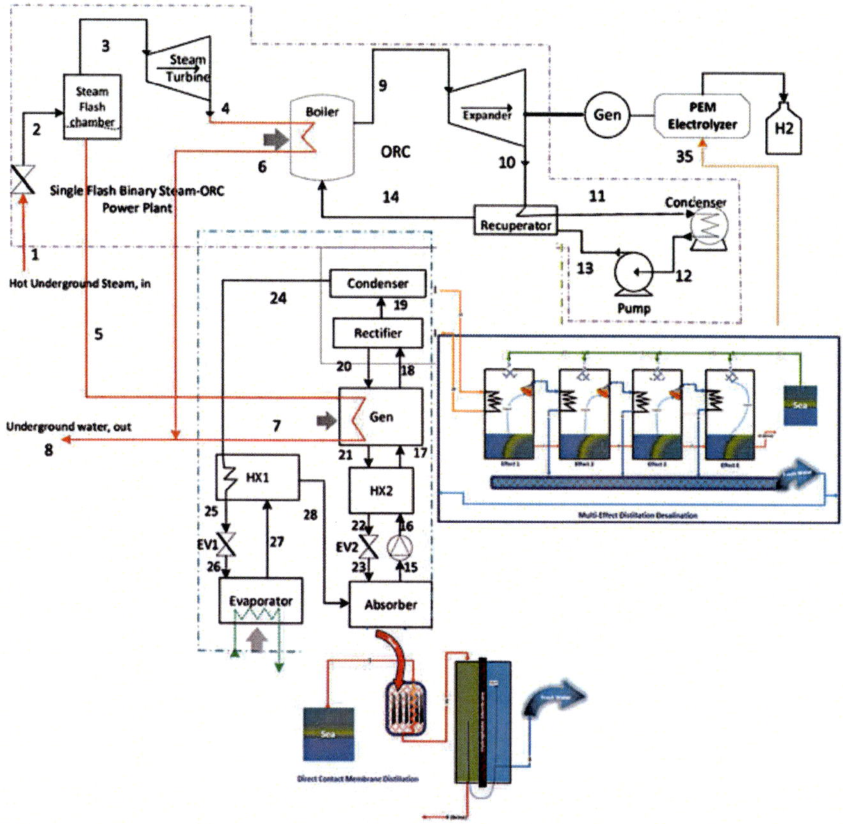

Fig. 8.28 The integrated system layout of case study 8 (modified from Ref. [9])

directed into the recuperator and then the condenser and pump. A six-effect multi-effect distillation system is used to exploit the waste heat from power generation and refrigeration cycles in order to generate fresh water. Figure 8.28 illustrates the integrated energy system in case study 8.

The overall exergy efficiency of the integrated energy system in case study 8 is between 21% and 24%. The overall energy and exergy change due to the reference temperature is given in Fig. 8.29. The hybrid desalination system can generate more than 165.5 tonnes of fresh water per day, which is suitable for a community with 1100 people. One of the main focuses of their system design is to address global concerns about fresh water by proposing a hybrid desalination system where heat and power are utilized. Besides the fresh water, power, cooling effect, and hydrogen are the other useful outputs, which are crucial commodities, especially for the communities with high cooling requirements. Moreover, hydrogen is a potential alternative fuel, which can improve sustainability when it is produced in an integrated fashion by using renewable sources.

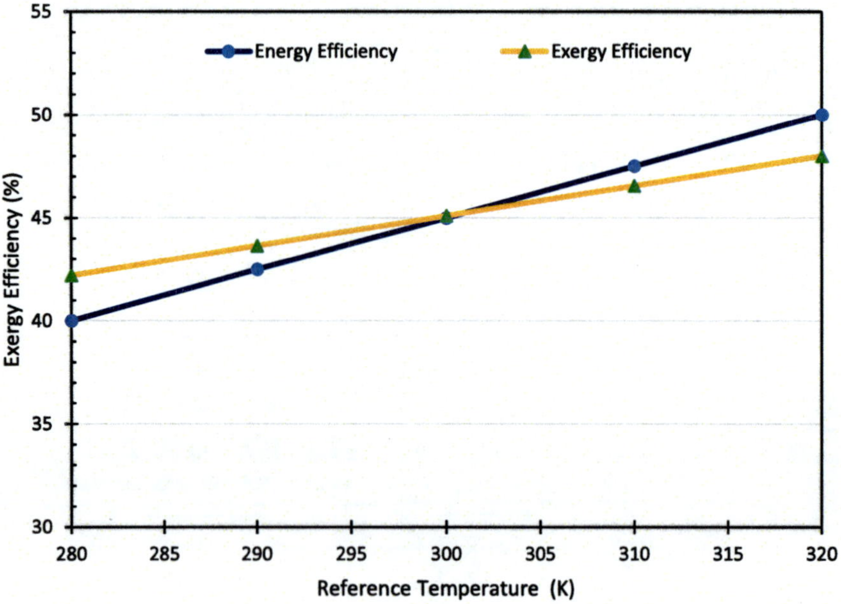

Fig. 8.29 The effect of reference temperature on overall system performance (data from Ref. [9])

8.3.9 *Case Study 9*

In case study 9, an integrated energy system developed by Sorgulu and Dincer [10] is considered. They proposed a biomass-based system with two different desalination systems in order to generate electricity and fresh water. In order to generate fresh water, they used a reverse osmosis desalination subsystem and a multi-effect desalination system in a hybrid manner, where the power generated by a Rankine cycle is employed in the reverse osmosis subsystem and the heat rejected from the condenser is used by the multi-effect distillation system. In the gasifier system, olive oils and municipal solid wastes are considered as biomass inputs to generate syngas and biogas. A digester is used to generate hydrogen and carbon dioxide, separately.

They considered the first and second laws of thermodynamics in their calculations in order to assess their proposed system from energy and exergy aspects. Although their proposed system generates heat, hydrogen, and other outputs, these are used within the system. Therefore, only useful outputs from the overall system are considered as power and fresh water. The system layout is presented in Fig. 8.30.

The energy and exergy efficiencies of their proposed system are considered by dividing the net power generation and net freshwater generation to the total biomass energy input along with the seawater input, as follows:

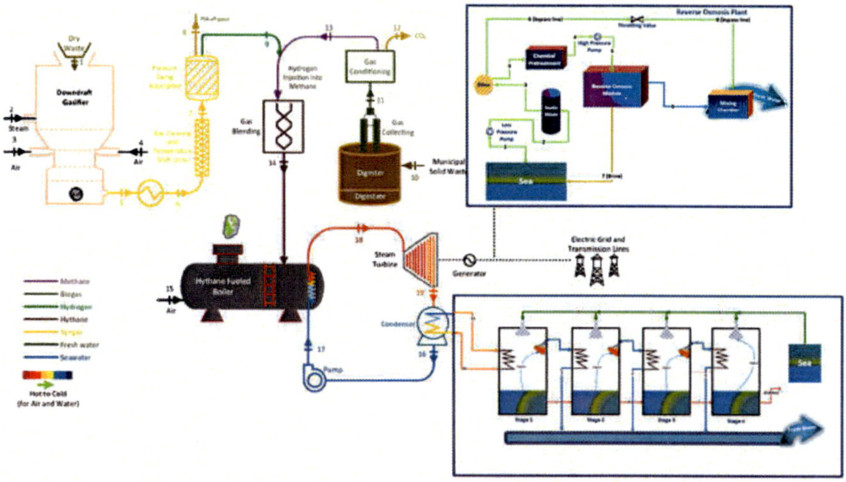

Fig. 8.30 The integrated system layout of case study 9 (modified from Ref. [4])

Table 8.4 A summary of the main parameters of the integrated system case study 9

Parameter	Value
Net power generation	4.5 MW_e
Power consumption of the reverse osmosis subsystem	0.89 MW_{th}
Heat consumption of the multi-effect distillation unit	5.6 MW_{th}
Hydrogen-methane mixture feed rate	0.28 kg/s
Olive oil waste feed rate	0.13 kg/s

$$\eta_{en,overall} = \frac{\dot{W}_{net} + \dot{m}_{fw,MED}h_{fw,MED} + \dot{m}_{fw,RO}h_{fw,RO}}{\dot{E}_{biomass} + \dot{m}_{sw,MED}h_{sw,MED} + \dot{m}_{sw,RO}h_{sw,RO}}$$

and

$$\eta_{ex,overall} = \frac{\dot{W}_{net} + \dot{m}_{fw,MED}ex_{fw,MED} + \dot{m}_{fw,RO}ex_{fw,RO}}{\dot{E}x_{biomass} + \dot{m}_{sw,MED}ex_{sw,MED} + \dot{m}_{sw,RO}ex_{sw,RO}}$$

In the boiler, hydrogen is mixed with methane with fractions up to 20% of the hydrogen ratio. The generated heat from the hydrogen-methane fuel mixture is used in a steam Rankine cycle. The generated power is directed to the reverse osmosis unit to generate fresh water (Table 8.4).

In their approach, seawater is also considered within the total input to the system. This approach can be considered especially when the seawater is warmer than the reference temperature.

The overall energy and exergy efficiencies of the integrated system are 37% and 19.8% in steady-state conditions. The overall energy and exergy efficiency changes

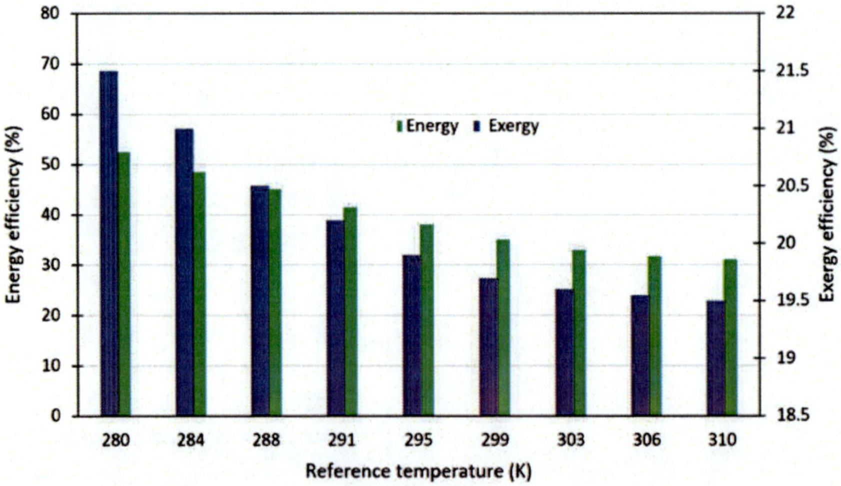

Fig. 8.31 The energy and exergy efficiencies of the overall integrated system by changing reference temperature (data from Ref. [10])

due to the reference temperature change are given in Fig. 8.31. According to the reference temperature change, the change in the energy and exergy efficiencies are shown in Fig. 8.28. The lower heating value of the hydrogen and methane mixture is considered as 52153 kJ/kg, where 0.284 kg/s feed rate leads a 5.43 MW of power generation via a steam Rankine cycle. The freshwater generation from the reverse osmosis system is found as 83.29 kg/s, where 897.6 kW of power is consumed. In the multi-effect distillation desalination system, 9.1 kW of power and 5.6 MW of heat are consumed in order to generate fresh water at 9 kg/s of mass flow rate.

8.3.10 Case Study 10

Temiz and Dincer [11] designed an integrated energy system, where they hybridized a wind, solar, and nuclear energy sources. They combined energy storage options with the nuclear energy source by using molten salt storage tanks. Moreover, they integrated subsystems in order to generate electric power, fresh water, cooling effect, heat, and hydrogen, as shown in Fig. 8.32.

A case study is carried out by considering the needs of a community in California, the United States. When we apply the principles to design an integrated energy system, the first step is to determine available resources. Wind and solar energy are two of the available resources in that location. Moreover, hybridization can also bring synergetic advantages which are not available for only renewable or nuclear energy systems. These synergetic advantages can be mentioned as the capability of nuclear energy to supply continuous energy, which can be used to meet the base

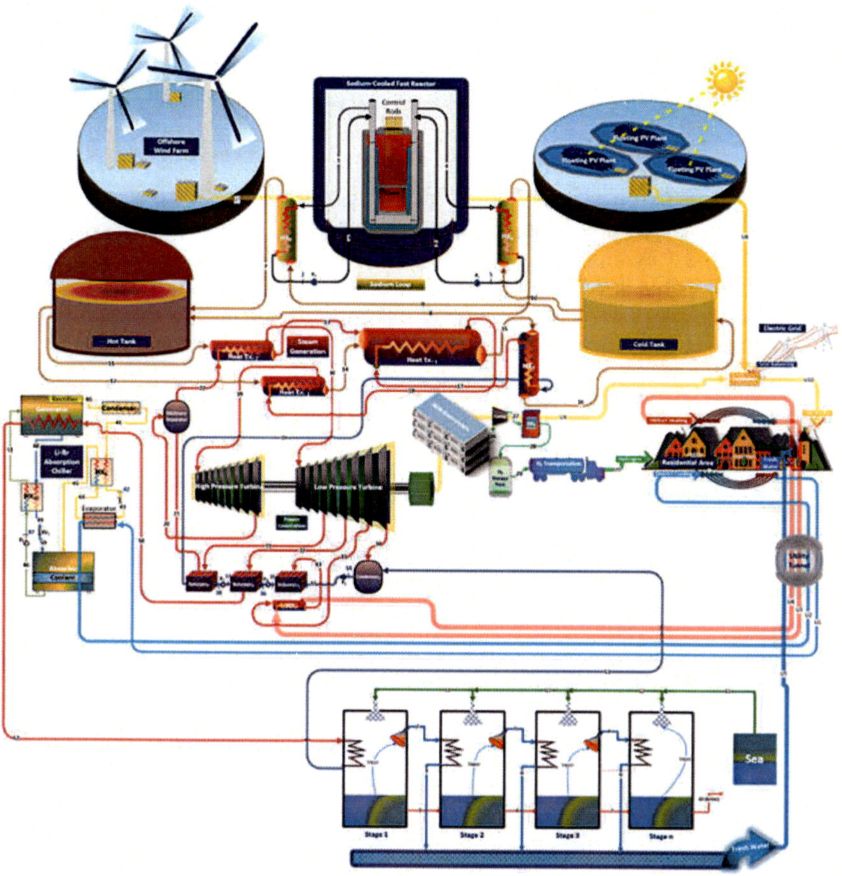

Fig. 8.32 The integrated system layout of case study 10 (modified from Ref. [11])

load. Moreover, less costly wind and solar energy systems can help to lower the costs of useful outputs. Combining the nuclear energy system with a storage option brings additional flexibility to the system in order to meet peak and charge during valley formations of the load. When we look at the forms of these energies, solar PV and wind farms utilize renewables to generate power. However, nuclear energy can bring process heat to the table, which can multiply the combinations of integrated energy systems that can improve the benefits.

California state is a progressive location where renewable energy technologies are highly supported and new technologies can be implemented with fast infrastructure changes. Hydrogen energy technologies are one of the new technologies which can be the key to complete renewables in order to achieve the ultimate target, sustainability. Therefore, hydrogen is considered as one of the needs in the selected location. Heat, cooling, power, and fresh water are the needs of every community anywhere on Earth. Therefore, these are also considered as the needs in the selected location. Therefore, the integrated energy system is developed by considering the above-mentioned factors.

In the proposed integrated energy system, energy input is made by utilizing three different energy sources: wind energy, solar energy, and nuclear energy. Wind energy is available in the form of kinetic energy, which can be utilized by wind turbines. Solar energy is in the form of photonic energy, which can be collected by using PV modules that generate DC power. The atomic energy, on the other hand, can be exploited by employing a nuclear reactor. A sodium-cooled fast reactor is suggested, which is also coupled with molten salt energy storage units.

They applied the principles of the first and second laws of thermodynamics in order to design and analyze their proposed system. When we look at the performance assessment of the system in case study 10, the outputs can be considered as the electric power, heat, cooling effect, fresh water, and hydrogen. On the other hand, solar energy, wind energy, and nuclear energy can be considered as the primary energy inputs. The overall energy and exergy efficiencies of the system can be calculated with:

$$\eta_{en,ov} = \frac{\dot{W}_{El,net} + \dot{m}_{H_2} LHV_{H_2} + \dot{m}_{fw} h_{fw} + \dot{Q}_{District\ Heating} + \dot{Q}_{District\ Cooling}}{\dot{Q}_{solar} + \dot{Q}_{SCFR} + \dot{E}_{wind}}$$

and

$$\eta_{ex,ov} = \frac{\dot{W}_{El,net} + \dot{m}_{H_2} ex_{H_2} + \dot{m}_{fw} ex_{fw} + \dot{E}x_{Q_{District\ Heating}} + \dot{E}x_{Q_{District\ Cooling}}}{\dot{E}x^{\dot{Q}_{solar}} + \dot{E}x^{\dot{Q}_{SCFR}} + \dot{E}x_{wind}}$$

where

$$\dot{E}x^{\dot{Q}_{solar}} = A_{PV} \times \dot{q}_{GHI,FPV} \times \left(1 - \frac{4}{3}\frac{T_o}{T_{sun}}\left(1 - \cos\delta\right)^{1/4} + \frac{1}{3}\left(\frac{T_o}{T_{sun}}\right)^4\right)$$

where $\dot{Q}_{solar}$expresses the total solar energy input, δ represents the half angle of the sun, and A represents the area of photovoltaic modules.

One of their objectives was to carry out their study in a realistic fashion. For this reason, commercially available components are selected with specific characteristics. Moreover, the load data is generated for a community by using the database of ResStock from the National Renewable Energy Laboratory as EUSS ResStock National TMY3 Release 2022.1 for California, the United States location. Moreover, annual water needs and the required hydrogen due to the driver statistics are scaled for a community with 275000 people. The electric, heat, and cooling needs of the community are shown in Fig. 8.33.

They designed an offshore wind farm where they used wind turbines from General Electric with a rotor diameter of 100 m and a hub height of 100 m, where

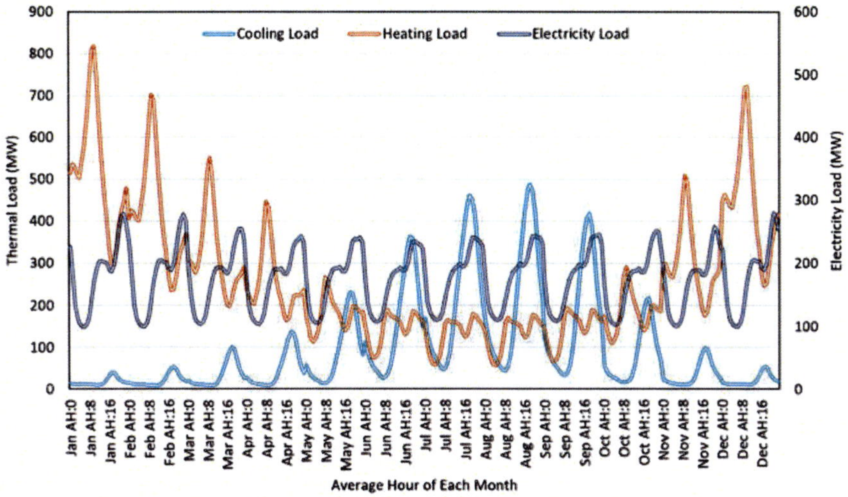

Fig. 8.33 The community load of case study 10 (data from Ref. [11])

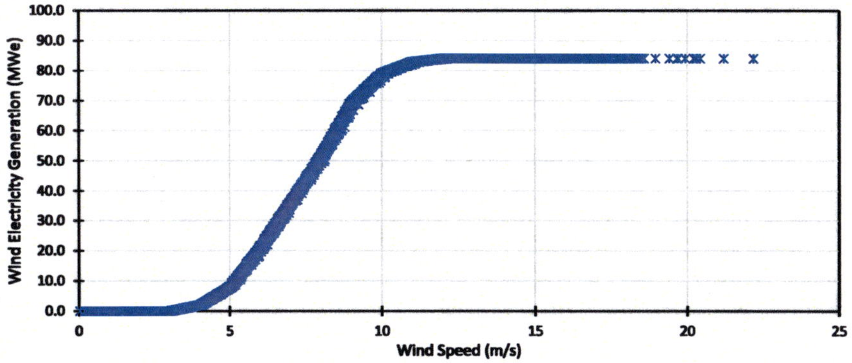

Fig. 8.34 The power curve of the selected wind turbine (data from Ref. [11])

each turbine can generate 1.6 MW. 63 of these wind turbines are used for an offshore wind farm for the selected location. The power curve of their offshore wind farm design is shown in Fig. 8.34.

One of the challenges in metropolitan areas is the land scarcity. Therefore, the authors considered to design the system on top of unutilized surfaces, specifically the ocean. Floating PV plants are designed in order to minimize the land footprint of these renewable power plants. A SunPower SPR-MAX3-400 PV modules are selected within the integrated system. 250000 units of PV modules are used to form the entire floating PV plant. The solar electricity generation from the floating PV plant at different solar intensities is shown in Fig. 8.35.

Moreover, a sodium-cooled fast reactor design is used where the sodium is heated up to 505°C in order to generate steam at 490°C and 137.3 bar. A molten salt storage

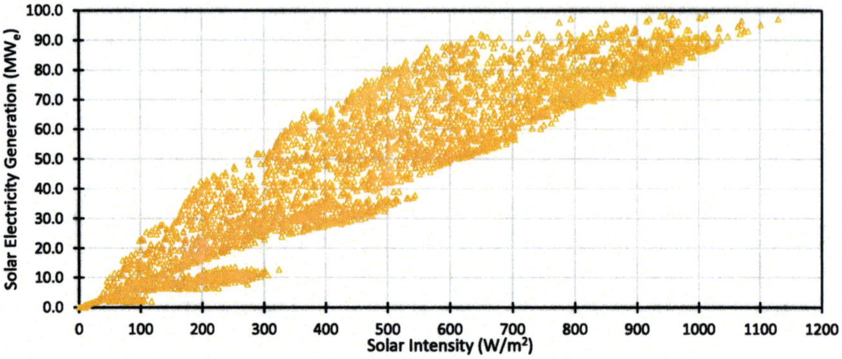

Fig. 8.35 The power generation from the floating PV plant (data from Ref. [11])

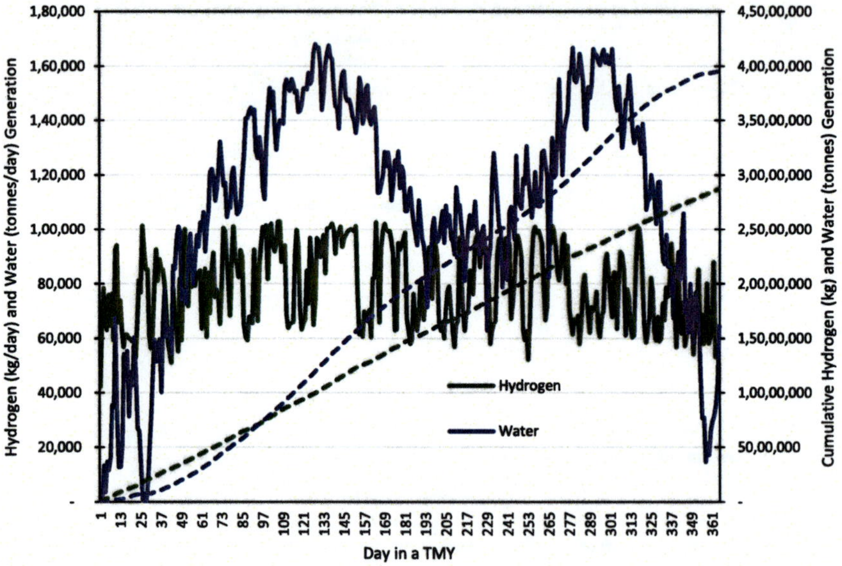

Fig. 8.36 The water and hydrogen generations from the integrated energy systems in a typical year (data from Ref. [11])

system is also combined in order to balance the imbalances between the demand of supply, which can store 5530 MWh of thermal energy with the molten salt energy storage medium.

Hydrogen and freshwater generations are shown in Fig. 8.36. An anion exchange membrane electrolysis unit is used to generate hydrogen by utilizing the excess power. On the other hand, a multi-effect distillation desalination cycle is used in order to exploit the excess heat from the system. In a typical year, 39575 kilo-tonnes of fresh water are generated. Moreover, 28713 tonnes of hydrogen is generated, in the same typical year.

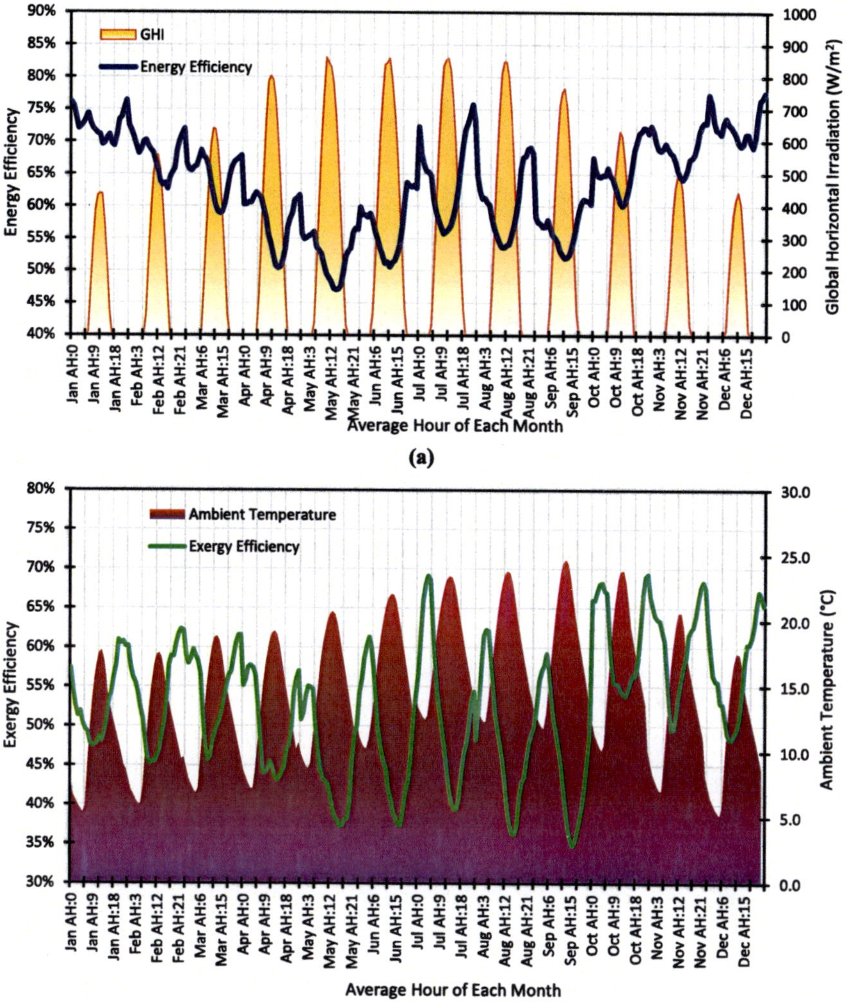

(a)

(b)

Fig. 8.37 The overall (**a**) energy and (**b**) exergy efficiency variations of the integrated energy system in case study 10 (data from Ref. [11])

The proposed system capacities were 345 MWe for the sodium-cooled fast reactor which can generate heat up to 1 GW; 100.8 MWe of the offshore wind farm; and a 100 MWp of floating PV plant, for 275000 people to meet their fresh water, hydrogen, electric power, heat, and cooling needs in a hybridized and integrated fashion.

The overall energy and exergy efficiencies were varying in a typical year as shown in Fig. 8.37. Among the months, the highest energy efficiency was in

December with 72.54% on average. The highest exergy efficiency was in October, with 61.11%, on average. The average energy and exergy efficiencies for the overall integrated energy systems are 63.36% and 52.42%, in a typical year for the selected location.

8.4 Closing Remarks

The ultimate target is to achieve sustainability. From this aspect, it is very critical to utilize the sources available on Earth, while ensuring to keep the sources available and suitable for future generations. When it comes to energy, it is very critical to design energy systems with a target to minimize wastes and maximize benefits. In energy systems, the integrated energy systems can achieve this while minimizing the wastes and maximizing the benefits in terms of useful outputs. The only limit to multiply the useful outputs is the principles of thermodynamics and the imagination of the designer. Those are extensively presented in this section with rich case studies, where the principles of thermodynamics are applied, especially from energy and exergy points of views.

Renewables and desalination are a key combination with the integrated energy systems. First, the renewables are the new energy systems; therefore, there is an opportunity to design them in an integrated manner. Second, desalination is a process where low-temperature waste heat or excess power can be utilized. Both waste heat and excess power are the common wastes from typical energy systems. Therefore, for each renewable energy system, there is room to integrate a desalination subsystem to exploit the unutilized wastes.

References

1. Dincer I, Bicer Y (2021) Integrated energy systems for multigeneration. Elsevier
2. El-Emam RS, Dincer I (2014) Thermodynamic and thermoeconomic analyses of seawater reverse osmosis desalination plant with energy recovery. Energy 64:154–163. https://doi.org/10.1016/j.energy.2013.11.037
3. Ghosh S, Dincer I (2014) Development and analysis of a new integrated solar-wind-geothermal energy system. Solar Energy 107:728–745. https://doi.org/10.1016/j.solener.2014.06.006
4. Ozlu S, Dincer I (2016) Performance assessment of a new solar energy-based multigeneration system. Energy 112:164–178. https://doi.org/10.1016/j.energy.2016.06.040
5. Demir ME, Dincer I (2017) Development of an integrated hybrid solar thermal power system with thermoelectric generator for desalination and power production. Desalination 404:59–71. https://doi.org/10.1016/j.desal.2016.10.016
6. El-Emam RS, Dincer I (2018) Investigation and assessment of a novel solar-driven integrated energy system. Energy Convers Manag 158:246–255. https://doi.org/10.1016/j.enconman.2017.12.062

7. Siddiqui O, Dincer I (2018) Examination of a new solar-based integrated system for desalination, electricity generation and hydrogen production. Solar Energy 163:224–234. https://doi.org/10.1016/j.solener.2018.01.077
8. Gevez Y, Dincer I (2023) A novel renewable energy system designed with Mg–Cl thermochemical cycle, desalination and heat storage options. Energy 283:129101. https://doi.org/10.1016/j.energy.2023.129101
9. Farsi A, Dincer I (2019) Development and evaluation of an integrated MED/membrane desalination system. Desalination 463:55–68. https://doi.org/10.1016/j.desal.2019.02.015
10. Sorgulu F, Dincer I (2021) Development and assessment of a biomass-based cogeneration system with desalination. Appl Therm Eng 185:116432. https://doi.org/10.1016/j.applthermaleng.2020.116432
11. Temiz M, Dincer I (2023) A new solar and wind integrated sodium fast reactor with molten salt to generate multiple useful outputs with hydrogen. Appl Therm Eng 232:120960. https://doi.org/10.1016/j.applthermaleng.2023.120960

Chapter 9
Future Directions

Nomenclature

BRICS	Brazil, Russia, India, China and South Africa
COVID-19	Coronavirus Disease 2019
C/H	carbon/hydrogen ratio
ICT	information and communication technologies
OECD	Organization for Economic Cooperation and Development

9.1 Introduction

There is no doubt that the reliance on desalination will continue to grow, not only for drinking water but also for many other purposes. Climate change is driven by extreme heat and prolonged drought; moreover, global population growth contributes to water scarcity and problems related to water quantity and quality globally.

The water withdrawals surpassed the critical levels, and renewable water resources are declining to unprecedented levels. Figure 9.1 depicts clearly how fast we are losing renewable water resources with historical data as well as future forecasts. In 1960, almost 14,000 m^3 of renewable water resources was available per capita. When it comes to the present (2023), more than half of the resources were lost. Our forecast for the future shows that only a quarter of renewable freshwater resources will remain by 2050, compared to the levels in 1960. This clearly shows that generating fresh water with desalination will be inevitable for our society in the future.

Figure 9.2 depicts the other side of the story: freshwater consumption globally. At the beginning of the twentieth century, global freshwater consumption was around 671 billion m^3. In the second half of the twentieth century, the post–World War II boom can be observed in parallel to freshwater consumption. The improved life

I. Dincer, M. Temiz, *Renewable Energy Options for Power Generation and Desalination*, https://doi.org/10.1007/978-3-031-53437-9_9

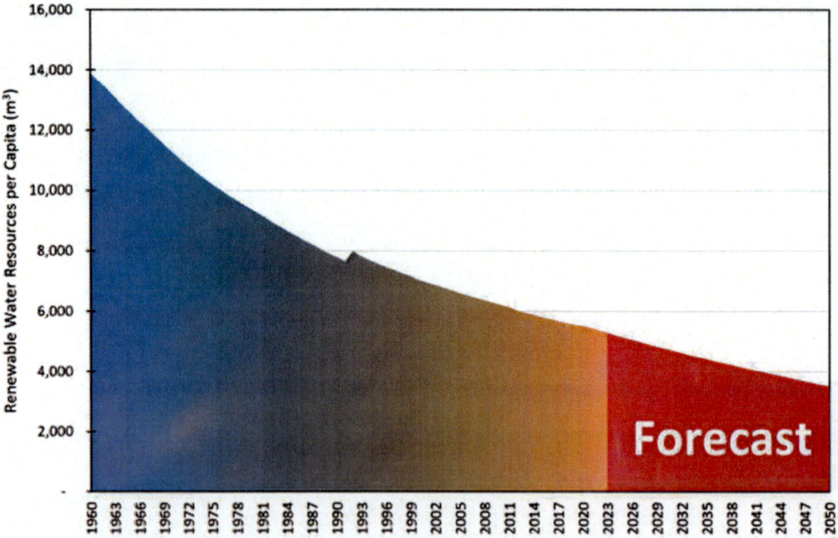

Fig. 9.1 Renewable water resources with forecast (historical data from Ref. [1])

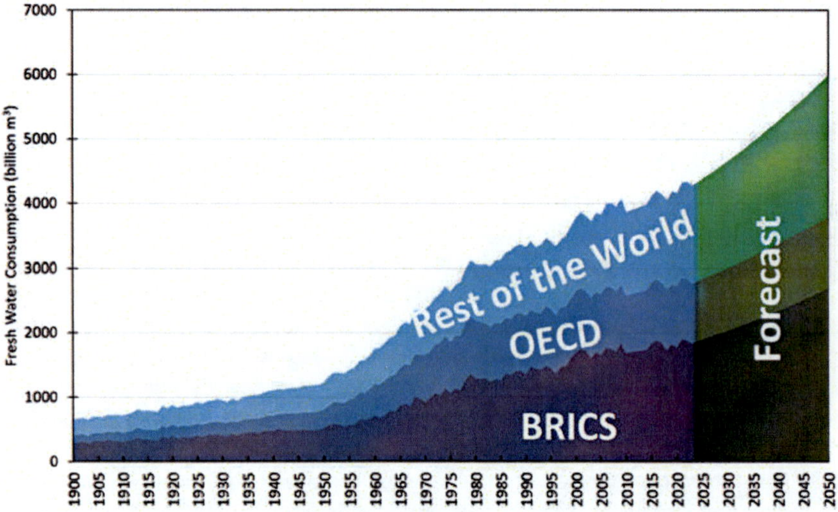

Fig. 9.2 Freshwater consumption by different regions with forecast (historical data from Ref. [1])

quality in modern society, the population growth as well as the modernization at any sector and industry led to growing freshwater consumption. Global freshwater consumption quintupled in the twentieth century. Presently, 4300 billion m^3 of fresh water is consumed annually. Our estimations show that the global freshwater consumption will become 6000 billion m^3 by 2050.

It is clearly shown in both charts that while renewable water resources are declining, water consumption is growing. More than half of the renewable water resources have been lost since 1960. This leads to an emerging problem of water quantity. There are numerous processes and sectors that require fresh water. However, the natural sources of fresh water are limited. If there will be no changes, the natural freshwater sources will be insufficient for the global society. It is also important to keep the other limited natural sources and not harm the environment while generating fresh water. From this aspect, sustainable freshwater generation is the ultimate target in order to achieve the global water challenge. Renewable desalination is the potential solution to generate fresh water in a sustainable manner.

9.2 Future Developments on Renewable Desalination

While determining the potential solutions for renewable energy and desalination technologies, it is critical to understand what the future demand will look like. Drinking water is one of the first things that comes to mind; however, it only represents a small portion of freshwater needs globally. Growing crops, manufacturing, energy and transportation sectors rely on fresh water. Municipal consumption including drinking water only represents 12% of global freshwater withdrawals. Agricultural activities are responsible for the majority of global freshwater withdrawals, as 72%. Industrial activities also use more fresh water than municipal activities. Approximately 16% of the global freshwater consumption is related to industrial activities. The sectoral distribution of global freshwater withdrawals is shown in Fig. 9.3.

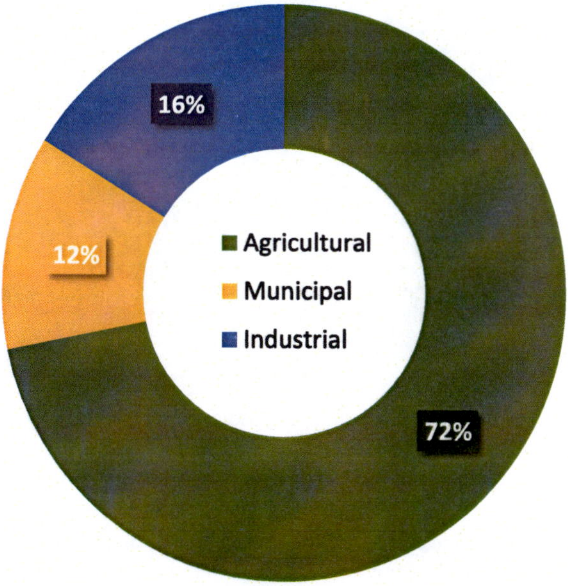

Fig. 9.3 The current global freshwater withdrawal shares by sectors (data from Ref. [1])

Many other reasons contribute to the consumption of water on Earth. However, humanity was careless when it comes to protect our vital freshwater resources. Mismanagement of the sources, pollution, and damaged and lost rivers and freshwater ways led to loosing many wetlands and renewable water resources. Freshwater scarcity is one of the contributors to deforestation, which is very critical. Healthy forests require fresh water; if water scarcity remains, deforestation can be accelerated. Global food security is another critical aspect that is highly sensitive to water scarcity, since agricultural activities rely on fresh water.

The future needs will not be similar to today's needs, due to the new developments and growth in different sectors. In future, there will be new sectors that rely on fresh water, as depicted in Fig. 9.4. Our daily lives have become more engaged with technology. Especially after the COVID-19 pandemic, the panoptical digital transformation has accelerated. This digital transformation is a new type of industrialization, which is clear when we look at the largest companies. Technology companies such as Apple, Microsoft, Amazon, Alphabet-Google, and Facebook-Meta are five of the top six largest companies in terms of market capitalization. Investments are

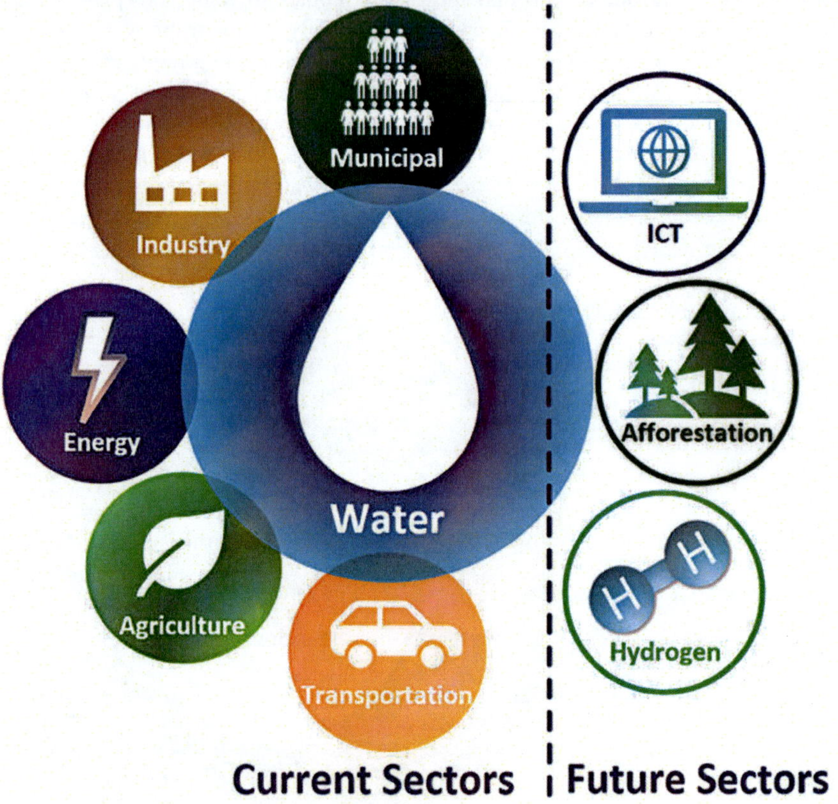

Fig. 9.4 The current and future sectors that rely on fresh water

also flowing through information and communication technologies (ICT), which will lead to the growing demands in power, cooling, and freshwater needs in these sectors. ICT companies will create energy consumption boomtowns at their centers, to provide their data-intensive services to the users. There will be a growing number of ICT centers, such as data centers in the future. This growth will create a hunger for power demand, cooling demand, as well as water demand. Currently, only 10% of the power consumption is used by the ICT sector. By 2030, 23% of the global power consumption will be used by ICT-related activities [2]. Data processing and other ICT-related processes cause heat generation. Almost 80% of the power consumed by the data processors is turned into heat. The generated heat by the data processors needs to be dissipated and rejected. This creates a secondary demand in the ICT sector, which is cooling. The current technologies and methods for cooling highly rely on water systems since evaporation of water can help reject unwanted heat much faster than air. Therefore, the panoptical digital transformation and the growth in the ICT sector lead to a new boom in water consumption.

The energy journey started for humankind with hydrocarbons. Wood was the first type of fuel that has been used by human society. The Industrial Revolution opened the coal era, where coal was the dominant fuel source for society. After World War I, the oil era started and oil became the dominant energy source. After 1980s, the natural gas era started. When we look at these developments from the carbon/hydrogen (C/H) ratio, it is clear that carbon has been declining from wood to natural gas. Especially the COVID-19 coronavirus affected the way we live, every sector and every community. As mentioned earlier, human beings need clean water, clean air, clean food, and clean energy. The necessity of cleaner water, cleaner food, and cleaner energy became more important after COVID-19. From this aspect, "COVID-19 coronavirus is closing carbon age but opening hydrogen age" [3]. Hydrogen age can help achieve many critical challenges, such as environmental, ecosystem, efficiency, source utilization, industrial development, economic development, synchronization with renewable energy technologies, health of the society, and sustainable development. Hydrogen society refers to a society where the main driver is hydrogen, instead of hydrocarbons. World recognized hydrogen as a potential solution, and in order to promote the hydrogen society, Hydrogen Council was launched by 13 global industry leaders in January 2017 in Davos, Switzerland, during the World Economic Forum [4]. In 2023, "Hydrogen 1.0" was introduced for the first time by Professor Dr. Ibrahim Dincer [5]. Hydrogen pioneers are expected to make great contributions in order to meet the needs of the hydrogen era with technology, innovation, and digitalization. Global energy and water equations will be reconstructed where hydrocarbon fuels will be replaced by hydrogen.

Afforestation is another establishment that will require fresh water with growing demands. Currently, jurisdictions and nongovernmental organizations established afforestation projects in order to create new forests. It is seen as a solid tool in order to fight against the climate change and minimize the impacts of the climate change. There are many benefits that can be mentioned regarding the afforestation. Trees improve the soil quality in the area and boost the organic carbon levels in the soil, which can prevent desertification. Desertification is very critical, especially in the regions where vulnerability is high. Africa as a continent has a majorly high vulnerability against desertification. Eastern European countries, Turkey, Australia, the Middle East region, and the west coast of the United States are other regions which are vulnerable to desertification. Many countries announced afforestation programs in order invest into planting trees to reduce emissions. Canada has a program called the "2 Billion Trees," which aims to motivate and support new tree-planting projects by 2031 [6]. They invested $3.2 billion for tree-planting projects via afforestation and reforestation. Afforestation projects will accelerate the demand for fresh water. Especially in the areas where there are no available freshwater resources, large-scale desalination needs to be coupled with afforestation projects. An estimation [7] indicates that if renewable desalination systems will be built for afforestation projects, there is a potential to sequestrate carbon dioxide 730 gigatons of carbon dioxide during the period between 2030 and 2100. The average cost of tonnes of carbon dioxide sequestration is estimated as €457 in 2030; moreover, a cost decline is also estimated which can drop to €100 per tonnes of carbon dioxide by 2100.

There are new technologies and developments in desalination technologies which improve the process of desalination in order to make it a more feasible process. In materials technology, new membranes can lower the need for energy to separate salt and water. One of the recent progresses in the membranes area is the achievements with graphene oxide–assisted membranes in desalination. Due to limited permeability, the reverse osmosis process becomes a costly option in terms of operational expenditures [8]. The Nobel Prize in Physics in 2010 was given to Andre Geim and Konstantin Novosolev regarding their pioneering research on graphene [9]. Graphene became a rising star in the industry and enabled numerous improvements in many processes, including desalination. Desalination is one of the areas where graphene can be exploited in order to enable benefits and improve the process to much more feasible levels. Innovative graphene oxide–assisted membranes can improve the desalination applications in different ways such as mechanical strength, antifouling properties, antimicrobial properties, selectivity, thermal properties, and water flux [10]. It is important to illustrate the current sectors and potentially the future sectors in terms of water demands, as shown in Fig. 9.4.

Figure 9.5 shows the development stages of desalination. As mentioned earlier in Chap. 6, historical development of desalination started with simple distillation methods. When we look at the currently available desalination technologies, they are majorly driven by fossil fuels via traditional energy systems. Cogeneration and trigeneration systems such as combined heat and power or combined cooling, heat, and power systems are also integrated with desalination systems. Nuclear desalination is also an important stage, where nuclear energy systems replaced the fossil fuels. It is a cleaner method in order to generate fresh water by using energy. Moreover, the nuclear power generation process is suitable for the integration of desalination. Many countries launched nuclear desalination programs. The United Arab Emirates initiated a program where they planned to have four nuclear power plants; each has a capacity to produce 1.5 GW of electricity. The United Arab Emirates considers different desalination technologies such as multi-stage flashing,

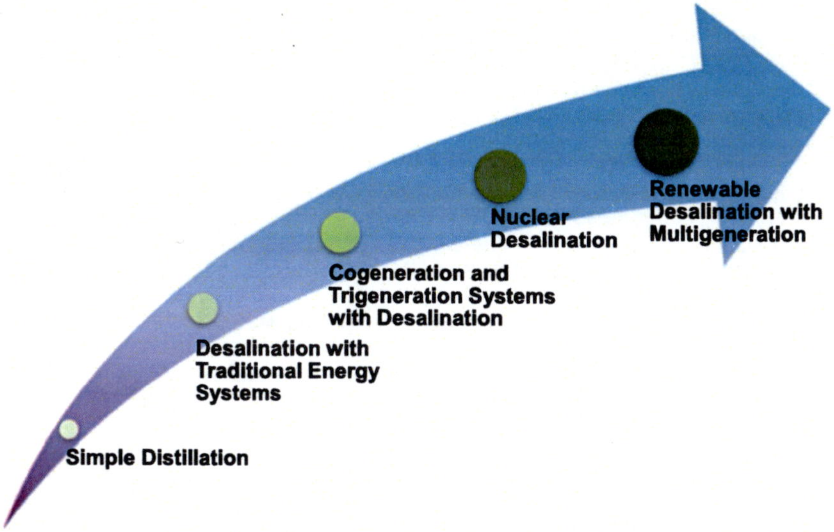

Fig. 9.5 Development stages of desalination

multi-effect desalination, and reverse osmosis in order to couple the nuclear power plants with desalination [11]. The currently available largest desalination plant is also in the United Arab Emirates [12]. Jebel Ali Power Plant and Water Desalination Complex operates the world's largest single-site desalination plant which can generate 2,227,587 m^3 of fresh water in a day. In the Jebel Ali Complex, the multi-stage flashing distillation is the main method, where the waste heat is exploited to generate fresh water. Additionally, a system is also built in the same site, where power is utilized via a reverse osmosis process in order to generate fresh water. Ultimately, the desalination process needs to be carried out in a sustainable fashion. From this aspect, renewable desalination is the potential solution. Especially when integrated energy systems are used, source utilization can be maximized. As the final development stage of desalination, renewable desalination carried out by integrated energy systems in order to generate multiple useful outputs can be mentioned, which is extensively presented throughout the book.

References

1. Ritchie H, Roser M (2023) Water use and stress. Our world in data. Retrieved from: https://ourworldindata.org/water-use-stress (Online Resource)
2. Andrae ASG, Edler T (2015) On global electricity usage of communication technology: trends to 2030. Challenges 6:117–157. https://doi.org/10.3390/challe6010117
3. Dincer I (2020) Covid-19 coronavirus: closing carbon age, but opening hydrogen age. Int J Energy Res 44:6093–6097. https://doi.org/10.1002/ER.5569

4. World Economic Forum (2017) World economic forum annual meeting 2017 responsive and responsible leadership. Davos-Klosters
5. Dincer I (2023) Hydrogen 1.0: a new age. Int J Hydrogen Energy 48:16143–16147. https://doi.org/10.1016/j.ijhydene.2023.01.124
6. Government of Canada (2023) 2 Billion trees commitment. https://www.canada.ca/en/campaign/2-billion-trees.html
7. Caldera U, Breyer C (2023) Afforesting arid land with renewable electricity and desalination to mitigate climate change. Nat Sustain 6:526–538. https://doi.org/10.1038/s41893-022-01056-7
8. Hughes ZE, Carrington LA, Raiteri P, Gale JD (2011) A computational investigation into the suitability of purely siliceous zeolites as reverse osmosis membranes. J Phys Chem C 115: 4063–4075. https://doi.org/10.1021/JP109591F/ASSET/IMAGES/LARGE/JP-2010-09591F_0009.JPEG
9. Geim AK, Novoselov KS (2007) The rise of graphene. Nat Materials 6:183–191. https://doi.org/10.1038/nmat1849
10. Hegab HM, Zou L (2015) Graphene oxide-assisted membranes: fabrication and potential applications in desalination and water purification. J Memb Sci 484:95–106. https://doi.org/10.1016/j.memsci.2015.03.011
11. Naizghi MS, Tesfay WG, Fath HES (2011) Nuclear desalination and its viability for UAE. In: The Sharjah International conference on nuclear and renewable energy energies for the 21st century (SHJ-NRE11)
12. Largest water desalination plant. Guinness World Records. https://www.guinnessworldrecords.com/world-records/425709-largest-water-desalination-plant. Accessed 21 Nov 2023

Index

I. Dincer, M. Temiz, *Renewable Energy Options for Power Generation and Desalination*, https://doi.org/10.1007/978-3-031-53437-9